W9-CFW-993

HAZARDOUS
CHEMICALS
IN THE
POLYMER
INDUSTRY

Environmental Science and Pollution Control Series

Additional Volumes in Preparation

RA
1242
.P66C48
1995
c.2

HAZARDOUS CHEMICALS IN THE POLYMER INDUSTRY

Nicholas P. Cheremisinoff

National Association of Safety and Health Professionals
Morganville, New Jersey

BUCKMAN LABORATORIES, INTL.
TECHNICAL INFORMATION CENTER

Marcel Dekker, Inc.   **New York • Basel • Hong Kong**

Library of Congress Cataloging-in-Publication Data

Cheremisinoff, Nicholas P.
    Hazardous chemicals in the polymer industry / Nicholas P. Cheremisinoff.
      p.   cm. — (Environmental science and pollution control series; 14)
    ISBN 0-8247-9273-4
    1. Plastics—Additives—Toxicology. 2. Polymers—Toxicology. 3. Plastics
industry and trade—Safety measures. I. Title. II. Series: Environmental science
and pollution control; 14.
    RA1242.P66C48 1994
    363.17'91—dc20   94-22884
                                                          CIP

The publisher offers discounts on this book when ordered in bulk quantities. For more information, write to Special Sales/Professional Marketing at the address below.

This book is printed on acid-free paper.

Copyright © 1995 by MARCEL DEKKER, INC. All Rights Reserved.

Neither this book nor any part may be reproduced or transmitted in any form or by any means, electronic or mechanical, including photocopying, micro-filming, and recording, or by any information storage and retrieval system, without permission in writing from the publisher.

MARCEL DEKKER, INC.
270 Madison Avenue, New York, New York 10016

Current printing (last digit):
10 9 8 7 6 5 4 3 2 1

PRINTED IN THE UNITED STATES OF AMERICA

PREFACE

Several hundred distinct chemicals are widely used in elastomer and plastics processing operations. Many of these are used almost exclusively in the rubber industry. Despite this widespread use, less than 10% of the more than 500 chemical additives have established threshold limit values. Considerable knowledge has and is being accumulated in the industry, particularly among chemical suppliers, in meeting right-to-know legislative requirements. This volume not only provides basic safety and toxicity data for industry users, but also points out some of the areas in which further information is needed.

The underlying aim of those involved with toxicity problems in the rubber and plastics industries is to prevent or eliminate the risk of health hazards to those handling or exposed to chemicals. This is a continuous learning process as new chemicals and new ways of using existing chemicals continue to be introduced. The science of relating mechanisms of the interaction of elastomeric chemical additives with human physiology is still relatively young. On the other hand, the knowledge base is rapidly expanding. It should be further noted that control of exposure to chemicals in this industry should be a routine part of normal plant management practices and should heavily involve the industrial hygienist and process safety engineer.

This volume is written as a practical reference for engineers, technicians, and elastomer/plastics compounders. Information on toxic and hazardous properties, symptoms of overexposure, safe handling and shipping, fire hazards, and spill responses for both generic and specific polymer chemicals are given. The author has made every effort to ensure that the information presented in this handbook is accurate and up-to-date. A wide range of sources was used in compiling this information. Despite this effort, no claims of assurance or liability for the information researched or presented is assumed by either the author or the publisher. On a personal note, the author wishes to express gratitude to Linda Jastrzebski for her assistance in the production of this work.

Nicholas P. Cheremisinoff, Ph.D.

iii

CONTENTS

1 HOW TO USE THE HANDBOOK

This handbook constitutes a database on the toxic and hazardous properties, as well as shipping, packaging, and emergency response procedures for polymeric additives. The chemicals covered in this volume are organized into nine categories.

Section Number	Class of Chemicals
2	Monomers of Synthetic Rubbers
3	Vulcanizing Agents
4	Accelerators and Activators
5	Mineral Fillers, Activators, and Dusting Agents
6	Antioxidants, Stabilizers, and Fire Retardants
7	Retarders
8	Blowing Agents
9	Oils, Plasticizers, Lubricants, Solvents, and Mold Release Agents
10	Amines

The information gathered for each chemical has been extracted from several sources including NIOSH, OSHA, RTECS, CHRIS Manual, DOT regulations, the *Federal Register*, ACGIH, and manufacturers' literature.

The user will find two general formats in which information is presented. The first format and most dominant presentation is an expanded information sheet. This format gives the following information on a chemical:

1. General Information
   Name
   Synonyms
   CAS Registration Number
   Chemical Formula
   Wisswesser Line Notation
   Chemical Class

2. Physical Properties

3. Regulations Governing the Chemical

4. Toxicity Data

5. Protection and First Aid

6. Initial Incident Response and Emergency Action

The second general format is an abbreviated Material Safety Data Sheet. In this case, either the specific chemical is proprietary in composition or only manufacturer's information could be obtained. This format provides basic health and safety information and limited properties data only.

The following is a list of the specific chemicals covered in the handbook. The last part of this first section provides a list of definitions and abbreviations used throughout the volume.

DEFINITION OF TERMS AND ABBREVIATIONS
USED THROUGHOUT THE BOOK

The following are definitions of the terms and abbreviations used throughout the handbook.

ACGIH - American Conference of Governmental Hygienists. Publications and recommendations regarding exposure to chemicals are quoted from this source.

Action - Refers to the OSHA action level concentration of the chemical.

Agents - The term as used herein refers to neutralizing agents.

AIT - Autoignition temperature in degrees Kelvin.

AITC - Autoignition temperature in degrees Celsius.

AITF - Autoignition temperature in degrees Fahrenheit.

ANAL.AGENCY - Refers to the agency (NIOSH or OSHA) that is responsible for the air sampling analysis method.

AQ.TOX - Abbreviation for the aquatic toxicity of the chemical.

BP - Boiling point.

CARC - Refers to carcinogen data; A "y" indicates that this chemical is a known or strongly suspected carcinogen by IARC, NTP, OSHA, ACGIH, or MAK.

CARC.ACGIH - Carcinogenicity coding: Al = confirmed human carcinogen. Substances associated with industrial processes, recognized to have carcinogenic potential. A2 = suspected human carcinogen. Chemical substances associated with industrial processes that are suspected of inducing cancer, based on either limited epidemiological evidence or demonstration of carcinogenesis in one or more animal species.

CARC.IARC - IARC carcinogenicity rating: 1 = carcinogen as defined by IARC as carcinogenic to humans, with sufficient epidemiological evidence. 2A = carcinogen defined by IARC to be probably carcinogenic to humans with (usually) at least human evidence. 2B = carcinogen defined by IARC to be possibly carcinogenic to humans, but having (usually) no human evidence. 3 = not classified as to human carcinogenicity or probably not carcinogenic to humans. 4 = the agent is probably not carcinogenic to humans. If a "No" is indicated, then the chemical is not listed by IARC.

CAS - Chemical Abstract Service Registry number.

CCA - Clean Air Act reference.

CEIL - Refers to Ceiling PEL. On January 19, 1989, OSHA published in the *Federal Register* the final rule (29CFR 1910) for air contaminants. This rule amended OSHA's existing air contaminants standard in 1910.100, including Tables Z-1, Z-2, and Z-3.

CERCLA - Comprehensive Environmental Resource Cleanup and Liability Act.

CHEM.REACT - Chemical reactivity.

Class - Refers to chemical class or classes.

CPC - chemical protective clothing.

CHRIS - Chemical Hazard Response Information System.

CTNR.EXC - Refers to 49 CFR Table 172.101. DOT cargo tank exceptions.

CTNR.SPEC - See 49 CFR Table 172.101. DOT cargo tank specifications.

CWA - Clean Water Act; means that the chemical is on one of two lists: Section 307, priority pollutants list, or Section 311, chemicals considered hazardous if spilled into navigable waterways. Section 311 chemicals have reportable quantities (RQs). The Section 311 list reviewed comes from 40 CFR 401.15, and the Section 311 list is from 40 CFR 116.4 (revised in 54 FR 33426, August 14, 1989).

Density - The weight per unit volume of the material in grams per cubic centimeter at a given temperature or ambient.

DOT - Department of Transportation.

DOT.CLASS - Refers to the DOT hazard class.

DOT.GUIDE - Refers to the DOT guide number.

DOT.ID - The U.S. Department of Transportation identification number. Those numbers preceded by a "UN" are associated with descriptions considered appropriate for international as well as domestic use. Those preceded by an "NA" are not recognized for international shipments except to and from Canada.

DOT.NAME - Refers to DOT shipping name from 49 CFR Table 172.101.

Duration - OSHA duration period for exposures where a maximum concentration is specified.

EPA.ID - EPA RCRA hazardous waste ID.

FIFRA - Refers to a chemical listed by EPA's Office of Pesticides Programs on April 24, 1989, as one of the 150 active ingredients subject to re-registration as required by 1988 FIFRA amendments [Section 4(c) (2) (B)].

FP - Refers to the flash point temperature of the material.

GCC - Refers to the EPA Generic Classification Codes from SARA Title III, Sec. 313(c). See 52 FR, p. 21168, June 4, 1987.

IDLH - NIOSH published concentration that is immediately dangerous to life and health.

LC50 - Lethal concentration for 50% of a test population (RETCS 1992).

LD50 - Lethal dose for 50 % of a test population of rats . Dose administered orally .

LD50.RAT - The numeric value (in ml/kg) of the oral LD50 for rats. From the 1992 RETCS.

LD50.SKN.RBT - LD50 value (in mg/kg) when applied to the skin of albino rabbits. From RETCS data (1992).

LEL - Lower explosion limit.

LTT - Long-term toxicity.

MAIL - Mailability information from the U.S. Postal Service's "Hazardous Materials Table Postal Mailability Guide," which is an Appendix to Publication 52, April, 1990.

MAX - OSHA maximum allowable concentration.

MAX.CARGO - Maximum net quantity in one package. From the Department of Transportation Hazardous Materials Table 172.101-Revised, October 27, 1987.

MCL - Refers to the maximum contaminant level, that is, the maximum permissible level of a contaminant in water that is delivered to the free-flowing outlet of the ultimate user of a public water system.

MCLG - Refers to maximum contaminant level goals. The Safe Drinking Water Act of 1986 requires EPA to publish MCLGs for contaminants that may have adverse health effects and are known or anticipated to occur in water systems. MCLG values allow for an adequate margin of safety.

MP - Melting point.

MUT - Mutagenic data from RTECS, 1992. Only positive mutation test results are reported in RETCS.

MW - Molecular weight.

NFPA - National Fire Protection Association.

NFPA.BLUE - NFPA health code (blue).

NFPA.RED - NFPA flammability code (red).

NFPA.SPECIAL - NFPA special code (usually radioactive or water reactive or for oxidizer).

NFPA.YELLOW - NFPA reactivity code (yellow).

NIOSH - National Institute of Occupational Safety and Health.

Odor - Lower odor detection limit. Note that these data are very subjective and probably vary from individual to individual.

OSHA - Occupational Safety and Health Administration.

PEL - Permissible Exposure Limit. On January 19, 1989 OSHA published in the *Federal Register* the final rule (29 CFR 1910) for air contaminants. This rule amended OSHA's existing Air Contaminants standard in 1910.1000, including Tables Z-1, Z-2, and Z-3. A new Table (Z-1-A) makes more protective 212 PEL values while adding 164 new substances. In addition, short-term exposure limits (STEL) were added to complement the 8-hour time weighted average (TWA) limits, skin designations were established, and ceiling limits were added for some substances.

POLY - Polymerization possibilities for the chemical.

RCRA.NO - EPA RCRA ID number.

RE - Routes of entry.

REL - NIOSH Recommended Exposure Limits. NIOSH develops and periodically revises recommendations for limiting exposure to potentially hazardous substances in the workplace. These are published in Criteria Documents, Current Intelligence Bulletins, Special Hazard Reviews, and Occupational Hazard Assessments.

REPRO - Reproductive toxicity from 1992 RTECS. Effects listed include paternal effects, maternal effects, effects on fertility, effects on embryo or fetus, specific developmental abnormalities, and tumorigenic effects.

RQ - Reportable quantity. CERCLA reportable quantity symbols: X, A, B, C, or D. RQs may be designated for a chemical under SARA Title III, Section 304 (Extremely Hazardous Substances-EHS) or under CERCLA.

RTECS - Registry of Toxic Effects of Chemical Substances.

SMALL.SPILL - Refers to small spill action as defined under DOT guidelines. Note that, if a "small" spill exceeds the reportable quantity (RQ) for a chemical, reporting of the spill may be required under EPA regulations.

SP.GR - Specific gravity. The ratio of the density of the substance at a given temperature to that of water at a given temperature.

SPILL.LEAK - Refers to the spill leak information from the DOT Guide P5800.4.

STCC - Standard tariff commodity code number.

STEL - ACGIH short-term exposure limit.

STOR - Refers to information on the general storage requirements for the chemical.

STT - Short-term toxicity.

TARGET.ORGANS - Target organs, that is, data derived primarily from NIOSH published sources and the CHRIS manual.

TC - Toxic characteristics identification number. Found in EPA's list of TC procedure, published in 55 FR 11798, March 29, 1990.

TERAT - Teratogenic data.

TLV - American Conference of Governmental and Industrial Hygienists established threshold limit value.

TO - Target organs.

TPQ - Threshold planning quantity. Based on Superfund Amendments and Reauthorization Act (SARA) of October 17, 1986. For solids, the lower TPQ applies only if the solid exists in powdered form and has a particle size less than 100 microns; or is handled in solution or in molten form; or meets the National Fire Protection Association criteria rating of 2, 3, or 4 for reactivity. If the solid does not meet any of these listed criteria, then the upper TPQ applies.

TQ - Threshold quantity, from 57 FR 6407, February 24, 1992. Appendix A to 1910.119, List of Highly Hazardous Chemicals, Toxics, and Reactives, and chemicals meeting the published requirements for flammable in 29 CFR 1910.1200. Generally, this includes liquids with flash points less than 100°F.

TWA - OSHA's 8-hour time weighted average limits for air contaminants.

UEL - Upper explosion limits.

VAPOR.DENSITY - Vapor density (air = 1); that is, the vapor's specific gravity.

VD - Vapor density.

VP - Vapor pressure.

WATER - Refers to water reactivity data.

WATER.SOL - Refers to water solubility data.

WEEL - Workplace Environmental Exposure Levels prepared by the American Industrial Hygiene Association (AIHA) for chemicals that have no current exposure guidelines published by other agencies. All WEELs are expressed as ceiling values. Different time periods are expressed, depending on the chemical. Ceiling values should not be exceeded at any time. "Skin" indicates that the chemical may be absorbed through the skin in amounts that may be toxicologically significant.

WLN - Refers to Wisswesser line notation. WLN is a line-formula notation that conveys information about the structural formula of a chemical compound. The WLN allows searching for special functional groups and constituents that are part of the molecule.

2 MONOMERS OF SYNTHETIC RUBBERS

Polymers used in the rubber industry generally do not cause health-related problems. The high molecular weight and chemical stability of these materials prevent significant absorption through skin contact, and they are stable and inert when in contact with fluids in medical applications such as catheter tubes and implants.

On the other hand, polymers may contain trace or residual impurities derived from the manufacturing process, and they may also have additives such as stabilizers. Both these classes of chemicals should be considered from the standpoint of toxic hazards.

This section describes the toxic and hazardous properties of some of the widely used monomers employed in synthesizing polymers. Traces of these monomers often remain in commercial polymer supplies at levels generally less than 1 ppm by weight. It is unusual, but not impossible for the TLV to be reached during normal processing and fabrication operations. As an example in some operations, the author has detected concentrations of acrylonitrile between 2 and 3 ppm on mill rolls during the milling of nitrile rubber. Acrylonitrile is a carcinogenic material, and compounders should proceed with caution when working with nitrile rubbers to ensure that operator exposure is well below the TLV.

Other polymers can contain even higher concentrations of monomers. Fabricators, compounders, and mill operators should always check to ensure that exposures to residual monomers do not exceed the TLV. Another example of a high-level exposure risk is polychloroprene latex, where several thousand ppm by weight of chloroprene monomer can occur during processing operations.

Two final examples of exposure risks are the processing of polyurethane rubbers and chlorosulfonated polyethylene. Polyurethane elastomers may contain free isocynate. Chlorosulfonated polyethylene can contain carbon tetrachloride, a well-known carcinogen.

--- IDENTIFIERS ---

Name: ACRYLONITRILE
Synonyms: Acrylnitril (German, Dutch); Acrylon; Acrylonitrile; Acrylonitrile
  Monomer; Akrylonitryl (Polish); Carbacryl; Cianuro DiVinile (Italian);
  Cyanoethylene; Cyanure De Vinyle (French); Ent 54; Fumigrain; Miller's
  Fumigrain; Nitrile Acrilico (Italian); Nitrile Acrylique (French);
  Propenenitrile; 2-Propenenitrile; TL 314; VCN; Ventox; Vinyl Cyanide;
  Vinyl Cyanide, Propenenitrile
CAS: 107-13-1; **RTECS:** AT5250000
Formula: C_3H_3N; **Mol Wt:** 53.06
WLN: NC1U1
Chemical Class: Nitrile

See other identifiers listed below under Regulations.

--- PROPERTIES ---

Physical Description: Clear, colorless to pale yellow liquid (NYDH)
Boiling Point: 350.38 K; 77.2°C; 171°F
Melting Point: 190.38 K; -82.8° C; -117°F
Flash Point: 273.15 K; 0°C; 32°F
Autoignition: 898 K; 624.8°C; 1156.7°F
Critical Temp: 536 K; 262.85°C; 505.13°F
Critical Press: 4.6 kN/m^2; 45.3 atm; 666 psia
Heat of Vap: 265 Btu/lb; 147.17 cal/g; 6.158x E5 J/kg
Heat of Comb: -14300 Btu/lb; -7950 cal/g; -332x E5 J/kg
Vapor Pressure: 83 mm
UEL: 17%
LEL: 3 %
Ionization Potential (eV): 10.91
Vapor Density: 1.8 (air = 1)
Evaporation Rate: 3.31(n-butyl acetate = 1)
Specific Gravity: 0.807 at 20°C
Density: 0.806
Water Solubility: 7.1%
Incompatibilities: Strong oxidizers, bromine, strong bases, copper, copper
  alloys, ammonia, amines

Reactivity with Water: No data on water reactivity
Reactivity with Common Materials: Attacks copper, copper alloys, and
  aluminum, penetrates leather

Stability during Transport: No data
Neutralizing Agents: May occur in absence of oxygen or exposure to light or heat
Polymerization Possibilities: No data

Toxic Fire Gases: None reported other than possible unburned vapors
Odor Detected at (ppm): 21.4 ppm (sense of smell fatigues rapidly)
Odor Description: Peach seed kernels (Source: CHRIS)
100% Odor Detection: No data

--- REGULATIONS ---

DOT hazard class: 3 FLAMMABLE LIQUID
DOT guide: 30
Identification number: UN1093
DOT shipping name: Acrylonitrile, inhibited
Packing group: I
Label(s) required: FLAMMABLE LIQUID, POISON
Special Provisions: B9
Packaging exceptions: 173.None
Nonbulk packaging: 173.201
Bulk packaging: 173.243
Quantity limitations:
  Passenger air/rail: Forbidden
  Cargo aircraft only: 30 L
  Vessel stowage: E
  Other stowage provisions: 40

STCC Number: 4906420

Clean Water Act Sect. 307: Yes
Clean Water Act Sect. 311: Yes
Clean Air Act: CAA '90 listed
EPA Waste Number: U009
CERCLA Ref: Y
RQ Designation: B 100 pounds (45.4 kg) CERCLA
SARA TPQ Value: 10,000 pounds
SARA Sect. 312 Categories:
  Acute toxicity: irritant.
  Chronic toxicity: carcinogen.
  Chronic toxicity: adverse effect to target organs after long periods of exposure.

Chronic toxicity: mutagen.
Chronic toxicity: reproductive toxin.
Fire hazard: flammable.
Listed in SARA Sect. 313: Yes
De Minimus Concentration: 0.1%

U.S. Postal Service Mailability:
Hazard class: Flammable liquid; mailable as ORM-D
Mailability: Domestic surface mail only
Max per parcel: 1 qt metal; 1 pt other

NFPA Codes:
Health Hazard (blue): (4) Full protection will not be adequate unless designed specifically for this chemical. DO NOT ENTER AREA!
Flammability (red): (3) This material can be ignited under almost all temperature conditions.
Reactivity (yellow): (2) Normally unstable and readily undergoes violent change, but does not detonate.
Special: Unspecified.

--- TOXICITY DATA ---

Short-term Toxicity: *Inhalation:* Levels of 16 to 100 ppm for 20 to 45 min have caused headache, eye, nose, and throat irritation, general irritability, and liver and kidney irritation. Higher levels may produce nausea, vomiting, diarrhea, tremors, unconsciousness and changes in the liver. *Skin:* Direct contact may cause irritation, redness, and blisters. Absorption is significant and may contribute to symptoms listed above. *Eyes:* May cause severe irritation and chemical burns. *Ingestion:* May cause severe irritation and chemical burns of the mouth, throat, and stomach, as well as other symptoms as listed under inhalation. (NYDH)

Long-term Toxicity: Levels of 5 to 20 ppm may produce headache, fatigue, nausea, weakness, anemia, and other blood abnormalities, mild liver damage and eye irritation. Acrylonitrile has been shown to cause cancer and birth defects in laboratory animals. It has been linked to an increase in cancer among occupationally exposed workers. (NYDH)

Target Organs: CVS, liver, kidneys, CNS, skin

Symptoms: Similar to those of hydrogen cyanide. Vapor inhalation may cause weakness, headache, sneezing, abdominal pain, and vomiting. Similar symptoms shown if large amounts of liquid are absorbed through the skin; lesser amounts cause stinging and sometimes blisters; contact with eyes causes severe irritation. Ingestion produces nausea, vomiting, and abdominal pain (Source: CHRIS).

Conc IDLH: 500 ppm

NIOSH REL: Potential occupational carcinogen 1 (skin). Time weighted averages for 8-hour exposure, 10 (skin). Ceiling exposures that shall at no time be exceeded.

ACGIH TLV: TLV = 2, A2 ppm (4.5, A2 mg/m^3) skin
ACGIH STEL: Not listed

OSHA PEL: Final rule limits: TWA = 2 ppm (skin); Ceiling = 10 ppm (skin); Consult 29 CFR 1910.1045.

MAK Information: Danger of cutaneous absorption; Carcinogenic working material without MAK; In the Commission's view, an animal carcinogen.

Carcinogen: Y; **Status:** See below
References: Human suspected IARC** 19, 73, 79; Animal positive IARC** 19,73,79; Human suspected IARC** 28, 151, 82; Animal suspected IARC** 28, 151, 82

Carcinogen Lists: *IARC*: Carcinogen defined by IARC to be probably carcinogenic to humans with (usually) at least limited human evidence; *MAK*: An animal carcinogen; *NIOSH*: Carcinogen defined by NIOSH with no further categorization; *NTP*: Carcinogen defined by NTP as reasonably anticipated to be carcinogenic, with limited evidence in humans or sufficient evidence in experimental animals; *ACGIH*: Carcinogen defined by ACGIH TLV Committee as a suspected carcinogen, based on either limited epidemiological evidence or demonstration of carcinogenicity in experimental animals; *OSHA*: Cancer hazard.

Human Toxicity Data: (Source: NIOSH RTECS)
ihl-hmn TCLo:16 ppm/20 M INMEAF 17, 199, 48
   Sense Organs
     Nose
       Other
     Eye
       Conjunctive irritation

Lungs, Thorax, or Respiration
Other changes

ihl-man LCLo: 1 gm/m³/1H ZAARAM 16, 1, 66
Behavioral
Somnolence (general depressed activity)
Gastrointestinal
Hypermotility, diarrhea
Nausea or vomiting

LD50 Value: orl-rat LD50:78 mg/kg

Other Species Toxicity Data: (Source: NIOSH RTECS 1992)
orl-rat LD50:78 mg/kg
ihl-rat LC50:425 ppm/4H
skn-rat LD50:148 mg/kg
ipr-rat LD50:65 mg/kg
scu-rat LD50:96 mg/kg
unr-rat LDLo:200 mg/kg
orl-mus LD50:27 mg/kg
ipr-mus LD50:46 mg/kg
scu-mus LD50:25 mg/kg
ihl-dog LCLo:110 ppm/4H
ivn-dog LDLo:200 mg/kg
ihl-cat LCLo:600 ppm/4H
ihl-rbt LCLo:260 ppm/4H
skn-rbt LD50:250 mg/kg
ivn-rbt LD50.69 mg/kg
orl-gpg LD50:50 mg/kg
ihl-gpg LCLo:575 ppm/4H
skn-gpg LD50:202 mg/kg
scu-gpg LD50:130 mg/kg

Irritation Data: (Source: NIOSH RTECS 1992)
skn-hmn 500 mg nse
skn-rbt 10 mg/24H
eye-rbt 20 mg SEV

Reproductive Toxicity (1992 RTECS): This chemical is a mammalian
reproductive toxin.

Reproductive Toxicity Data (1992 RTECS):
orl-rat TDLo:650 mg/kg (6-15D preg) DOWCC* 03NOV76
  Effects on Fertility
    Female fertility index
  Effects on Embryo or Fetus
    Fetotoxicity (except death, e.g., stunted fetus)
  Specific Developmental Abnormalities
    Musculoskeletal system

orl-rat TDLo:650 mg/kg (6-15D preg) TJADAB 17, 50A, 78
  Specific Developmental Abnormalities
    Musculoskeletal system
    Cardiovascular (circulatory) system

ihl-rat TCLo:40 ppm/6H (6-15D preg) FCTXAV 16, 547, 78
  Maternal Effects
    Other effects on female
  Nutritional and Gross Metabolic
    See also Biochemical
    Weight loss or decreased weight gain

ihl-rat TCLo:80 ppm/6H (6-15D preg) FCTXAV 16, 547, 78
  Specific Developmental Abnormalities
    Musculoskeletal system

ipr-mus TDLo:32 mg/kg (5D preg) ZHYGAM 26, 564, 80
  Effects on Fertility
    Postimplantation mortality

ipr-ham TDLo:641 mg/kg (8D preg) TJADAB 23, 317, 81
  Effects on Fertility
    Postimplantation mortality
  Specific Developmental Abnormalities
    Central nervous system

ipr-ham TDLo:641 mg/kg (8D preg) TJADAB 23, 325, 81
  Effects on Embryo or Fetus
    Extra embryonic features (e.g., placenta, umbilical cord)
    Cytological changes (including somatic cell genetic material)
    Fetotoxicity (except death, e.g., stunted fetus)

No Significant Risk Level (Ca P65): No. 7 mg/day

---PROTECTION AND FIRST AID ---

Protection Suggested from the CHRIS Manual: Air-supplies mask, industrial chemical type, with approved canister for acrylonitrile in low (less than 25%) concentrations; rubber or plastic gloves; cover goggles or face mask; rubber boots; slicker suit; safety helmet.

NIOSH Pocket Guide to Chemical Hazards:
  **Wear Appropriate Equipment to Prevent: Repeated or prolonged skin contact.

  **Wear Eye Protection to Prevent: Reasonable probability of eye contact.

  **Exposed Personnel Should Wash: Immediately when skin becomes wet.

  **Remove Clothing: Immediately remove any clothing that becomes wet to avoid any flammability.

  **The Following Equipment Should Be Made Available: Eyewash, quick drench.

  **Reference: NIOSH

Recommended Respiration Protection Source: NIOSH Pocket Guide (85-114) OSHA (Acrylonitrile) -
Less than or equal to 20 ppm: (1) Chemical cartridge respirator with organic vapor cartridge(s) and half-mask facepiece; or (2) supplied air respirator with half-mask facepiece.
Less than or equal to 100 ppm or maximum use concentration (MUC) of cartridges or canisters, whichever is lower: (1) Full facepiece respirator with (A) organic vapor cartridges, (B) organic vapor gas mask chin-style, or (C) organic vapor gas mask canister, front- or back-mounted; (2) supplied air respirator with full facepiece; or (3) self-contained breathing apparatus with full facepiece.
Less than or equal to 4000 ppm: Supplied air respirator operated in the positive pressure mode with full facepiece, helmet, suit, or hood.
Greater than 4000 ppm or unknown concentration: (1) Supplied air and auxiliary self-contained breathing apparatus with full facepiece in positive pressure mode; or (2) self-contained breathing apparatus with full facepiece in positive pressure mode.

Escape: (1) Any organic vapor respirator; or (2) any self-contained breathing apparatus.
Firefighting: Self-contained breathing apparatus with full facepiece in positive pressure mode.

First Aid Source: CHRIS Manual 1991.
Skilled medical treatment is necessary; call physician for all cases of exposure.
Inhalation: Remove victim to fresh air. (Wear an oxygen or fresh-air-supplied mask when entering contaminated area.)
Ingestion: Induce vomiting by administering strong solution of salt water, but only if victim is conscious.
Skin: Remove contaminated clothing and wash affected area thoroughly with soap and water.
Eyes: Hold eyelids apart and wash with continuous gentle stream of water for at least 15 min. If victim is not breathing, give artificial respiration until physician arrives. If he is unconscious, crush an amyl nitrite ampule in a cloth and hold it under his nose for 15 sec every minute. Do not interrupt artificial respiration while doing this. Replace ampule when its strength is spent and continue treatment until condition improves or physician arrives.

First Aid Source: DOT Emergency Response Guide 1993.
Move victim to fresh air and call emergency medical care; if not breathing, give artificial respiration; if breathing is difficult, give oxygen. In case of contact with material, immediately flush skin or eyes with running water for at least 15 min. Remove and isolate contaminated clothing and shoes at the site. Keep victim quiet and maintain normal body temperature. Effects may be delayed; keep victim under observation.

--- INITIAL INCIDENT RESPONSE ---

Fire Extinguishment: Dry chemical, alcohol foam, carbon dioxide.
Note: Water or foam may cause frothing (CHRIS 91).

U.S. Department of Transportation Guide to Hazardous Materials Transport Information - Publication DOT 5800.5 (1990).
DOT Shipping Name: Acrylonitrile, inhibited
DOT ID Number: UN1093

* POTENTIAL HAZARDS *

*Health Hazards
Poisonous; may be fatal if inhaled, swallowed, or absorbed through skin.
Contact may cause burns to skin and eyes.
Runoff from fire control or dilution water may cause pollution.

*Fire or Explosion
Extremely flammable; may be ignited by heat, sparks, and flames.
Vapors may travel to a source of ignition and flash back.
Container may explode in heat of fire.
Vapor explosion and poison hazard indoors, outdoors, or in sewers.
Runoff to sewer may create fire or explosion hazard.

* EMERGENCY ACTION *

Keep unnecessary people away; isolate hazard area and deny entry.
Stay upwind; keep out of low areas.
Positive-pressure self-contained breathing apparatus (SCBA) and chemical
  protective clothing that is specifically recommended by the shipper or
  manufacturer may be worn. It may provide little or no thermal protection.
Structural firefighter's protective clothing is not effective for these materials.
Isolate the leak or spill area immediately for at least 150 ft. in all directions.
See the Table of Initial Isolation and Protective Action Distances. If you find
  the ID Number and the name of the material there, begin protective action.
Isolate for 1/2 mile in all directions if tank, rail car, or tank truck is involved
  in fire.
**CALL CHEMTREC AT 1-800-424-9300 FOR EMERGENCY ASSIS-
TANCE.**

*Fire
Small fires: Dry chemical, CO_2, water spray or regular foam.
Large fires: Water spray, fog, or regular foam.
Do not get water inside container.
Apply cooling water to sides of containers that are exposed to flames until
  well after fire is out. Stay away from ends of tanks.
For massive fire in cargo area, use unmanned hose holder or monitor nozzles;
  if this is impossible, withdraw from area and let fire burn.
Withdraw immediately in case of rising sound from venting safety device or
  any discoloration of tank due to fire.

*Spill or Leak

Shut off ignition sources; no flares, smoking, or flames in hazard area.

Fully encapsulating, vapor-protective clothing should be worn for spills and leaks with no fire.

Do not touch or walk through spilled material.

Small spills: Flush area with flooding amounts of water.

Large spills: Dike far ahead of liquid spill for later disposal.

*First Aid

Move victim to fresh air and call emergency medical care; if not breathing, give artificial respiration; if breathing is difficult, give oxygen.

In case of contact with material, immediately flush skin or eyes with running water for at least 15 minutes.

Remove and isolate contaminated clothing and shoes at the site.

Keep victim quiet and maintain normal body temperature.

Effects may be delayed; keep victim under observation.

▼   ▼   ▼   ▼   ▼   ▼   ▼   ▼   ▼   ▼   ▼   ▼   ▼   ▼

--- IDENTIFIERS ---

Name: 1,3-BUTADIENE
Synonyms: Biethylene; Bivinyl; Butadieen (Dutch); Buta-1,2-Dieen (Dutch); Butadien (Polish); Buta-1,3-Dien (German); Butadiene; Buta-1,3-Diene; alpha-gamma-Butadiene; Divinyl; Erythrene; NCI-C50602; Pyrrolylene; Vinylethylene
CAS: 106-99-0; **RTECS:** EI9275000
Formula: C_4H_6; **Mol Wt:** 54.09
WLN: 1U2U1
Chemical Class: Olefin

See other identifiers listed below under Regulations.

--- PROPERTIES ---

Physical Description: Colorless liquified compressed gas with gasolinelike odor.
Boiling Point: 268.71 K; -4.5°C; 24°F
Melting Point: 164.27 K; -108.9°C; -164°F
Flash Point: 196.89 K; -76.3°C; -105.3°F
Autoignition: 693 K; 419.8°C; 787.7°F
Critical Temp: 425 K; 151.85°C; 305.33°F

Critical Press: 4.32 kN/M^2; 42.5 atm; 625 psia
Heat of Vap: 180 Btu/lb; 99.96 cal/g; 4.182x E5 J/kg
Heat of Comb: -19008 Btu/lb; -10567 cal/g; -442x E5 J/kg
Vapor Pressure: 1840 mm at 21°C
UEL: 11.5%
LEL: 2%
Ionization Potential (eV): 9.07
Vapor Density: 1.87 (air = 1)
Specific Gravity: 0.6212 0°C
Density: 0.621 at 20°C
Water Solubility: 0.05%
Incompatibilities: Strong oxidizers, copper, copper alloys

Reactivity with Water: No data on water reactivity
Reactivity with Common Materials: Powerful oxidizers (Source: SAX)
Stability during Transport: Explosive decomposition when contaminated with peroxides formed by reaction with air
Neutralizing Agents: No data
Polymerization Possibilities: Stable when inhibitors present

Toxic Fire Gases: Acrid fumes
Odor Detected at (ppm): 4 mg/m^3
Odor Description: Mildly aromatic (Source: CHRIS)
100% Odor Detection: 1.3 ppm

--- REGULATIONS ---

DOT hazard class: 2.1 FLAMMABLE GAS
DOT guide: 17
Identification number: UN1010
DOT shipping name: Butadienes, inhibited
Label(s) required: FLAMMABLE GAS
Packaging exceptions: 173.306
Nonbulk packaging: 173.304
Bulk packaging: 173.314, 315
Quantity limitations:
  Passenger air/rail: Forbidden
  Cargo aircraft only: 150 kg
  Vessel stowage: B
  Other stowage provisions: 40

STCC Number: 4905705, 4905704, 4905703

Clean Water Act Sect. 307: No
Clean Water Act Sect. 311: No
Clean Air Act: CAA '90 listed
EPA Waste Number: None
CERCLA Ref: Not listed
RQ Designation: Not listed
SARA TPQ Value: Not listed
SARA Sect. 312 Categories:
   Acute toxicity: adverse effect to target organs.
   Chronic toxicity: carcinogen.
   Chronic toxicity: mutagen.
   Chronic toxicity: reproductive toxin.
   Fire hazard: flammable.
   Sudden pressure: compressed gases.
   Reactive hazard: unstable/reactive.
Listed in SARA Sect 313: Yes
De Minimus Concentration: 0.1%

U.S. Postal Service Mailability:
   Hazard class: Not given
   Mailability: Nonmailable
   Max per parcel: 0

NFPA Codes:
   Health Hazard (blue): (2) Hazardous to health. Area may be entered with self-contained breathing apparatus.
   Flammability (red): (4) This material forms readily ignitable mixtures in air.
   Reactivity (yellow): (2) Normally unstable and readily undergoes violent change, but does not detonate.
   Special: Unspecified.

--- TOXICITY DATA ---

Short-term Toxicity: *Inhalation:* May cause irritation to the nose, throat, and lungs. Concentrations above 8000 ppm may cause narcotic effects, dizziness, headache, drowsiness, and loss of consciousness. *Skin:* The gas may cause irritation; the liquid may cause freezing and frostbite. *Eyes:* May cause irritation. *Ingestion:* The liquid may cause freezing and frostbite to the mouth and throat. (NYDH)

Long-term Toxicity: Exposures to levels below 5000 ppm over long periods has caused no ill effects. Butadiene has been shown to cause cancer and birth defects in laboratory animals. Whether it does so in humans is unknown. NIOSH recommends butadiene be considered a potential occupational carcinogen, teratogen, and a possible reproductive hazard. (NYDH)

Target Organs: Skin, mucous membrane, eyes

Symptoms: Slight anesthetic effect at high concentration; causes "frostbite" from skin contact; slight irritation to eyes and nose at high concentrations (Source: CHRIS).

Conc IDLH: 20,000 ppm

NIOSH Rel: Potential occupational carcinogen; Lowest Feasible (LOQ 0.19 ppm)

ACGIH TLV: TLV = 10 ppm. Suspected human carcinogen (A2)
ACGIH STEL: Not listed

OSHA PEL: Transitional limits: PEL = 1000 PPM (2200 mg/m^3); Final rule limits: TWA = 1000 ppm (2200 mg/m^3)

MAK Information: Carcinogenic working material without MAK. In the Commission's view, an animal carcinogen.

Carcinogen: Y; **Status:** See below

Carcinogen Lists: *IARC:* Carcinogen defined by IARC to be possibly carcinogenic to humans, but having (usually) no human evidence; *MAK:* An animal carcinogen; *NIOSH:* Carcinogen defined by NIOSH with no further categorization; *NTP:* Not listed; *ACGIH:* Carcinogen defined by ACGIH TLV Committee as a suspected carcinogen, based on either limited epidemiological evidence or demonstration of carcinogenicity in experimental animals; *OSHA:* Not listed.

Human Toxicity Data: (Source: NIOSH RTECS)
ihl-hmn TCLo:2000 ppm/7H JIHTAB 26, 69, 44
  Sense Organs
    Eye
      Other

Behavioral
Hallucinations, distorted perceptions

LD50 Value: orl-rat LD50:5480 mg/kg

Other Species Toxicity Data: (Source: NIOSH RTECS 1992)
orl-rat LD50:5480 mg/kg
ihl-rat LC50:285 g/m^3/4H
ihl-mus LC50:259 g/m^3
ihl-rbt LCLo:25 pph/23M

Reproductive Toxicity (1992 RTECS): This chemical is a mammalian reproductive toxin.

Reproductive Toxicity Data (1992 RTECS):
ihl-rat TCLo:8000 ppm/6H (6-15D preg) EPASR* 8 EHQ-0382-0441
   Specific Developmental Abnormalities
      Musculoskeletal system

ihl-mus TCLo:1000 ppm/6H (6-15D preg) EVHPAZ 86, 79, 90
   Effects on Fertility
      Postimplantation mortality
   Effects on Embryo or Fetus
      Extra embryonic features e.g., placenta, umbilical cord
      Fetotoxicity (except death, e.g., stunted fetus)

No Significant Risk Level (Ca P65): 0.4 mg/day

--- PROTECTION AND FIRST AID ---

Protection Suggested from the CHRIS Manual: Chemical-type safety goggles; rescue harness and safe line for those entering a tank or enclosed storage space; hose mask with hose inlet in vapor-free atmosphere; self-contained breathing apparatus; rubber suit.

NIOSH Pocket Guide to Chemical Hazards:
   **Wear Appropriate Equipment to Prevent: Prevent skin freezing.

   ** Wear Eye Protection to Prevent: Reasonable probability of eye contact.

   **Exposed Personnel Should Wash: Immediately when skin becomes wet.

**Remove Clothing: Immediately remove any clothing that becomes wet to avoid any flammability.

**Reference: NIOSH

Recommended Respiration Protection Source: NIOSH Pocket Guide (85-114); NIOSH (1,3-Butadiene) -
Greater at any detectable concentration: Any self-contained breathing apparatus with full facepiece and operated in a pressure-demand or other positive-pressure mode. Any supplied-air respirator with a full facepiece and operated in pressure-demand or other positive-pressure mode in combination with an auxiliary self-contained breathing apparatus operated in pressure-demand or other positive-pressure mode.
Escape: Any air-purifying full facepiece respirator (gas mask) with a chin-style or front- or back-mounted canister providing protection against the compound of concern. Any appropriate escape-type self-contained breathing apparatus.

First Aid Source: CHRIS Manual 1991.
Remove from exposure immediately. Call a physician.
Inhalation: If breathing is irregular or stopped, start resuscitation, administer oxygen.
Skin contact: Remove contaminated clothing and wash affected skin area.
Eye contact: Irrigate with water for 15 min.

First Aid Source: DOT Emergency Response Guide 1993.
Move victim to fresh air and call emergency medical care; if not breathing, give artificial respiration; if breathing is difficult, give oxygen. In case of frostbite, thaw frosted parts with water. Keep victim quiet and maintain normal body temperature.

--- INITIAL INCIDENT RESPONSE ---

Fire Extinguishment: Stop flow of gas (CHRIS 91).

U.S. Department of Transportation Guide to Hazardous Materials Transport Information - Publication DOT 5800.5 (1990).
DOT Shipping Name: Butadienes, inhibited
DOT ID Number: UN1010

* POTENTIAL HAZARDS *

*Fire or Explosion
Extremely flammable.
May be ignited by heat, sparks, and flames.
Vapors may travel to a source of ignition and flash back.
Container may explode violently in heat of fire.
Vapor explosion hazard indoors, outdoors, or in sewers.

*Health Hazards
May be poisonous if inhaled.
Contact may cause burns to skin and eyes.
Vapors may cause dizziness or suffocation.
Contact with liquid may cause frostbite.
Fire may produce irritating or poisonous gases.

* EMERGENCY ACTION *

Keep unnecessary people away; isolate hazard area and deny entry.
Stay upwind, out of low areas, and ventilate closed spaces before entering.
Positive-pressure self-contained breathing apparatus (SCBA) and structural
   firefighters' protective clothing will provide limited protection.
Isolate for 1/2 mile in all directions if tank, rail car, or tank truck is involved
   in fire.
CALL CHEMTREC AT 1-800-424-9300 AS SOON AS POSSIBLE,
especially if there is no local hazardous team available.

*Fire
Let tank, tank car, or tank truck burn unless leak can be stopped; with smaller
   tanks or cylinders, extinguish or isolate from other flammables.
Small fires: Dry chemical or CO_2
Large fires: Water spray, fog, or regular foam.
Move container from fire area if you can do it without risk.
For massive fire in cargo area, use unmanned hose holder or monitor nozzles;
   if this is impossible, withdraw from area and let fire burn.
Withdraw immediately in case of rising sound from venting safety device or
   any discoloration of tank due to fire.
Cool container with water using unmanned device until well after fire is out.

*Spill or Leak
Shut off ignition sources; no flares, smoking, or flames in hazard area.
Stop leak if you can do it without risk.
Water spray may reduce vapors; but it may not prevent ignition in closed
    spaces.
Isolate area until gas has dispersed.

*First Aid
Move victim to fresh air and call emergency medical care; if not breathing,
    give artificial respiration; if breathing is difficult, give oxygen.
In case of frostbite, thaw frosted parts with water.
Keep victim quiet and maintain normal body temperature.

▼   ▼   ▼   ▼   ▼   ▼   ▼   ▼   ▼   ▼   ▼   ▼   ▼   ▼

--- IDENTIFIERS ---

Name: CHLOROPRENE
Synonyms: 2-Chloor-1,3-Butadieen (Dutch); 2-Chlor-1,3-Butadien (German);
    Chlorobutadiene; 2-Chlorobuta-1,3-Diene; 2-Chloro-1,3-Butadiene;
    Chloropreen (Dutch); Chloropren (German, Polish); Chloroprene; beta-
    Chloroprene; 2-Cloro-1,3-Butadiene (Italian); Cloroprene (Italian); Neoprene
CAS: 126-99-8; **RTECS:** EI9625000
Formula: C_4H_5Cl; **Mol Wt:** 88.54
WLN: 1UYG1U1
Chemical Class: Halogenated h-carbon

See other identifiers listed below under Regulations.

--- PROPERTIES ---

Physical Description: Colorless liquid with an etherlike odor
Boiling Point: 332.04 K; 58.8°C; 138°F
Melting Point: 143.16 K; -130°C; -202°F
Flash Point: 253 K; -20.2°C; -4.3°F
Autoignition: NA
Vapor Pressure: 179 mm
UEL: 20%
LEL: 4%
Ionization Potential (eV): 8.83
Vapor Density: 3.0 (air = 1)
Specific Gravity: .958 at 29°C

Density: .958 g/cc or 8.9094 lb/gal
Water Solubility: INSOL
Incompatibilities: Peroxides, other oxidizers, fluorine

Reactivity with Water: No data on water reactivity
Reactivity with Common Materials: No data
Stability during Transport: No data
Neutralizing Agents: No data
Polymerization Possibilities: No data

Toxic Fire Gases: Hydrogen chloride upon decomp by heat
Odor Detected at (ppm): Unknown
Odor Description: No data
100% Odor Detection: No data

--- REGULATIONS ---

DOT hazard class: 3 FLAMMABLE LIQUID
DOT guide: 30
Identification number: UN1991
DOT shipping name: Chloropropene, inhibited
Packing group: I
Label(s) required: FLAMMABLE LIQUID, POISON
Special provisions: B57, T15
Packaging exceptions: 173.None
Nonbulk packaging: 173.201
Bulk packaging: 173.243
Quantity limitations:
  Passenger air/rail: Forbidden
  Cargo aircraft only: 30 L
  Vessel stowage: D
  Other stowage provisions: 40, M2

STCC Number: 4921414

Clean Water Act Sect. 307: No
Clean Water Act Sect. 311: No
Clean Air Act: CAA '90 listed
EPA Waste Number: None
CERCLA Ref: Not listed
RQ Designation: Not listed
SARA TPQ Value: Not listed

SARA Sect. 312 Categories:
  Acute toxicity: adverse effect to target organs.
  Chronic toxicity: mutagen.
  Chronic toxicity: reproductive toxin.
  Fire hazard: flammable.
Listed in SARA Sect. 313: Yes
De Minimus Concentration: 1.0%

U.S. Postal Service Mailability:
  Hazard class: Flammable liquid; mailable as ORM-D
  Mailability: Domestic surface mail only
  Max per parcel: 1 qt metal; 1 qt other

NFPA Codes:
  Health Hazard (blue): (2) Hazardous to health. Area may be entered with
  self-contained breathing apparatus.
  Flammability (red): (3) This material can be ignited under almost all
  temperature conditions.
  Reactivity (yellow): (0) Stable even under fire conditions.
  Special: Unspecified.

--- TOXICITY DATA ---

Short-term Toxicity: Unknown

Long-term Toxicity: Unknown

Target Organs: Respiratory system, skin, eyes, lungs

Symptoms: Irritated eyes and respiratory system, nervous, irritable; dermatitis,
  alopecia (Source: NIOSH)

Conc IDLH: 400 ppm

NIOSH REL: Potential occupational carcinogen at 1 ppm. Ceiling exposures
that shall at no time be exceeded, 3.6 mg/m^3.

ACGIH TLV: TLV = 10 ppm (45 mg/m^3), skin
ACGIH STEL: Skin

OSHA PEL: Transitional limits: PEL - 25 ppm (90 mg/m^3) (skin); Final rule limits: TWA = 10 ppm (35 mg/m^3) (skin).

MAK Information: 10 ppm; 36 mg/m^3; Substance with systemic effects, onset of effect less than or equal to 2 hr: Peak = 2 x MAK for 30 min, 4 times per shift of 8 hr; danger of cutaneous absorption.

Carcinogen: N; **Status:** See below
References: Animal indefinite IARC** 19, 131, 79; Carcinogenic determination

Carcinogen Lists: *IARC*: Not classified as to human carcinogenicity or probably not carcinogenic to humans; *MAK*: Not listed; *NIOSH*: Carcinogen defined by NIOSH with no further categorization; *NTP*: Not listed; *ACGIH*: Not listed; *OSHA*: Not listed.

LD50 Value: orl-rat LD50:450 mg/kg

Other Species Toxicity Data: (Source: NIOSH RTECS 1992)
orl-rat LD50:450 mg/kg
ihl-rat LC50:11800 mg/m^3/4H
scu-rat LDLo:500 mg/kg
orl-mus LD50:146 mg/kg
ihl-mus LC50:2300 mg/m^3
scu-mus LDLo:1 g/kg
ihl-cat LCLo:1290 mg/m^3/8H
scu-cat LDLo:100 mg/kg
ihl-rbt LCLo:3870 mg/m^3/8H
ivn-rbt LDLo:96 mg/kg
scu-pgn LDLo:13 g/kg

Reproductive Toxicity (1992 RTECS): This chemical is a mammalian reproductive toxin.

Reproductive Toxicity Data (1992 RTECS):
orl-rat TDLo:1 mg/kg (3-4D preg) GTPZAB 19(3), 30, 75
  Effects on Embryo or Fetus
    Fetal death

orl-rat TDLo:1 mg/kg (11-12D preg) GTPZAB 19(3), 30, 75
  Specific Developmental Abnormalities
    Central nervous system

orl-rat TDLo:1 mg/kg (11-12D preg) GTPZAB 19(7), 30, 75
   Effects on Embryo or Fetus
      Fetotoxicity (except death, e.g., stunted fetus)

orl-rat TDLo:1 mg/kg (9-10D preg) GTPZAB 19(7), 30, 75
   Specific Developmental Abnormalities
      Other developmental abnormalities

ihl-rat TCLo:4 mg/m^3/24H (3-4D preg) GTPZAB 19(3), 30, 75
   Effects on Embryo or Fetus
      Fetal death

ihl-rat TCLo:4 mg/m^3/24H (11-12D preg) GTPZAB 19(3), 30, 75
   Specific Developmental Abnormalities
      Central nervous system

ihl-rat TCLo:10 ppm/4H (3-20D preg) TXAPA9 44, 81, 78
   Effects on Fertility
      Postimplantation mortality

ihl-rat TCLo:150 mg/m^3/24H (19W male) EVHPAZ 17, 85, 76
   Paternal Effects
      Spermatogenesis

ihl-rat TCLo:500 mg/m^3/5H (17W pre) BEBMAE 60(7), 107, 65
   Maternal Effects
      Menstrual cycle changes or disorders

--- PROTECTION AND FIRST AID ---

NIOSH Pocket Guide to Chemical Hazards:
   **Wear Appropriate Equipment to Prevent: Any possibility of skin contact.

   **Wear Eye Protection to Prevent: Reasonable probability of eye contact.

   **Exposed Personnel Should Wash: Promptly when skin becomes contaminated.

   **Remove Clothing: Immediately remove any clothing that becomes wet to avoid any flammability.

**Reference: NIOSH

First Aid Source: CSDS.
Inhalation: Leave contaminated area immediately; if difficult breathing or any other symptoms develop, seek medical attention at once, even if symptoms develop many hours after exposure.
Ingestion: If convulsions are not present, give a glass or two of water or milk to dilute the substance. Assure that the person's airway is unobstructed; contact a hospital or poison center immediately for advice on whether or not to induce vomiting.
Skin: Flood contacted areas with water. Remove contaminated clothing under skin. Use soap. Isolate contaminated clothing when removed to prevent contact by others.
Eyes: Remove contacts. Flush well with water or normal saline for 20 to 30 minutes. Seek medical attention.

First Aid Source: DOT Emergency Response Guide 1993.
Move victim to fresh air and call emergency medical care; if not breathing, give artificial respiration; if breathing is difficult, give oxygen. In case of contact with material, immediately flush skin or eyes with running water for at least 15 min. Remove and isolate contaminated clothing and shoes at the site. Keep victim quiet and maintain normal body temperature. Effects may be delayed; keep victim under observation.

--- INITIAL INCIDENT RESPONSE ---

Fire Extinguishment: Alcohol foam recommended for fires. Water may be ineffective but should be used to keep fire-exposed containers cool (Note: CHRIS 91).

U.S. Department of Transportation Guide to Hazardous Materials Transport Information - Publication DOT 5800.5 (1990).
DOT Shipping Name: Chloropropene, inhibited
DOT ID Number: UN1991

ERG90   GUIDE 30
* POTENTIAL HAZARDS *

*Health Hazards
Poisonous; may be fatal if inhaled, swallowed, or absorbed through skin. Contact may cause burns to skin and eyes.

*Fire or Explosion

Extremely flammable; may be ignited by heat, sparks, and flames.

Vapors may travel to a source of ignition and flash back.

Container may explode in heat of fire.

Vapor explosion and poison hazard indoors, outdoors, or in sewers.

Runoff to sewer may create fire or explosion hazard.

* EMERGENCY ACTION *

Keep unnecessary people away; isolate hazard area and deny entry.

Stay upwind; keep out of low areas.

Positive-pressure self-contained breathing apparatus (SCBA) and chemical protective clothing that is specifically recommended by the shipper or manufacturer may be worn. It may provide little or no thermal protection.

Structural firefighters' protective clothing is not effective for these materials.

Isolate the leak or spill area immediately for at least 150 ft in all directions.

See the Table of Initial Isolation and Protective Action Distances. If you find the ID number and the name of the material there, begin protective action.

Isolate for 1/2 mile in all directions if tank, rail car, or tank truck is involved in fire.

CALL CHEMTREC AT 1-800-424-9300 FOR EMERGENCY ASSISTANCE.

*Fire

Small fires: Dry chemical, CO_2, water spray, or regular foam.

Large fires: Water spray, fog or regular foam.

Do not get water inside container.

Apply cooling water to sides of containers that are exposed to flames until well after fire is out. Stay away from ends of tanks.

For massive fire in cargo area, use unmanned hose holder or monitor nozzles; if this is impossible, withdraw from area and let fire burn.

Withdraw immediately in case of rising sound from venting safety device or any discoloration of tank due to fire.

*Spill or Leak

Shut off ignition sources; no flares, smoking, or flames in hazard area.

Fully encapsulating, vapor-protective clothing should be worn for spills and leaks with no fire.

Do not touch or walk through spilled material.

Small spills: Flush area with flooding amounts of water.

Large spills: Dike far ahead of liquid spill for later disposal.

*First Aid
   Move victim to fresh air and call emergency medical care; if not breathing,
      give artificial respiration; if breathing is difficult, give oxygen.
   In case of contact with material, immediately flush skin or eyes with running
      water for at least 15 min.
   Remove and isolate contaminated clothing and shoes at the site.
   Keep victim quiet and maintain normal body temperature.
   Effects may be delayed; keep victim under observation.

▼   ▼   ▼   ▼   ▼   ▼   ▼   ▼   ▼   ▼   ▼   ▼   ▼

--- IDENTIFIERS ---

Name: ETHYLIDENE NORBORNENE
Synonyms: 5-Ethylidenebicyclo (2.2.1)Hept-2-Ene; 5-Ethylidene-2-Norbornene;
   5-Ethylidenebichcyl (2.2.1)Hept-2-Ene; 2-Norbornene, 5-Ethylidene-
CAS: 16219-75-3; **RTECS:** RB9450000
Formula: C_9H_{12}; **Mol Wt:** 120.21
WLN: L55 A CY FUTJ CU2

See other identifiers listed below under Regulations.

--- PROPERTIES ---

Physical Description: White liquid with a turpentinelike odor
Boiling Point: 420.75 K; 147.6°C; 297.6°F
Melting Point: 353.15 K; 80°C; 176°F
Flash Point: 309.81 K; 36.6°C; 97.9°F
Autoignition: NA
UEL: NA
LEL: NA
Vapor Density: 4.1 (air = 1)
Specific Gravity: 0.896 at 20°C
Density: 0.893
Incompatibilities: Reacts with oxygen

Reactivity with Water: N/R; floats on water
Reactivity with Common Materials: No data
Stability during Transport: No data
Neutralizing Agents: No data
Polymerization Possibilities: No data

BUCKMAN LABORATORIES, INTL.
TECHNICAL INFORMATION CENTER

Toxic Fire Gases: Acrid smoke and fumes
Odor Detected at (ppm): 0.007 ppm
Odor Description: Like turpentine (Source: CHRIS)
100% Odor Detection: No data

--- REGULATIONS ---

DOT hazard class: 3 FLAMMABLE LIQUID
DOT guide: 27
Identification number: UN1993
DOT shipping name: FLAMMABLE LIQUID, N.O.S.
Packing group: III
Label(s) required: FLAMMABLE LIQUID
Special Provisions: B1, B52, T7, T30
Packaging exceptions: 173.150
Nonbulk packaging: 173.203
Bulk packaging: 173.242
Quantity limitations:
  Passenger air/rail: 60 L
  Cargo aircraft only: 220L
  Vessel stowage: A

STCC Number: Not listed

Clean Water Act Sect. 307: No
Clean Water Act Sect. 311: No
Clean Air Act: Not listed
EPA Waste Number: None
CERCLA Ref: Not listed
RQ Designation: Not listed
SARA TPQ Value: Not listed
SARA Sect. 312 Categories:
  Acute toxicity: irritant.
  Acute toxicity: adverse effect to target organs.

U.S. Postal Service Mailability: Not given

NFPA Codes:
  Health Hazard (blue): Unspecified
  Flammability (red): Unspecified
  Reactivity (yellow): Unspecified
  Special: Unspecifed

--- TOXICITY DATA ---

Short-term Toxicity: Unknown

Long-term Toxicity: Causes kidney lesions and gain in kidney and liver weights in rats (Source: HCDB).

Target Organs: Respiratory system, digestive system, eyes, skin

Symptoms: Inhalation of vapors causes headache, confusion, and respiratory distress. Ingestion causes irritation of entire digestive system. Aspiration causes severe pneumonia. Contact with liquid causes irritation of eyes and skin (Source: CHRIS).

Conc IDLH: Unknown

ACGIH TLV: TLV = 5 ppm (25 mg/m^3) C
ACGIH STEL: Not listed

OSHA PEL: Final Rule Limits: Ceiling = 5 ppm (25 mg/m^3); enforcement of limit indefinitely stayed.

MAK Information: Not listed

Carcinogen: N; **Status:** See below

Carcinogen Lists: *IARC*: Not listed; *MAK*: Not listed; *NIOSH*: Not listed; *NTP*: Not listed; *ACGIH*: Not listed; *OSHA*: Not listed.

Human Toxicity Data: (Source: NIOSH RTECS)
ihl-hmn TCLo:6 ppm/30M TXAPA9 20, 250, 71
   Sense Organs
     Nose
      Other
     Eye
      Conjunctive irritation
     Taste
      Change in function

LD50 Value: orl-rat LD50:2527 mg/kg

Other Species Toxicity Data: (Source: NIOSH RTECS 1992)
orl-rat LD50:2527 mg/kg
ihl-rat LC50:1246 ppm/4H
orl-mus LD50:3250 mg/kg
ihl-mus LC50:732 ppm/4H
ihl-rbt LC50:3104 ppm/4H
skn-rbt LD50:8189 mg/kg
ihl-gpg LC50:2896 ppm/4H

Irritation Data: (Source: NIOSH RTECS 1992)
skn-rbt 445 mg open MLD

Reproductive Toxicity (1992 RTECS): This chemical has no known mammalian reproductive toxicity.

--- PROTECTION AND FIRST AID ---

Protection Suggested from the CHRIS Manual: Organic canister or air-supplied mask; goggles or face shield; rubber gloves.

First Aid Source: CHRIS Manual 1991.
Inhalation: Remove victim to fresh air; administer artificial respiration and oxygen if required; call a doctor.
Ingestion: Give large amount of water and induce vomiting; get medical attention at once.
Skin: Wipe off, wash with soap and water.
Eyes: Flush with water for at least 15 min.

First Aid Source: DOT Emergency Response Guide 1993.
Move victim to fresh air and call emergency medical care; if not breathing, give artificial respiration; if breathing is difficult, give oxygen. In case of contact with material, immediately flush eyes with running water for at least 15 min. Wash skin with soap and water. Remove and isolate contaminated clothing and shoes at the site.

--- INITIAL INCIDENT RESPONSE ---

Fire Extinguishment: Dry chemical, foam, carbon dioxide. *Note*: Water may be ineffective (CHRIS 91).

U.S. Department of Transportation Guide to Hazardous Materials Transport Information - Publication DOT 5800.5 (1990).
DOT Shipping Name: FLAMMABLE LIQUID, N.O.S.
DOT ID Number: UN1993

ERG90   GUIDE 27
* POTENTIAL HAZARDS *

*Health Hazards
   May be poisonous if inhaled or absorbed through skin.
   Vapors may cause dizziness or suffocation.
   Contact may irritate or burn skin and eyes.
   Fire may produce irritating or poisonous gases.
   Runoff from fire control or dilution water may cause pollution.

*Fire or Explosion
   Flammable/combustible material; may be ignited by heat, sparks, or flames.
   Vapors may travel to a source of ignition and flash back.
   Container may explode in heat of fire.
   Vapor explosion hazard indoors, outdoors, or in sewers.
   Runoff to sewer may create fire or explosion hazard.

* EMERGENCY ACTION *

Keep unnecessary people away; isolate hazard area and deny entry.
Stay upwind; keep out of low areas.
Positive-pressure self-contained breathing apparatus (SCBA) and structural
   firefighters' protective clothing will provide limited protection.
Isolate for 1/2 mile in all directions if tank, rail car, or tank truck is involved
   in fire.
**CALL CHEMTREC AT 1-800-424-9300 FOR EMERGENCY ASSIS-
TANCE.** If water pollution occurs, notify the appropriate authorities.

*Fire
   Small fires: Dry chemical, CO_2, water spray, or regular foam.
   Large fires: Water spray, fog or regular foam.
   Move container from fire area if you can do it without risk.
   Apply cooling water to sides of containers that are exposed to flames until
      well after fire is out. Stay away from ends of tanks.
   For massive fire in cargo area, use unmanned hose holder or monitor nozzles;
      if this is impossible, withdraw from area and let fire burn.

Withdraw immediately in case of rising sound from venting safety device or any discoloration of tank due to fire.

*Spill or Leak

Shut off ignition sources; no flares, smoking, or flames in hazard area.

Stop leak if you can do it without risk.

Water spray may reduce vapor; but it may not prevent ignition in closed spaces.

Small spills: Take up with sand or other noncombustible absorbent material and place into containers for later disposal.

Large spills: Dike far ahead of liquid spill for later disposal.

*First Aid

Move victim to fresh air and call emergency medical care; if not breathing, give artificial respiration; if breathing is difficult, give oxygen.

In case of contact with material, immediately flush skin or eyes with running water for at least 15 min. Wash skin with soap and water.

Remove and isolate contaminated clothing and shoes at the site.

▼   ▼   ▼   ▼   ▼   ▼   ▼   ▼   ▼   ▼   ▼   ▼   ▼   ▼

--- IDENTIFIERS ---

Name: ISOPRENE

Synonyms: 1,3-Butadiene, 2-Methyl-; Isoprene (DOT); beta-Methylbivinyl; 2-Methylbutadiene; 2-Methyl-1,3-Butadiene

CAS: 78-79-5; **RTECS:** NT4037000

Formula: C_5H_8; **Mol Wt:** 68.11

WLN: 1UY1

Chemical Class: Olefin

See other identifiers listed below under Regulations.

--- PROPERTIES ---

Physical Description: Colorless water liquid with a mild odor

Boiling Point: 307 K; 33.8°C; 92.9°F

Melting Point: 153 K; -120.2°C; -184.3°F

Flash Point: 220 K; -53.2°C; -63.7°F

Autoignition: 493 K; 219.8°C; 427.7°F

Critical Temp: 484.3 K; 211.15°C; 412.07°F

Critical Press: 3.79 kN/m^2; 37.3 atm; 549 psia

Heat of Vap: 150 Btu/lb; 83.3 cal/g; 3.485x E5 J/kg
Heat of Comb: -18,848 Btu/lb; -10,479 cal/g; -438x E5 J/kg
UEL: NA
LEL: NA
Ionization Potential (eV): 5.7
Vapor Density: 2.3 (air = 1)
Specific Gravity: 0.681 at 20°C
Density: 0.681

Reactivity with Water: No data on water reactivity
Reactivity with Common Materials: No data
Stability during Transport: No data
Neutralizing Agents: No data
Polymerization Possibilities: Accelerated by heat and oxygen, rusty iron. Iron should be treated with reducing agent, such as sodium nitrite, before placing in isoprene service.

Toxic Fire Gases: None reported other than possible unburned vapors
Odor Detected at (ppm): 0.005 ppm
Odor Description: Mild, aromatic (Source: CHRIS)
100% Odor Detection: No data

--- REGULATIONS ---

DOT hazard class: 3 FLAMMABLE LIQUID
DOT guide: 27
Identification number: UN1218
DOT shipping name: Isoprene, inhibited
Packing group: I
Label(s) required: FLAMMABLE LIQUID
Special Provisions: T20
Packaging exceptions: 173.150
Nonbulk packaging: 173.201
Bulk packaging: 173.243
Quantity limitations:
  Passenger air/rail: 1 L
  Cargo aircraft only: 30 L
  Vessel stowage: E

STCC Number: 4907230

Clean Water Act Sect. 307: No
Clean Water Act Sect. 311: Yes
Clean Air Act: Not listed
EPA Waste Number: None
CERCLA Ref: Y
RQ Designation: B; 100 lb (45.4 kg) CERCLA
SARA TPQ Value: Not listed
SARA Sect. 312 Categories:
  Acute toxicity: adverse effect to target organs.
  Chronic toxicity: mutagen.
  Fire hazard: flammable.
  Reactive hazard: unstable/reactive.

U.S. Postal Service Mailability:
  Hazard class: Not given
  Mailability: Nonmailable
  Max per parcel: 0

NFPA Codes:
  Health Hazard (blue): (2) Hazardous to health. Area may be entered with self-contained breathing apparatus.
  Flammability (red): (4) This material forms readily ignitable mixtures in air.
  Reactivity (yellow): (2) Normally unstable and readily undergoes violent change, but does not detonate.
  Special: Unspecified.

--- TOXICITY DATA ---

Short-term Toxicity: Unknown

Long-term Toxicity: Unknown

Target Organs: Irritator of mucous membranes of eyes, nose, and upper respiratory tract.

Symptoms: Vapor produces no effects other than slight irritation of the eyes and upper respiratory tract. Liquid may irritate eyes, like gasoline (Source: CHRIS).

Conc IDLH: Unknown

NIOSH REL: Not given

ACGIH TLV: Not listed
ACGIH STEL: Not listed
WEEL (ACGIH): 50 ppm (8-hr TWA)

OSHA PEL: Not in Table Z-1-A

MAK Information: Not listed

Carcinogen: N; **Status:** See below

Carcinogen Lists: *IARC*: Not listed; *MAK*: Not listed; *NIOSH*: Not listed; *NTP*: Not Listed; *ACGIH*: Not listed; *OSHA*: Not listed.

LD50 Value: No LD50 in RTECS 1992

Other Species Toxicity Data: (Source: NIOSH RTECS 1992)
ihl-rat LC50:180 $g/m^3/4H$
ihl-mus LC50:139 $g/m^3/2H$

Reproductive Toxicity (1992 RTECS): This chemical has no known mammalian reproductive toxicity.

--- PROTECTION AND FIRST AID ---

Protection Suggested from the CHRIS Manual: Vapor-proof goggles; self-contained breathing apparatus; leather or rubber shoes; rubber gloves.

First Aid Source: CHRIS Manual 1991.
Inhalation: Remove victim promptly from irritating or asphyxiating atmosphere; if symptoms of asphyxiation persist, administer artificial respiration and oxygen; treat symptomatically thereafter; call a physician.
Eyes: Flush with water for at least 15 min.

First Aid Source: DOT Emergency Response Guide 1993.
Move victim to fresh air and call emergency medical care; if not breathing, give artificial respiration; if breathing is difficult, give oxygen.
In case of contact with material, immediately flush skin or eyes with running water for at least 15 min. Wash skin with soap and water.
Remove and isolate contaminated clothing and shoes at the site.

--- INITIAL INCIDENT RESPONSE ---

Fire Extinguishment: Dry chemical, foam, or carbon dioxide. *Note:* Water
may be ineffective (CHRIS 91).

U.S. Department of Transportation Guide to Hazardous Materials Transport
Information - Publication DOT 5800.5 (1990).
DOT Shipping Name: Isoprene, inhibited
DOT ID Number: UN1218

ERG90   GUIDE 27
* POTENTIAL HAZARDS *

*Health Hazards
   May be poisonous if inhaled or absorbed through skin.
   Vapors may cause dizziness or suffocation.
   Contact may irritate or burn skin and eyes.
   Fire may produce irritating or poisonous gases.
   Runoff from fire control or dilution water may cause pollution.

*Fire or Explosion
   Flammable/combustible material; may be ignited by heat, sparks, or flames.
   Vapors may travel to a source of ignition and flash back.
   Container may explode in heat of fire.
   Vapor explosion hazard indoors, outdoors, or in sewers.
   Runoff to sewer may create fire or explosion hazard.

* EMERGENCY ACTION *

Keep unnecessary people away; isolate hazard area and deny entry.
Stay upwind; keep out of low areas.
Positive-pressure self-contained breathing apparatus (SCBA) and structural
   firefighters' protective clothing will provide limited protection.
Isolate for 1/2 mile in all directions if tank, rail car, or tank truck is involved
   in fire.
**CALL CHEMTREC AT 1-800-424-9300 FOR EMERGENCY ASSIS-
TANCE.** If water pollution occurs, notify the appropriate authorities.

*Fire
   Small fires: Dry chemical, CO_2, water spray, or regular foam.
   Large fires: Water spray, fog, or regular foam.
   Move container from fire area if you can do it without risk.

Apply cooling water to sides of containers that are exposed to flames until well after fire is out. Stay away from ends of tanks.

For massive fire in cargo area, use unmanned hose holder or monitor nozzles; if this is impossible, withdraw from area and let fire burn.

Withdraw immediately in case of rising sound from venting safety device or any discoloration of tank due to fire.

*Spill or Leak

Shut off ignition sources; no flares, smoking, or flames in hazard area.

Stop leak if you can do it without risk.

Water spray may reduce vapor; but it may not prevent ignition in closed spaces.

Small spills: Take up with sand or other noncombustible absorbent material and place into containers for later disposal.

Large spills: Dike far ahead of liquid spill for later disposal.

*First Aid

Move victim to fresh air and call emergency medical care; if not breathing, give artificial respiration; if breathing is difficult, give oxygen.

In case of contact with material, immediately flush skin or eyes with running water for at least 15 min. Wash skin with soap and water.

Remove and isolate contaminated clothing and shoes at the site.

▼   ▼   ▼   ▼   ▼   ▼   ▼   ▼   ▼   ▼   ▼   ▼   ▼

--- IDENTIFIERS ---

Name: VINYL CHLORIDE

Synonyms: Chloroethylene; Vinyl Chloroide; Chloroethen; Chlorure De Vinyle (French); Chloro Di Vinyle (Italian); Ethylene Monochloride; Monochloroethene; Monochloroethylene (DOT); Vinyl Chloride Monomer; Vinyl C Monomer; Winylu Chlored (Polish); VCM; VCL

CAS: 75-01-4; **RTECS:** KU9625000

Formula: C_2H_3Cl; **Mol Wt:** 62.50

WLN: G1U1

Chemical Class: Vinyl halide

See other identifiers listed below under Regulations.

--- PROPERTIES ---

Physical Description: Colorless liquified compressed gas with a sweet odor

Boiling Point: 259.4 K; -13.8°C; 7.2°F
Melting Point: 119.4 K; -153.8°C; -244.8°F
Flash Point: 194 K; -179.2°C; -110.5°F
Autoignition: 745 K; 471.8°C; 881.3°F
Critical Temp: 431.6 K; 158.45°C; 317.21°F
Critical Press: 5.34 kN/m^2; 52.6 atm; 773 psia
Heat of Vap: 160 Btu/lb; 88.85 cal/g; 3.718x E5 J/kg
Heat of Comb: -8136 Btu/lb; -4523 cal/g; -189x E5 J/kg
Vapor Pressure: 2600 mm at 25°C
UEL: 33%
LEL: 3.6%
Ionization Potential (eV): 7.57
Vapor Density: 2.2 (air = 1) (air = 1)
Specific Gravity: 0.969 at -13°C
Density: 0.969 g/cc or 9.0117 lb/gal
Water Solubility: INSOL

Reactivity with Water: No data on water reactivity
Reactivity with Common Materials: No data
Stability during Transport: No data
Neutralizing Agents: No data
Polymerization Possibilities: Polymerizes in presence of air, sunlight, or heat
unless stabilized by inhibitors.

Toxic Fire Gases: HCL and unburned toxic vapors
Odor Detected at (ppm): 260 ppm
Odor Description: Pleasant, sweet (Source: CHRIS)
100% Odor Detection: No data

--- REGULATIONS ---

DOT hazard class: 2.1 FLAMMABLE GAS
DOT guide: 17
Identification number: UN1086
DOT shipping name: Vinyl chloride, inhibited
Label(s) required: FLAMMABLE GAS
Special Provisions: B44
Packaging exceptions: 173.306
Nonbulk packaging: 173.304
Bulk packaging: 173.314, 315
Quantity limitations:
Passenger air/rail: Forbidden

Cargo aircraft only: 150 kg
  Vessel stowage: B
  Other stowage provisions: 40

STCC Number: 4905792

Clean Water Act Sect. 307: Yes
Clean Water Act Sect. 311: No
  National Primary Drinking Water Regulations
    Maximum contaminant levels (MCL): 0.002 mg/L$>>$ (01/09/89)
    Maximum contaminant level goals (MCLG): 0 mg/L$>>$ (01/09/89)
Clean Air Act: CAA '90 listed
EPA Waste Number: U043, D043
CERCLA Ref: Not listed
RQ Designation: X; 1 lb (0.454 kg) CERCLA
SARA TPQ Value: Not listed
SARA Sect. 312 Categories:
  Acute toxicity: adverse effect to target organs.
  Chronic toxicity: carcinogen.
  Chronic toxicity: adverse effect to target organs after long periods of exposure.
  Chronic toxicity: mutagen.
  Chronic toxicity: reproductive toxin.
  Fire hazard: flammable.
  Sudden pressure: compressed gases.
  Reactive hazard: unstable/reactive.
Listed in SARA Sect. 313: Yes
De Minimus Concentration: 0.1%

U.S. Postal Service Mailability:
  Hazard class: Not given
  Mailability: Nonmailable
  Max per parcel: 0

NFPA Codes:
  Health Hazard (blue): (2) Hazardous to health. Area may be entered with self-contained breathing apparatus.
  Flammability (red): (4) This material forms readily ignitable mixtures in air.
  Reactivity (yellow): (1) Normally stable, but may become unstable at elevated temperature and pressures.
  Special: Unspecified.

--- TOXICITY DATA ---

Short-term Toxicity: *Inhalation:* Exposure at 8000 ppm for 5 min can cause a feeling of intoxication, tiredness, drowsiness, abdominal pain, numbness, and tingling in fingers and toes, pains in joints, coughing, sneezing, irritability, and loss of appetite and weight. *Skin:* Contact with liquid may cause frostbite; contact with vapor may cause irritation and rash. Absorption is possible through the skin. *Eyes:* Can cause severe and immediate irritation. *Ingestion:* None found. (NYDH)

Long-term Toxicity: May cause clublike swelling and shortening of fingertips. Skin may become thickened and stiff with coarse, whitish patches. Bones and joints of arms and legs may suffer damage. Liver and spleen damage may occur. Not all symptoms disappear after exposure stops. Vinyl chloride has caused liver cancer in occupationally exposed individuals. (NYDH)

Target Organs: Skin, eyes, mucous membranes, nervous system, liver, kidneys.

Symptoms: *Inhalation:* High concentrations cause dizziness, anesthesia, lung irritation. *Skin:* May cause frostbite; phenol inhibitor may be absorbed through skin if large amounts of liquid evaporate (Source: CHRIS).

Conc IDLH: Unknown

NIOSH REL: Potential occupational carcinogen (use 1910.1017)

ACGIH TLV: TLV = 5 ppm; confirmed human carcinogen (A1)
ACGIH STEL: Not listed

OSHA PEL: Final Rule Limits: TWA = 1 ppm; Ceiling = 5 ppm; Consult 29 CFR 1910.1017.

MAK Information: Carcinogenic working material without MAK; Capable of inducing malignant tumors as shown by experience with humans.

Carcinogen: Y; **Status:** See below
References: Human positive IARC** 19, 377, 79; Animal positive IARC** 7, 291, 74; Human suspected IARC** 7, 291, 74; Animal positive IARC** 19, 377, 79; Human positive IARC** 28, 151, 82

Carcinogen Lists: *IARC*: Carcinogen as defined by IARC as carcinogenic to humans, with sufficient epidemiological evidence; *MAK*: Capable of inducing malignant tumors as shown by experience in humans; *NIOSH*: Carcinogen defined by NIOSH with no further categorization; *NTP*: Carcinogen defined by NTP as known to be carcinogenic, with evidence from human studies; *ACGIH*: Carcinogen defined by ACGIH TLV Committee as a confirmed human carcinogen, recognized to have carcinogenic or cocarcinogenic potential; *OSHA*: Cancer suspect.

LD50 Value: orl-rat LD50:500 mg/kg

Other Species Toxicity Data: (Source: NIOSH RTECS 1992)
orl-rat LD50:500 mg/kg
ihl-rat LC50:18 pph/15M
ihl-mam LCLo:200 ppm/18M

Reproductive Toxicity (1992 RTECS): This chemical is a mammalian reproductive toxin.

Reproductive Toxicity Data (1992 RTECS):
ihl-man TCLo:30 mg/m^3 (5Y male) GTPZAB 24(5), 28, 80
    Paternal Effects
        Spermatogenesis

ihl-rat TCLo:100 ppm/6H (26W male) EESADV 10, 281, 85
    Paternal Effects
        Testes, epididymis, sperm duct

ihl-rat TCLo:500 ppm/7H (6-15D preg) TXAPA9 33, 134, 75
    Effects on Embryo or Fetus
        Fetotoxicity (except death, e.g., stunted fetus)

ihl-rat TCLo:1500 ppm/24H (1-9D preg) TXCYAC 11, 45, 78
    Effects on Fertility
        Postimplantation mortality

ihl-rat TCLo:500 ppm/7H (6-15D preg) EVHPAZ 41, 171, 81
    Effects on Embryo or Fetus
        Fetotoxicity (except death, e.g., stunted fetus)
    Specific Developmental Abnormalities
        Musculoskeletal system

ihl-rat TCLo:250 ppm/6H (55D preg) JTEHD6 3, 965, 77
  Effects on Fertility
    Female fertility index

ihl-mus TCLo:30000 ppm/6H (5D male) EVHPAZ 21, 71, 77
  Effects on Fertility
    Preimplantation mortality

ihl-mus TCLo:500 ppm/7H (6-15D preg) EVHPAZ 41, 171, 81
  Effects on Embryo or Fetus
    Fetotoxicity (except death, e.g., stunted fetus)
  Specific Developmental Abnormalities
    Musculoskeletal system

No Significant Risk Level (Ca P65): 0.3 mg/day

--- PROTECTION AND FIRST AID ---

Recommended Respiration Protection Source: NIOSH Pocket Guide (85-114)
OSHA (Vinyl Chloride) -
Unknown, or above 3600 ppm: Open-circuit, self-contained breathing apparatus, pressure-demand type, with full facepiece.
Not over 3600 ppm: Combination type C supplied-air respirator, pressure-demand type, with full or half facepiece, and auxiliary self-contained air supply; or
Not over 1000 ppm: Combination type, supplied-air respirator, continuous-flow type, with full or half facepiece, and auxiliary self-contained air supply. Type C, supplied-air respirator, continuous-flow type, with full or half facepiece, helmet or hood.
Not over 100 ppm: (A) Combination type C supplied-air respirator, demand type; with full facepiece and auxiliary self-contained air supply; or (B) open-circuit self-contained breathing apparatus with full facepiece, in demand mode; or (C) type C supplied air respirator, demand type, with full facepiece.
Not over 25 ppm: (A) A powered air-purifying respirator with hood, helmet, full or half facepiece, and a canister that provides a service life of at least 4 hr for concentrations of vinyl chloride up to 25 ppm, or (B) gas mask, front- or back-mounted canister that provides a service life of at least 4 hr for concentration of vinyl chloride up to 25 ppm.
Not over 10 ppm: (A) Combination type C supplied-air respirator, demand type, with half facepiece, and auxiliary self-contained air supply; or (B) type C supplied-air respirator, demand type, with half facepiece; or (C) any

chemical cartridge respirator with an organic vapor cartridge that provides a service life of at least 1 hr for concentrations of vinyl chloride up to 10 ppm.

First Aid Source: CT HCDB.
Inhalation: Move to fresh air, keep quiet and warm, call doctor, artificial respirator.
Ingestion: None given
Skin: None given
Eyes: None given

First Aid Source: CHRIS Manual 1991.
Inhalation: Remove patient to fresh air and keep him quiet and warm; call a doctor; give artificial respiration if breathing stops.
Eyes and Skin: Flush with plenty of water for at least 15 min; for eyes, get medical attention; remove contaminated clothing.

First Aid Source: DOT Emergency Response Guide 1993.
Move victim to fresh air and call emergency medical care; if not breathing, give artificial respiration; if breathing is difficult, give oxygen. In case of frostbite, thaw frosted parts with water. Keep victim quiet and maintain normal body temperature.

--- INITIAL INCIDENT RESPONSE ---

Fire Extinguishment: For small fires use dry chemical or carbon dioxide. For large fires stop flow of gas. Cool exposed containers with water (CHRIS 91).

U.S. Department of Transportation Guide to Hazardous Materials Transport Information - Publication DOT 5800.5 (1990).
DOT Shipping Name: Vinyl chloride, inhibited
DOT ID Number: UN1086

ERG90   GUIDE 17
* POTENTIAL HAZARDS *

*Health Hazards
May be poisonous if inhaled.
Contact may cause burns to skin and eyes.
Vapors may cause dizziness or suffocation.
Contact with liquid may cause frostbite.
Fire may produce irritating or poisonous gases.

*Fire or Explosion
  Extremely flammable.
  May be ignited by heat, sparks, and flames.
  Vapors may travel to a source of ignition and flash back.
  Container may explode in heat of fire.
  Vapor explosion hazard indoors, outdoors, or in sewers.

* EMERGENCY ACTION *

Keep unnecessary people away; isolate hazard area and deny entry.
Stay upwind, out of low areas, and ventilate closed spaces before entering.
Positive-pressure self-contained breathing apparatus (SCBA) and structural
  firefighters' protective clothing will provide limited protection.
Isolate for 1/2 mile in all directions if tank, rail car, or tank truck is involved
  in fire.
CALL CHEMTREC AT 1-800-424-9300 AS SOON AS POSSIBLE,
especially if there is no local hazardous team available.

*Fire
  Let tank, tank car, or tank truck burn unless leak can be stopped; with smaller
    tanks or cylinders, extinguish/isolate from other flammables.
  Small fires: Dry chemical or CO_2.
  Large fires: Water spray, fog, or regular foam.
  Move container from fire area if you can do it without risk.
  For massive fire in cargo area, use unmanned hose holder or monitor nozzles;
    if this is impossible, withdraw from area and let fire burn.
  Withdraw immediately in case of rising sound from venting safety device or
    any discoloration of tank due to fire.
  Cool container with water using unmanned device until well after fire is out.

*Spill or Leak
  Shut off ignition sources; no flares, smoking, or flames in hazard area.
  Stop leak if you can do it without risk.
  Water spray may reduce vapor; but it may not prevent ignition in closed
    spaces.
  Isolate area until gas has dispersed.

*First Aid
  Move victim to fresh air and call emergency medical care; if not breathing,
    give artificial respiration; if breathing is difficult, give oxygen.
  In case of frostbite, thaw frosted parts with water.
  Keep victim quiet and maintain normal body temperature.

▼   ▼   ▼   ▼   ▼   ▼   ▼   ▼   ▼   ▼   ▼   ▼   ▼   ▼

--- IDENTIFIERS ---

Name: POLYETHYLENE
Synonyms: AC 8; AC 394; AC 1220; AC GA; ACP 6; AC 8 (Polymer); Acroart; Agilene; Alathon; Alathon 14; Alathon 15; Alathon 1560; Alathon 6600; Alathon 7026; Alathon 7040; Alathon 7050; Alathon 7140; Alathon 7511; Alathon 5B; Alathon 71XHN; Alcowax 6; Aldyl A; Alithon 7050; Alkathene; Alkathene 17/04/00; Alkathene 22 300; Alkathene 200; Alkathene ARN 60; Alkathene WJG 11; Alkathene WNG 14; Alkathene XDG 33; Alkathene XJK 25; Allied PE 617; Alphex FIT 221; Ambythene; Amoco 610A4; A60-20R; A60-70R; Bakelite DFD 330; Bakelite DHDA 4080; Bakelite DYNH; Bareco Polywax 2000; Bareco Wax C 7500; Bicolene C; BPE-I; Bralen KB 2-11; Bralen RB 03-23; Bulen A; Bulen A 30; Carlona 58-030; Carlona 900; Carlona 18020 FA; Carlona PXB; Chemplex 3006; CIPE; Coathylene HA 1671; Courlene-X3; CPE; CPE 16; CPE 25; Cryopolythene; Cry-O-Vac L; Daisolac; Daplen; Daplen 1810 H; DFD 0173; DFD 0188; DFD 2005; DFD 6005; DFD 6032; DFD 6040; DFDJ 5505; DGNB 3825; Diothene; Dixopak; DMDJ 4309; DMDJ 5140; DMDJ 7008; DQDA 1868; DQWA 0355; DXM 100; Dylan; Dylan Super; Dylan WPD 205; DYNH; DYNK 2; Eltex; Eltex 6037; Eltex A 1050; Epolene C; Epolene C 10; Epolene C 11; Epolene E; Epolene E 10; Epolene E 12; Epolene N; Ethene Polymer; Etherin; Etherol E; Ethylene Homopolymer; Ethylene Polymer; Ethylene Polymers (8CI); 23F203; Fabritone PE; FB 217; Fertene; Flamolin MF 15711; Flothene; FM 510; Fortiflex 6015; Fortiflex A 60/500; FP 4; 2100 GP; Grex; Grex PP 60-002; Grisolen; HFDB 4201; Hi-Fax; Hi-Fax 1900; Hi-Fax 4401; Hi-Fax 4601; Hizex; Hizex 5000; Hizex 5100; Hizex 3000B; Hizex 3300F; Hizex 7000F; Hizex 7300F; Hizex 1091J; Hizex 1291J; Hizex 1300J; Hizex 2100J; Hizex 2200J; Hizex 2100LP; Hizex 5100LP; Hizex 6100P; Hizex 3000S; Hizex 3300S; Hizex 5000S; Hoechst PA 190; Hoechst Wax PA 520; Hostalen; Hostalen GD 620; Hostalen GD 6250; Hostalen GF 4760; Hostalen GF 5750; Hostalen GM 5010; Hostalen GUR; Hostalen HDPE; Irax; Irrathene R; Lacqten 1020; LD 400; LD 600; LDPE 4; Lupolen 4261A; Lupolen 6042D; Lupolen 1010H; Lupolen 1800H; Lupolen 1810H; Lupolen 6011H; Lupolen KR 1032; Lupolen KR 1051; Lupolen KR 1257; Lupolen 6011L; Lupolen L 6041D; Lupolen N; Lupolen 1800S; Manolene 6050; Marlex 9; Marlex 50; Marlex 60; Marlex 960; Marlex 6003; Marlex 6009; Marlex 6015; Marlex 6050; Marlex 6060; Marlex EHM 6001; Marlex M 309; Marlex TR 704; Marlex TR 880; Marlex TR 885; Marlex TR 906; Microthene; Microthene 510; Microthene 704; Microthene 710; Mecrothene F; Microthene FN 500; Microthene FN 510; Microthene MN 754-18; Mirason 9; Mirason 16; Mirason M 15; Mirason M 50; Mirason M 68; Mirason NEO 23H; Mirathen; Mirathen 1313; Mirathen

1350; Moplen RO-QG 6015; Neopolin; Neopolin 30N; Neozex 4010B; Neozex 45150; Nopol (Polymer); Novatec JUO 80; Novatec JVO 80; NVC 9025; Okiten G 23; Orizon; Orizon 805; 6020P; PA 130; PA 190; PA 520; PA 560; PAD 522; P 2010B; PE 512; PE 617; PEN 100; PEP 211; PES 100; PES 200; Petrothene; Petrothene LB 861; Petrothene LC 731; Petrothene LC 941; Petrothene NA 219; Petrothene NA 227; Petrothene XL 6301; P 4007EU; P 4070L; Planium; Plaskon PP 60-002; Plastazote X 1016; Plastronga; Plastylene MA 2003; Plastylene MA 7007; Politen; Politen I 020; Polyaethylen (German); Poly-em 12; Poly-em 40; Poly-em 41; Polyethylene AS; Polymist A12; Polymul CS 81; Polysion N 22; Polythene; Polywax 1000; Porolen; P 2070P; PPE 2; Procene UF 1.5; Profax A 60-008; P 2020T; P2050T; P4007T; PTS 2; PVP 8T; PY 100; RCH 1000; Repoc; Rigidex; Rigidex 35; Rigidex 50; Rigidex Type 2; Ropol; Ropothene OB.03-110; Sanwax 161P; Sclair 59; Sclair 2911; Sclair 19A; Sclair 96A; Sclair 59C; Sclair 79D; Sclair 11K; Sclair 19X6; SDP 640; Sholex 5003; Sholex 5100; Sholex 6000; Sholex 6002; Sholex F 171; Sholex F 6050C; Sholex F 6080C; Sholex 4250HM; Sholex L 131; Sholex S 6008; Sholex Super; Sholex XMO 314; Socarex; SRM 1475; SRM 1476; Staflen E 650; Stamylan 900; Stamylan 1000; Stamylan 1700; Stamylan 8200; Stamylan 8400; Sumikathene; Sumikathene F 101-1; Sumikathene F 210-3; Sumikathene F 702; Sumikathene G 201; Sumikathene G 202; Sumikathene G 701; Sumikathene G 801; Sumikathene G 806; Sumikathene Hard 2052; Sunwax 151; Super Dylan; Suprathen; Suprathen C 100; Takathene; Takathene P 3; Takathene P 12; Telcotene; Telecothene; Tenaplas; Tenite 800; Tenite 1811; Tenite 2910; Tenite 2918; Tenite 3300; Tenite 3340; Trovidur PE; Tyrin; Tyvek; Unifos DYOB S; Unifos EFD 0118; Valeron; Valspex 155-53; Velustral KPA; Vestolen; Vestolen A 616; Vestolen A 6016; Wax LE; WJG 11; WNF 15; WVG 23; XL 335-1; XL 1246; XNM 68; XO 440; Yukalon EH 30; Yukalon HE 60; Yukalon K 3212; Yukalon LK 30; Yukalon MS 30; Yukalon PS 30; Yukalon YK 30; ZF 36

CAS: 9002-88-4; **RTECS:** TQ3325000
Formula: $(C_2H_4)_n$
Chemical Class: Polymer

See other identifiers listed below under Regulations.

--- PROPERTIES ---

Boiling Point: NA
Melting Point: 403.16 to 418.16 K; 130 to 145°C; 266 to 293°F
Flash Point: NA
Autoignition: NA

UEL: NA
LEL: NA
Vapor Density: No data
Specific Gravity: No data

Reactivity with Water: No data on water reactivity
Reactivity with Common Materials: No data
Stability during Transport: No data
Neutralizing Agents: No data
Polymerization Possibilities: No data

Toxic Fire Gases: None reported other than possible unburned vapors
Odor Detected at (ppm): Unknown
Odor Description: No data
100% Odor Detection: No data

--- REGULATIONS ---

DOT hazard class: Not given
Packaging exceptions: 173.
Nonbulk packaging: 173.
Bulk packaging: 173.

STCC Number: Not listed

Clean Water Act Sect. 307: No
Clean Water Act Sect. 311: No
Clean Air Act: Not listed
EPA Waste Number: None
CERCLA Ref: Not listed
RQ Designation: Not listed
SARA TPQ Value: Not listed
SARA Sect. 312 Categories: No category

U.S. Postal Service Mailability: Not given

--- TOXICITY DATA ---

Short-term Toxicity: Unknown

Long-term Toxicity: Unknown

Conc IDLH: Unknown

NIOSH REL: Not given

ACGIH TLV: Not listed
ACGIH STEL: Not listed

OSHA PEL: Not in Table Z-1-A

MAK Information: Not listed

Carcinogen: N; **Status:** See below
References: Animal positive IARC** 19, 157, 79; Human indefinite IARC**
19, 157, 79

Carcinogen Lists: *IARC:* Not classified as to human carcinogenicity or
probably not carcinogenic to humans; *MAK:* Not listed; *NIOSH:* Not listed;
NTP: Not listed; *ACGIH:* Not listed; *OSHA:* Not listed.

LD50 Value: No LD50 in RTECS 1992

Other Species Toxicity Data: (Source: NIOSH RTECS 1992)
orl-rat LD : >3 g/kg

Reproductive Toxicity (1992 RTECS): This chemical has no known mammalian reproductive toxicity.

--- INITIAL INCIDENT RESPONSE ---

No DOT Guide information for this compound.

▼   ▼   ▼   ▼   ▼   ▼   ▼   ▼   ▼   ▼   ▼   ▼   ▼   ▼

--- IDENTIFIERS ---

**Chemical Name and Synonyms: TRIMETHYLOLPROPANE
TRIMETHYLACRYLATE; TMPTMA**

Trade Name and Synonyms: Prespersion PLB-5405

Chemical Family: Acrylic Monomer

Manufacturer: Synthetic Products Co., Cleveland, Ohio

--- INGREDIENTS AND COMPOSITION DATA ---

Material	Cas No.	%	TLV
Trimethylolpropane trimethacrylate	3290-92-4	75	25 ppm (skin)
Calcium silicate carrier	1344-95-2	25	15 mg/m^3

--- PHYSICAL DATA ---

Boiling Point (°F): NA

Specific Gravity (H$_2$O = 1): NA

Vapor Pressure (mm Hg): NA

Percent Volatile by Volume (%): NA

Vapor Density (air = 1): NA

Evaporation Rate: NA

Solubility in Water: Negligible

Appearance/Color/Odor: Tan wetted powder.

--- FIRE AND EXPLOSION HAZARDS ---

Flash Point: 200°F

Flammability Limits: Unknown

Extinguishing Media: Water fog, carbon dioxide, dry chemicals, foam.

Special Fire Fighting Procedures: Firefighters should wear self-contained breathing apparatus and protective clothing when fighting fires involving chemicals.

Unusual Fire and Explosion Hazards: Excessive heat may cause polymerization.

--- HEALTH HAZARD DATA ---

Effects of Overexposure: Prolonged or repeated skin contact may cause irritation and possible sensitization. Dust may irritate eyes, nose, and lungs.

Emergency Response and First Aid:
Skin: Wash with soap and water.
Eyes: Irrigate with water, consult physician if irritation persists.
Inhalation: Remove victim to fresh air and consult physician if breathing difficulties arise.
Ingestion: Gastric lavage (stomach wash), consult physician.

--- REACTIVITY DATA ---

Stability and Conditions to Avoid: Material is stable; however, you should avoid elevated temperatures.

Materials to Avoid (Incompatibility): Strong oxidizing and reducing agents.

Hazard Decomposition Products: Oxides of carbon, acrid smoke, and fumes.

Hazardous Polymerization: Will not occur.

Conditions to Avoid: Polymerization catalysts, oxygen-free atmospheres, oxidizing and reducing agents, alkalines, amines, acids, UV radiation.

--- SPILL AND EMERGENCY RESPONSE ---

For Spills: Sweep up or vacuum.

Waste Disposal Method: Dispose of in accordance with federal, state, and local regulations.

--- SPECIAL PROTECTION AND PRECAUTIONS ---

Respiratory Protection: Use OSHA-approved respirator for dust.

Type of Ventilation Required: General mechanical.

Eye Protection Requirements: Safety goggles.

Protective Gloves and Other: Wear impervious rubber gloves. Wear protective clothing to minimize skin contact.

Storage and Handling Precautions: Store in cool, dry location away from all sources of heat, spark, and open flames. Keep partial containers closed to prevent contamination.

Other Precautions: Use all standard practices for good personal hygiene.

▼ ▼ ▼ ▼ ▼ ▼ ▼ ▼ ▼ ▼ ▼ ▼ ▼ ▼

--- IDENTIFIERS ---

Chemical Name and Synonyms: POLYBUTENE POLYMER, $(C_4H_8)_n$

Trade Name and Synonyms: Indopol H-300

Chemical Family: Polybutene

Manufacturer: Amoco Chemicals Co., Chicago, Illinois

--- INGREDIENTS AND COMPOSITION DATA ---

Material	Cas No.	%	TLV
Polybutene	9003-29-6	100	No exposure limit established

--- PHYSICAL DATA ---

Boiling Point (°F): NA

Specific Gravity (H_2O = 1): 0.9

Vapor Pressure (mm Hg): NA

Percent Volatile by Volume (%): NA

Vapor Density (air = 1): NA

Evaporation Rate: NA

Solubility in Water: Negligible, below 0.1%

Viscosity: 635 to 690 cSt at 210°F (99°C)

Pour Point: 35°F (1.7°C)

Appearance/Color/Odor: Clear, bright liquid.

--- FIRE AND EXPLOSION HAZARDS ---

Flash Point: 356°F, (185°C) ASTM D93;

Flammability Limits: Unknown

Extinguishing Media: Agents approved for Class B hazards (e.g., dry chemical, carbon dioxide, halogenated agents, foam steam) or water fog.

Special Fire Fighting Procedures: Firefighters should wear self-contained breathing apparatus and protective clothing when fighting fires involving chemicals.

Unusual Fire and Explosion Hazards: None reported

--- HEALTH HAZARD DATA ---

Effects of Overexposure: None reported by manufacturer.

Emergency Response and First Aid:
Skin: No effects reported under normal conditions of use. No first aid data reported by manufacturer.   A similar product has a percutaneous LD50 greater than 10.25 g/kg (rabbits).
Eyes: No significant irritation reported. Flush eyes with plenty of water should discomfort occur. A similar product produced an irritation score of 20.0/110.0 with complete disappearance of effects in 72 hr (rabbits).
Inhalation: No effects expected under normal conditions of use. No specific first aid requirements. No special protection required. Inhalation of a similar product for 4 hr at 17 mg/1 produced no deaths or untoward behavioral reactions in rats.
Ingestion: This material is expected to be relatively nontoxic. No special first aid required. A similar product has an oral LD50 greater than 34.6 g/kg

(rats). A two-year rat and dog toxicity feeding study and a three-generation reproduction study with rats were conducted on polybutene and revealed no untoward effects due to feeding of polybutene in any of the test animals.

No component of this product is identified as a carcinogen by NTP, IARC or OSHA.

--- REACTIVITY DATA ---

Stability and Conditions to Avoid: Material is stable. No dangerous reactions are reported by the manufacturer.

Materials to Avoid (Incompatibility): None reported by the manufacturer.

Hazard Decomposition Products: Polymerization will not occur. Incomplete burning can produce carbon monoxide, carbon dioxide, and other harmful products.

Hazardous Polymerization: Will not occur.

Conditions to Avoid: None reported by the manufacturer.

--- SPILL AND EMERGENCY RESPONSE ---

For Spills: Remove mechanically or contain on an absorbent material. Treat as an oil spill.

Waste Disposal Method: Disposal must be in accordance with federal, state, and local regulations. Determine waste classification at time of disposal. Conditions to use may render the spent product as hazardous waste. Enclosed controlled incineration is recommended unless directed otherwise by applicable ordinances.

Empty Containers: The container for this product can present explosion or fire hazards, even when emptied. To avoid the risk of injury, do not cut, puncture, or weld on or near this container.

--- SPECIAL PROTECTION AND PRECAUTIONS ---

Respiratory Protection: None recommended by manufacturer.

Type of Ventilation Required: General mechanical.

Eye Protection Requirements: Safety goggles.

Protective Gloves and Other: None recommended by manufacturer.

Storage and Handling Precautions: Store in cool, dry location away from all sources of heat, spark, and open flames. Keep partial containers closed to prevent contamination.

Other Precautions: Use all standard practices for good personal hygiene.

--- REGULATORY INFORMATION ---

OSHA Hazard Communication Standard: Not hazardous per 29 CFR 1910.1200 (d).

DOT Proper Shipping Name (Bulk, Land): Not regulated.

Miscellaneous: Various grades of this product meet FDA and USDA regulations. Information concerning compliance with a specific FDA regulation or USDA approval can be obtained directly from the manufacturer.

3 VULCANIZING AGENTS

Vulcanization is a chemical reaction between rubber and a functional group, usually brought about by heat. The result is a product that is stronger, more elastic, more resilient, and less sensitive to temperature changes and the action of solvents than the original polymer.

The first accelerators used were inorganic basic materials. They included basic lead carbonate, lime, magnesia, and litharge. Inorganic accelerators remained the most important type for many years. Today they have largely been superseded by organic accelerators because the latter are more powerful and produce vulcanizates of much greater strength and higher quality and durability. Organic accelerators are most effective when combined with zinc oxide.

The chemicals listed in this section are organized alphabetically by chemical family name. The following is a list of these chemicals along with trade names and manufacturer's names.

- Chemical Family - Alkaline Earth Oxides
  Chemicals - Magnesium Oxide and Zinc Oxide
  Trade names - Liquispers MBZ
  Manufacturer - Basic Chemical Division, Basic Incorporated

- Chemical Family - Aliphatic Amines
  Chemicals - Hexamethylene Diamine
  Trade names - Numerous
  Manufacturers - Numerous

- Chemical Family - Metal Oxides
  Chemicals - Proprietary
  Trade names - Maglite D and K
  Manufacturer - Merck & Co., Inc.

- Chemical Family - Organic Peroxide
    Chemicals - *t*-Butyl Perbenzoate, Dicumyl Peroxide,
        Calcium Carbonate, Silica, *t*-Butyl Peroxymaleic Acid,
        t-Butylperoxy Valerate

    Trade names - *t*-Butyl Perbenzoate, Di-Cup R, Di-Cup 40C, Di-
        Cup 40KE, Luperco 130-XL, Luperco 101-XL, Luperox PMA,
        Luperco 230-XL
    Manufacturers - Pennwalt Corp., Hercules, Inc.

- Chemical Family - Nonmetals and Sulfenamides
    Chemicals - Sulfur, *N*-Oxydiethylenebenzothiazole-2-Sulfenamide,
        Insoluble Sulfur Oil treated
    Trade names - Amax, Crystex
    Manufacturers - R. T. Vanderbilt Co., Akzo Corp.

- Chemical Family - Miscellaneous
    Chemicals - Dibenzoyl *p*-Quinonedioxime, Hexamethylenediamine
        Carbamate, *n,n*'-Dicinnamylidene-1,6-Hexanediamine, Lead
        Oxides
    Trade names - Dibenzo GMF, Diak No. 1, Diak No. 3, Poly-
        Dispersion
    Manufacturers - Uniroyal Chemical, Inc., E. I. Du Pont de
        Nemours & Co., Wyrough and Loser, Inc., American
        Cyanamid Co.

--- IDENTIFIERS ---

Name: MAGNESIUM OXIDE
Synonyms: Calcined Brucite; Calcined Magnesia; Calcined Magnesite;
  Granmag; Magcal; Maglite; Magnesia; Magnesia USTA; Magnezu Tlenek
  (Polish); Magox; Pericalse; Seawater Magnesia
CAS: 1309-48-4; **RTECS:** OM3850000
Formula: MgO; **Mol Wt:** 40.31
WLN: MG O
Chemical Class: Metal oxide

See other identifiers listed below under Regulations.

--- PROPERTIES ---

Physical Description: White fume
Boiling Point: 3855.34 K; 3582.1°C; 6479.9°F
Melting Point: 3073.13 K; 2799.9°C;5071.9°F
Flash Point: NA
Autoignition: NA

Vapor Pressure: About 0 mm
UEL: NA
LEL: NA
Vapor Density: No data
Specific Gravity: No data
Density: 2.580
Water Solubility: INSOL
Incompatibilities: Chlorine trifluoride

Reactivity with Water: No data on water reactivity
Reactivity with Common Materials: No data
Stability during Transport: No data
Neutralizing Agents: No data
Polymerization Possibilities: No data

Toxic Fire Gases: None reported other than possible unburned vapors
Odor Detected at (ppm): Unknown
Odor Description: No data
100% Odor Detection: No data

--- REGULATIONS ---

DOT hazard class: Not given
Packaging exceptions: 173.
Nonbulk packaging: 173.
Bulk packaging: 173.

STCC Number: Not listed

Clean Water Act Sect. 307: No
Clean Water Act Sect. 311: No
Clean Air Act: Not listed
EPA Waste Number: None
CERCLA Ref: Not listed

RQ Designation: Not listed
SARA TPQ Value: Not listed
SARA Sect. 312 Categories:
  Acute toxicity: adverse effect to target organs.
U.S. Postal Service Mailability: Not given

NFPA Codes:
  Health Hazard (blue): Unspecified
  Flammability (red): Unspecified
  Reactivity (yellow): Unspecified
  Special: Unspecified

--- TOXICITY DATA ---

Short-term Toxicity: Metal fume fever, irritation of eyes and nose, flulike symptoms: fever, cough, chest pains, muscle weakness, nausea, vomiting, leucocytosis.

Long-term Toxicity: Unknown

Target Organs: Respiratory system, lungs, eyes

Symptoms: Irritated eyes, nose; flulike fever, cough, chest pain (metal fume fever). (Source: NIOSH)

Conc IDLH: Unknown

ACGIH TLV: TLV = 10 mg/m^3 fume, as Mg
ACGIH STEL: Fume, as Mg

OSHA PEL: Transitional Limits: PEL = total dust 15mg/m^3; Final Rule Limits: TWA = total dust 10 mg/m^3.

MAK Information: 6 calculated as fine dust, mg/m^3

Carcinogen: N; **Status:** See below

Carcinogen Lists: *IARC*: Not listed; *MAK*: Not listed; *NIOSH*: Not listed; *NTP*: Not listed; *ACGIH*: Not Listed; *OSHA*: Not listed.

Human Toxicity Data: (Source: NIOSH RTECS)
  * ihl-hmn TCLo:400 mg/m^3 38MKAJ 2A, 1745, 81

LD50 Value: No LD50 in RTECS 1992

Reproductive Toxicity (1992 RTECS): This chemical has no known mammalian reproductive toxicity.

--- PROTECTION AND FIRST AID ---

Recommended Respiration Protection Source: NIOSH Pocket Guide (85-114) ACGIH (Magnesium Oxide) -
100 mg/m³: Any dust, mist, and fume respirator with a full facepiece. Any supplied-air respirator. Any self-contained breathing apparatus.
250 mg/m³: Any supplied-air respirator operated in a continuous-flow mode. Any powered air-purifying respirator with a dust, mist, and fume filter.
500 mg/m³: Any air-purifying full facepiece respirator with a high-efficiency particulate filter. Any self-contained breathing apparatus with a full facepiece. Any supplied-air respirator with a full facepiece. Any powered air-purifying respirator with a tight-fitting facepiece and a high-efficiency particulate filter. Substance reported to cause eye irritation or damage; may require eye protection.
7500 mg/m³: Any supplied-air respirator with a full facepiece and operated in a pressure-demand or other positive-pressure mode.
Emergency or Planned Entry in Unknown Concentrations or IDLH Conditions: Any self-contained breathing apparatus with a full facepiece and operated in a pressure-demand or other positive-pressure mode. Any supplied-air respirator with a full facepiece and operated in a pressure-demand or other positive-pressure mode in combination with an auxiliary self-contained breathing apparatus operated in a pressure-demand or other positive-pressure mode.
Escape: Any air-purifying full facepiece respirator with a high-efficiency particulate filter. Any appropriate escape-type self-contained breathing apparatus.

First Aid Source: NIOSH
Inhalation: Artificial Respiration
Ingestion: None given
Skin: None given
Eyes: None given

--- INITIAL INCIDENT RESPONSE ---

No DOT Guide Information for this compound.

▼   ▼   ▼   ▼   ▼   ▼   ▼   ▼   ▼   ▼   ▼   ▼   ▼

--- IDENTIFIERS ---

Name: ZINC OXIDE
Synonyms: Flowers of Zinc; Philosopher's Wool; Zinc White; Chinese White;
  Zinc Oxide Fume
CAS: 1314-132; **RTECS:** ZH4810000
Formula: OZn; **Mol Wt:** 81.37
WLN: ZN O
Chemical Class: Metal oxide

See other identifiers listed below under Regulations.

--- PROPERTIES ---

Physical Description: Odorless, white or yellow powder
Boiling Point: NA
Melting Point: >2073.15 K
Flash Point: NA
Autoignition: NA
UEL: NA
LEL: NA
Vapor Density:5.47 (air = 1)
Specific Gravity: No data
Water Solubility: INSOL

Reactivity with Water: No data on water reactivity
Reactivity with Common Materials: No data
Stability during Transport: No data
Neutralizing Agents: No data
Polymerization Possibilities: No data

Toxic Fire Gases: None reported other than possible unburned vapors
Odor Detected at (ppm): Unknown
Odor Description: No data
100% Odor Detection: No data

--- REGULATIONS ---

DOT hazard class: Not given
Packaging exceptions: 173.
Nonbulk packaging: 173.
Bulk packaging: 173.

STCC Number: Not listed

Clean Water Act Sect. 307: Yes
Clean Water Act Sect. 311: No
Clean Air Act: Not listed
EPA Waste Number: None
CERCLA Ref: Not listed
RQ Designation: Not listed
SARA TPQ Value: Not listed
SARA Sect. 312 Categories:
  Acute toxicity: irritant.
  Acute toxicity: adverse effect to target organs.
  Chronic toxicity: reproductive toxin.
Listed in SARA Sect. 313: Yes
De Minimus Concentration: 1.0%

U.S. Postal Service Mailability: Not given

NFPA Codes:
  Health Hazard (blue): Unspecified
  Flammability (red): Unspecified
  Reactivity (yellow): Unspecified
  Special: Unspecified

--- TOXICITY DATA ---

Short-term Toxicity: Unknown

Long-term Toxicity: Unknown

Target Organs: Skin, eye, respiratory system, lungs

Symptoms: May cause metal fume fever with chills, fever, tightness in chest, cough, and leukocytes; metallic taste in mouth, pains in muscles and joints, marked thirst, occasionally bronchitis or pneumonia, cyanosis. (Source: MI, SAX; THIC)

Conc IDLH: Unknown

NIOSH REL: 5 mg/m^3 time-weighted averages for 8-hr exposure; 15 mg/m^3 ceiling exposures, which shall at no time be exceeded.

ACGIH TLV: TLV = fume, 5 mg/m³; dust, 10 mg/m³; total dust containing no asbestos and less than 1% crystalline silica
ACGIH STEL: STEL = Fume, 10 mg/m³ > > total dust containing no asbestos

OSHA PEL: Transitional Limits: PEL = total dust 15 mg/m³; respirable fraction 5 mg/m³; Final Rule Limits: TWA = total dust 10 mg/m³; respirable fraction 5 mg/m³.

MAK Information: 5 calculated as fine dust mg/m³; Substance with systematic effects, onset of effect over 2 hr: Peak = 10xMAK for 30 min, once per shift of 8 hr.

Carcinogen: N; **Status:** See below

Carcinogen Lists: *IARC*: Not listed; *MAK*: Not listed; *NIOSH*: Not listed; *NTP*: Not listed; *ACGIH*: Not listed; *OSHA*: Not listed.

Human Toxicity Data: (Source: NIOSH RTECS)
  * orl-hmn LDLo:500 mg/kg YAKUD5 22, 291, 80

LD50 Value: No LD50 in RTECS 1992

Other Species Toxicity Data: (Source: NIOSH RTECS 1992)
  orl-rat LD : > 8437 mg/kg
  ipr-rat LD50:240 mg/kg
  orl-mus LD50:7950 mg/kg
  ihl-mus LC50:2500 mg/m³

Reproductive Toxicity (1992 RTECS): This chemical is a mammalian reproductive toxin.

Reproductive Toxicity Data (1992 RTECS):
  orl-rat TDLo:6846 mg/kg (1-22D preg) JONUAI 98, 303, 69
    Specific Developmental Abnormalities
      Homeostasis
    Effects on Newborn
      Stillbirth
      Growth statistics (e.g., reduced weight gain)

--- PROTECTION AND FIRST AID ---

Recommended Respiration Protection Source: NIOSH Pocket Guide (85-114)
NIOSH (Zinc Oxide) -
50 mg/m³: Any dust, mist, and fume respirator with a full facepiece. Any supplied-air respirator. Any self-contained breathing apparatus.
125 mg/m³: Any powered air-purifying respirator with a dust, mist, and fume filter. Any supplied-air respirator operated in a continuous-flow mode.
250 mg/m³: Any air-purifying full facepiece respirator with a high-efficiency particulate filter. Any powered air-purifying respirator with a tight-fitting facepiece and a high-efficiency particulate filter. Any self-contained breathing apparatus with a full facepiece. Any supplied-air respirator with a full facepiece. Any supplied-air respirator with a tight-fitting facepiece operated in a continuous-flow mode.
2500 mg/m³: Any supplied-air respirator with a half-mask and operated in a pressure-demand or other positive-pressure mode.
Emergency or Planned Entry in Unknown Concentrations or IDLH Conditions: Any self-contained breathing apparatus with a full facepiece and operated in a pressure-demand or other positive-pressure mode. Any supplied-air respirator with a full facepiece and operated in a pressure-demand or other positive-pressure mode in combination with an auxiliary self-contained breathing apparatus operated in a pressure-demand or other positive-pressure mode.
Escape: Any air-purifying full facepiece respirator with a high-efficiency particulate filter. Any appropriate escape-type self-contained breathing apparatus.

--- INITIAL INCIDENT RESPONSE ---

No DOT Guide information for this compound.

▼   ▼   ▼   ▼   ▼   ▼   ▼   ▼   ▼   ▼   ▼   ▼   ▼   ▼

--- IDENTIFIERS ---

Name: HEXAMETHYLENE DIAMINE
Synonyms: 1,6-Hexanediamine; 1,6-Diaminohexane; Hexamethylenediamine; 1,6-Hexamethylenediamine; HMDA
CAS: 124-09-4; **RTECS:** MO1180000
Formula: $C_6H_{16}N_2$; **Mol Wt:** 116.24
WLN: Z6Z

Chemical Class: Aliphatic amine
See other identifiers listed below under Regulations.

--- PROPERTIES ---

Physical Description: Colorless solid or watery liquid with a weak ammonia
    odor
Boiling Point: 478 K; 204.8°C; 400.7°F
Melting Point: 315.15 K; 42°C; 104°F
Flash Point: 354 K; 80.8°C; 177.5°F
Autoignition: NA
UEL: NA
LEL: NA
Vapor Density: No data
Specific Gravity: 0.799 60°C
Density: 0.799 g/cc or 7.4307 lb/gal
Water Solubility: Freely soluble

Reactivity with Water: No data on water reactivity
Reactivity with Common Materials: Absorbs moisture and carbonic acid gas
    in air; corrosive, can react with oxidizing materials (Source: SAX); this can
    react with oxidizing materials (Source: SAQXX)
Stability during Transport: No data
Neutralizing Agents: Flush with water
Polymerization Possibilities: No data

Toxic Fire Gases: None reported other than possible unburned vapors
Odor Detected at (ppm): 0.0041
Odor Description: Weak, fishy (Source: CHRIS)
100% Odor Detection: .0032 mg/m^3 ppm

--- REGULATIONS ---

DOT hazard class: 8 CORROSIVE
DOT guide: 60
Identification number: UN2280
DOT shipping name: Hexamethylenediamine, solid
Packing group: III
Label(s) required: CORROSIVE
Packaging exceptions: 173.154
Nonbulk packaging: 173.213
Bulk packaging: 173.242

Quantity limitations:
  Passenger air/rail: 25 kg
  Cargo aircraft only: 100 kg
  Vessel stowage: A
  Other stowage provisions: 12, 48

STCC Number: 4935640

Clean Water Act Sect. 307: No
Clean Water Act Sect. 311: No
Clean Air Act: Not listed
EPA Waste Number: None
CERCLA Ref: Not listed
RQ Designation: Not listed
SARA TPQ Value: Not listed
SARA Sect. 312 Categories:
  Acute toxicity: corrosive.
  Chronic toxicity: reproductive toxin.

U.S. Postal Service Mailability:
  Hazard class: Corrosive material - mailable as ORM-D
  Mailability: Domestic service and air transportation shipper's declaration
  Max per parcel: 25 lb 15% mixture

NFPA Codes:
  Health Hazard (blue): Unspecified
  Flammability (red): Unspecified
  Reactivity (yellow): Unspecified
  Special: Unspecified

--- TOXICITY DATA ---

Short-term Toxicity: Unknown

Long-term Toxicity: Unknown

Target Organs: Skin, eyes, mucous membranes

Symptoms: Vapors cause irritation of eyes and respiratory tract. Liquid irritates eyes and skin, may cause dermatitis; edema, hepatitis. (Source: THIC)

Conc IDLH: Unknown

NIOSH REL: Not given

ACGIH TLV: Not listed
ACGIH STEL: Not listed
WEEL (ACGIH): 5 mg/m^3 (8-hr TWA)

OSHA PEL: Not in Table Z-1-A

MAK Information: Not listed

Carcinogen: N; **Status:** See below

Carcinogen Lists: *IARC*: Not listed; *MAK*: Not listed; *NIOSH*: Not listed; *NTP*: Not listed; *ACGIH*: Not listed; *OSHA*: Not listed.

LD50 Value: orl-rat LD50:750 mg/kg

Other Species Toxicity Data: (Source: NIOSH RTECS 1992)
  orl-rat LD50:750 mg/kg
  ihl-rat LC : >950 mg/m^3/4H
  ihl-mus LCLo:750 mg/m^3/10M
  ipr-mus LD50:320 mg/kg
  scu-mus LD50:1300 mg/kg
  ivn-mus LD50:180 mg/kg
  skn-rbt LD50:1110 mg/kg

Reproductive Toxicity (1992 RTECS): This chemical is a mammalian reproductive toxin.

Reproductive Toxicity Data (1992 RTECS):
  orl-rat TDLo:3 g/kg (6-15D preg) JJATDK 7, 259, 87
    Effects on Embryo or Fetus
      Fetotoxicity (except death, e.g., stunted fetus)
    Specific Developmental Abnormalities
      Hepatobilinary system

  orl-rat TDLo:1840 mg/kg (6-15D preg) JJATDK 7, 259, 87
    Specific Developmental Abnormalities
      Urogenital system

orl-rat TDLo:77 g/kg (multigeneration) FAATDF 16, 490, 91
Effects on Newborn
Growth statistics (e.g., reduced weight gain)

orl-rat TDLo:66500 mg/kg (multigeneration) FAATDF 16, 490, 91
Effects on Fertility
Litter size (number of fetuses per litter; measured before birth)

--- PROTECTION AND FIRST AID ---

Protection Suggested from the CHRIS Manual: Protective clothing; eye protection.

First Aid Source: DOT Emergency Response Guide 1993.
Move victim to fresh air; call emergency medical care. In case of contact with material, immediately flush skin or eyes with running water for at least 15 min. Remove and isolate contaminated clothing and shoes at the site. Keep victim quiet and maintain normal body temperature.

--- INITIAL INCIDENT RESPONSE ---

U.S. Department of Transportation Guide to Hazardous Materials Transport Information - Publication DOT 5800.5 (1990).
DOT Shipping Name: Hexamethylenediamine, solid
DOT ID Number: UN2280

ERG90   GUIDE 60
* POTENTIAL HAZARDS *

*Health Hazards
Contact causes burns to skin and eyes.
If inhaled, may be harmful.
Fire may produce irritating or poisonous gases.
Runoff from fire control or dilution water may cause pollution.

*Fire or Explosion
Some of these materials may burn, but none of them ignites readily.
Flammable/poisonous gases may accumulate in tanks and hopper cars.
Some of these materials may ignite combustibles (wood, paper, oil, etc.).

* EMERGENCY ACTION *

Keep unnecessary people away; isolate hazard area and deny entry.
Stay upwind; keep out of low areas.
Positive-pressure self-contained breathing apparatus (SCBA) and structural
firefighters' protective clothing will provide limited protection.

**CALL CHEMTREC AT 1-800-424-9300 FOR EMERGENCY ASSIS-
TANCE.** If water pollution occurs, notify the appropriate authorities.

*Fire
Some of these materials may react violently with water.
Small fires: Dry chemical, CO_2, water spray, or regular foam.
Large fires: Water spray, fog, or regular foam.
Move container from fire area if you can do it without risk.
Apply cooling water to sides of containers that are exposed to flames until
well after fire is out. Stay away from ends of tanks.

*Spill or Leak
Do not touch spilled material; stop leak if you can do it without risk.
Small spills: Take up with sand or other noncombustible absorbent material
and place into containers for later disposal.
Small dry spills: With clean shovel, place material into clean, dry container
and cover loosely; move containers from spill area.
Large spills: Dike far ahead of liquid spill for later disposal.

*First Aid
Move victim to fresh air; call emergency medical care.
In case of contact with material, immediately flush skin or eyes with running
water for at least 15 min.
Remove and isolate contaminated clothing and shoes at the site.
Keep victim quiet and maintain normal body temperature.

▼   ▼   ▼   ▼   ▼   ▼   ▼   ▼   ▼   ▼   ▼   ▼   ▼   ▼

--- IDENTIFIERS ---

Name: TERT-BUTYL PEROXYBENZOATE
Synonyms: *t*-Butyl Perbenzoate; Tert-Butylperbenzoan (Czech); *t*-Butyl Peroxy
Benzoate; Perbenzoate De Butyle Tertiaire (French)
CAS: 614-45-9; **RTECS:** SD450000
Formula: $C_{11}H_{14}O_3$; **Mol Wt:** 194.23

Chemical Class: Organic peroxide

See other identifiers listed below under Regulations.

--- PROPERTIES ---

Physical Description: Colorless to slightly yellow liquid
Boiling Point: -349.16 K at 0.2 mm; 75 to 76°C at 0.2 mm; 168.8°F at 0.2 mm
Melting Point: 281.15 K; 8°C; 46.4°F
Flash Point: 360.92 K; 87.7°C; 189.9°F
Autoignition: 366.15 K; 93°C; 199.4°F
Vapor Pressure: 0.33 mm hG at 50°C
UEL: NA
LEL: NA
Vapor Density: No data
Specific Gravity: 1.0
Density: 1.02 g/ml
Water Solubility: INSOL

Reactivity with Water: No data on water reactivity
Reactivity with Common Materials: Organic matter (Source: SAX)
Stability during Transport: No data
Neutralizing Agents: No data
Polymerization Possibilities: No data

Toxic Fire Gases: None reported other than possible unburned vapors
Odor Detected at (ppm): Unknown
Odor Description: Aromatic odor ■CSDS (Source: Unspecified)
100% Odor Detection: No data

--- REGULATIONS ---

DOT hazard class: 5.2
DOT guide: 48
Identification number: UN3103
DOT shipping name: ORGANIC PEROXIDE TYPE C, LIQUID
Packing group: II
Label(s) required: ORGANIC PEROXIDE
Packaging exceptions: 173.152
Nonbulk packaging: 173.225
Bulk packaging: 173.None

Quantity limitations:
 Passenger air/rail: 5 L
 Cargo aircraft only: 10 L
 Vessel stowage: D
 Other stowage provisions: 12, 40

STCC Number: 4919151, 4919150, 4919149

Clean Water Act Sect. 307: No
Clean Water Act Sect. 311: No
Clean Air Act: Not listed
EPA Waste Number: None
CERCLA Ref: Not listed
RQ Designation: Not listed
SARA TPQ Value: Not listed
SARA Sect. 312 Categories:
 Acute toxicity: irritant.
 Acute toxicity: adverse effect to target organs.
 Chronic toxicity: mutagen.
 Reactive hazard: organic peroxide.

U.S. Postal Service Mailability:
 Hazard class: Not given
 Mailability: Nonmailable
 Max per parcel: 0

NFPA Codes:
 Health Hazard (blue): (1) Slightly hazardous to health. As a precaution, wear self-contained breathing apparatus.
 Flammability (red): (3) This material can be ignited under almost all temperature conditions.
 Reactivity (yellow): (4) Capable of detonation or explosive decomposition at normal temperatures.
 Special: Unspecified.

--- TOXICITY DATA ---

Short-term Toxicity: Unknown

Long-term Toxicity: Unknown

Target Organs: Skin, eye irritant

Symptoms: Irritation of eyes, nose, and mucous membranes. (Source: CSDS).

Conc IDLH: Unknown

NIOSH REL: Not given

ACGIH TLV: Not listed
ACGIH STEL: Not listed

OSHA PEL: Not in Table Z-1-A

MAK Information: Not listed

Carcinogen: N; **Status:** See below

Carcinogen Lists: *IARC*: Not listed; *MAK*: Not listed; *NIOSH*: Not listed; *NTP*: Not listed; *ACGIH*: Not listed; *OSHA*: Not listed.

LD50 Value: orl-rat LD50:1012 mg/kg

Other Species Toxicity Data: (Source: NIOSH RTECS 1992)
orl-rat LD50:1012 mg/kg
orl-mus LD50:914 mg/kg

Reproductive Toxicity (1992 RTECS): This chemical has no known mammalian reproductive toxicity.

--- PROTECTION AND FIRST AID ---

First Aid Source: DOT Emergency Response Guide 1993.
Move victim to fresh air; call emergency medical care. In case of contact with material, immediately flush skin or eyes with running water for at least 15 min. Wash skin with soap and water. Remove and isolate contaminated clothing and shoes at the site. Keep victim quiet and maintain normal body temperature.

--- INITIAL INCIDENT RESPONSE ---

U.S. Department of Transportation Guide to Hazardous Materials Transport Information - Publication DOT 5800.5 (1990).
DOT Shipping Name: Organic Peroxide Type C, Liquid

DOT ID Number: UN3103

ERG90   GUIDE 48
* POTENTIAL HAZARDS *

*Health Hazards
Contact causes burns to skin and eyes.
Fire may produce irritating or poisonous gases.
Runoff from fire control or dilution water may cause pollution.

*Fire or Explosion
May be ignited by heat, sparks, and flames.
Container may explode in heat of fire.
May explode from heat or contamination
Runoff to sewer may create fire or explosion hazard.

* EMERGENCY ACTION *

Keep unnecessary people away; isolate hazard area and deny entry.
Stay upwind; keep out of low areas.
Positive-pressure self-contained breathing apparatus (SCBA) and structural firefighters' protective clothing will provide limited protection.
CALL CHEMTREC AT 1-800-424-9300 FOR EMERGENCY ASSISTANCE. If water pollution occurs, notify the appropriate authorities.

*Fire
Small fires: Dry chemical, CO_2, water spray, or regular foam.
Large fires: Flood fire area with water.
Apply cooling water to sides of containers that are exposed to flames until well after fire is out. Stay away from ends of tanks.
For massive fire in cargo area, use unmanned hose holder or monitor nozzles; if this is impossible, withdraw from area and let fire burn.

*Spill or Leak
Shut off ignition sources; no flares, smoking, or flames in hazard area.
Do not touch or walk through spilled material; stop leak if you can do it without risk.
Small spills: Take up with inert, damp noncombustible material; move containers from spill area.
Large spills: Wet down with water and dike for later disposal.

*First Aid
Move victim to fresh air; call emergency medical care.
In case of contact with material, immediately flush eyes with running water
for at least 15 min. Wash skin with soap and water.
Remove and isolate contaminated clothing and shoes at the site.
Keep victim quiet and maintain normal body temperature.

▼   ▼   ▼   ▼   ▼   ▼   ▼   ▼   ▼   ▼   ▼   ▼   ▼   ▼

--- IDENTIFIERS ---

Chemical Name and Synonyms: ORGANIC PEROXIDE MIXTURE

Trade Name and Synonyms: Di-Cup R Organic Peroxide

Chemical Family: Organic Peroxide

Manufacturer: Hercules Inc., Wilmington, Delaware

UN ID Number: 2121

--- INGREDIENTS AND COMPOSITION DATA ---

Material	Cas No.	%	TLV
Dicumyl Peroxide [peroxide, bis (alpha alpha-dimethylbenzyl)]	80-43-3	99	Not established*

* This material is stated by manufacturer as not being expected to cause
physiological impairment at low concentration.   Until a specific TLV is
adopted by the ACGIH or an OSHA standard is issued, the manufacturer
recommends that the product be treated as a NUISANCE DUST OR
PARTICULATE in accordance with the recommendations of the ACGIH.

--- PHYSICAL DATA ---

Boiling Point (°F): NA

Specific Gravity (H$_2$O = 1): 0.9

Vapor Pressure (mm Hg): 0.004 mm Hg at 50°C

Percent Volatile by Volume (%): Negligible at 20°C

Vapor Density (air = 1): NA

Evaporation Rate: Slower than Butyl Acetate

Solubility in Water: Negligible

Appearance/Color/Odor: Yellow, acrid color. Material is a granular solid at room temperature.

--- FIRE AND EXPLOSION HAZARDS ---

Flash Point: 127°C (260°F) SETA-CC

Flammability Limits: Unknown

Autoignition: NA

Extinguishing Media: Water spray, dry chemical, carbon dioxide.

Special Fire Fighting Procedures: Use water to keep fire-exposed containers cool. Use self-contained breathing apparatus.

Unusual Fire and Explosion Hazards: Exposure of containers to fire results in rapid product decomposition, container pressure buildup, and failure, followed by vigorous burning with flare effect. Cleanup should not be attempted until all the product has cooled completely. Refer to Reactivity, Incompatibility, and Combustion Data Sections.

Hazardous Product of Combustion: Combustion products depend on other materials present in the fire and on fire conditions. In most cases, these will be carbon monoxide and carbon dioxide. Usually, carbon dioxide will be the principal combustion product. In some cases, irritating partial combustion products will be formed.

--- HEALTH HAZARD DATA ---

Effects of Overexposure: May cause mild, temporary eye, skin, and respiratory irritation. Decomposition products may cause eye, skin, and respiratory irritation and may be skin sensitizers. Decomposition products and fumes from vulcanizing and crosslinking operations can cause eye, skin, and respiratory irritation and may be skin sensitizers.

Additional Information on Effects: Slight primary irritation but no evidence of sensitization was seen in a human patch test with DI-CUP T, which is dicumyl peroxide slightly less refined than DI-CUP R. Increased numbers of blood vessels in the nasal septum have been noted in workers exposed to dicumyl peroxide as compared to other groups examined.

Emergency Response and First Aid:
Skin: Wash thoroughly with soap and running water. Remove contaminated clothing and wash before reuse.
Eyes: In case of contact, immediately flush with plenty of low-pressure water for at least 15 min. Remove any contact lenses to ensure thorough flushing. Call a physician.
Inhalation: Remove victim to fresh air and, if indicated, give artificial respiration. If breathing is difficult, give oxygen. Call a physician. Treat immediately.
Ingestion: If conscious, the person should immediately drink large quantities of liquid to dilute this product. *Never* give liquids to an unconscious person. Call a physician.

Cancer Information: Not listed as a carcinogen by NTP; not regulated as a carcinogen by OSHA; not evaluated by IARC.

--- TOXICITY DATA ---

Animal Effects: *Rat oral LD50:* 4100 mg/kg (administered as 20% solution in corn oil). Mild conjunctivitis but no corneal effects were seen in testing of a 50% solution in corn oil. Eyes returned to normal during the next 48 hr. *Guinea pig sensitization:* Not a sensitizer; no primary irritation was obtained from repeated intradermal injection of 0.1% solution of DI-CUP R. *Acute inhalation study* (study conducted with DI-CUP RR at 40°C, another product composed of DI-CUP (40%) absorbed onto an inert calcium carbonate carrier): No noticeable effects in animals exposed at 21 to 224 mg dust/m^3 for 6 hr. No observable signs of toxicity during the subsequent 14-day postexposure period.

Other: Dicumyl peroxide has shown negative in the Ames test. Dicumyl peroxide has been reported to cause embryotoxicity (death and/or malformations) in 3-day-old chick embryos with an ED50 of 1.5 micromole(s) (weight) per egg.

--- REACTIVITY DATA ---

Stability and Conditions to Avoid: Stable at room temperatures. Avoid temperatures above 55°C (131°F) or contact with materials listed below. Higher temperatures or contact with listed materials promote exothermic decomposition.

Materials to Avoid (Incompatibility): Acids and acidic-type materials such as Friedel-Crafts catalysts, strong oxidizing catalysts, copper, copper alloys, lead, and iron. Glass and stainless steel are preferred structural materials for handling this product because corroded metals and acid clays may be catalytically active in causing decomposition of the peroxide.

Hazard Decomposition Products: Acetophenone, cumyl alcohol, alpha-methyl styrene, methane, and ethane. Under acidic decomposition conditions, phenols may also be formed.

Hazardous Polymerization: Will not occur.

--- SPILL AND EMERGENCY RESPONSE ---

For Spills: If material is not contaminated, scoop it into clean containers for use. If it is contaminated, scoop it into containers for disposal.

Waste Disposal Method: An acceptable method of disposal would be to incinerate after diluting to 5% concentration in No. 2 fuel oil. Use an RCRA-approved incinerator. *Do Not* attempt to burn product in containers. *Do not* mix concentrations or solutions of product with other chemical wastes. *Do not* put solutions containing product into sewer systems. Refer to Regulations Section for specific federal environmental and regulatory data regarding use or disposal of this product.

Empty Containers: The container for this product can present explosion or fire hazards, even when emptied. To avoid the risk of injury, do not cut, puncture, or weld on or near this container.

--- SPECIAL PROTECTION AND PRECAUTIONS ---

Respiratory Protection: Appropriate respirator selected and used in accordance with OSHA Subpart I (29 CFR 1910.134) and manufacturer's recommendations is required if airborne dust is not adequately controlled or if excessive fumes are generated from product decomposition. Self-contained breathing apparatus should be worn in high vapor and dust concentrations.

Type of Ventilation Required: General mechanical. Avoid breathing dusts and vapors. Adequate ventilation should be provided to keep mist and dust concentrations below acceptable exposure limits. Discharge from the ventilation system should comply with applicable air-pollution regulations.

Eye Protection Requirements: Chemical goggles. Eyewash fountains and safety showers should be easily accessible.

Protective Gloves and Other: Impervious gloves and clothing should be worn.

Storage and Handling Precautions: Store in cool, dry location away from all sources of heat, spark, and open flames. Keep partial containers closed to prevent contamination.

Other Precautions: Use all standard practices for good personal hygiene. Completely isolate and thoroughly clean all equipment, piping, or vessels before beginning maintenance or repairs. Keep area clean. Product will burn.

--- REGULATORY INFORMATION ---

SARA TITLE III (see footnotes):

Component No.	SEC. 304 EHS RQ (lb)	SEC.302 EHS TPQ (lb)	SEC.311/312 Hazard Category	SEC.313 Toxic Chemical (Yes, No)
P	NA	NA	HC-1, HC-3, HC-4, HC-5	NA
1	NA	NA	NHH	NO

CERCLA (40 CFR 302.4 Hazardous Substance and Reportable Quantities): This product may contain hazardous substances listed under 40 CFR 302.4. However, they will be present only in trace or residual quantities. Under reasonably anticipated conditions of shipment or handling, it is highly unlikely that an accidental release would exceed the reportable quantity of the listed chemical.

RCRA Information: This product exhibits the characteristic of reactivity (D003) as defined in hazardous waste regulations 40 CFR 261 Subpart C. Therefore, disposal of unused product must comply with hazardous waste regulations.

Other: This product may release phenol upon decomposition, which is listed as a "Toxic Pollutant" under Section 307 of the Clean Water Act, and specific discharge limitations on wastewaters containing it may apply. Refer to the Effluent Guidelines for your industry (40 CFR 401 through 469).

Footnotes to Regulatory Data:

SEC. 302 - Threshold Planning Quantity, Extremely Hazardous Substance. (EHS) (40 CFR 355 Emergency Planning and Notification Regulations).

NA - This chemical is not an EHS.   Therefore, there is no Threshold Planning Quantity (TPQ).

SEC. 304 - Reportable Quantity for Releases of an EHS (40 CFR 355, Appendix A).

SEC. 311/312 - 40 CFR 370 Hazardous Chemical Reporting Requirements. The "hazard categories" are:

HC-1 Immediate (acute) health hazard
HC-2 Delayed (chronic) health hazard
HC-3 Fire hazard
HC-4 Sudden release of pressure hazard
HC-5 Reactive hazard
NHH Not a health hazard
NPH Not a physical hazard

˙SEC. 313 - 40 CFR 372 Toxic Chemical Release Reporting Requirements: NO: This component is *not* subject to the reporting requirements of Section 313 of Title III of the Superfund Amendments and Reauthorization Act of 1986 and 40 CFR Part 372 Toxic Chemical Reporting requirements.

YES: This component is subject to the reporting requirements of Section 313 of Title III of the Superfund Amendments and Reauthorization Act of 1986 and 40 CFR Part 372 Toxic Chemical Reporting requirements. Percent composition (or estimated range) is listed above.

NA: This product is a mixture. As such, it is not listed as a Toxic Chemical under 40 CFR 372, Section 313 reporting requirements. Reportable constituents are listed individually where they exceed threshold concentration limits.

▼ ▼ ▼ ▼ ▼ ▼ ▼ ▼ ▼ ▼ ▼ ▼ ▼ ▼

--- IDENTIFIERS ---

Chemical Name and Synonyms: ORGANIC PEROXIDE MIXTURE

Trade Name and Synonyms: DI-CUP C Organic Peroxide

Chemical Family: Organic Peroxide, supported on precipitated calcium carbonate

Manufacturer: Hercules Inc., Wilmington, Delaware

UN ID Number: NA

--- INGREDIENTS AND COMPOSITION DATA ---

Material	Cas No.	%	TLV
Dicumyl Peroxide [(peroxide, bis (alpha, alpha-dimethylbenzyl)]	80-43-3	38.8	Not established*
Calcium carbonate	1317-65-3	60	10 mg/m^3
Cumene	98-82-8	1.2	50 ppm (skin)

*This material is stated by manufacturer as not being expected to cause physiological impairment at low concentration. Until a specific TLV is adopted by the ACGIH or an OSHA standard is issued, the manufacturer recommends that the product be treated as a NUISANCE DUST OR PARTICULATE in accordance with the recommendations of the ACGIH.

Other Data: Recommended OSHA PEL is:

Dicumyl peroxide   PEL not established

Calcium carbonate   PEL is 15 mg/m^3

Cumene   PEL is 50 ppm (skin)

Additional Data: 10 ppm [Gage, J.C., *British Journal of Industrial Medicine*
27:1-18 (1970)]

--- PHYSICAL DATA ---

Boiling Point (°F): NA, Decomposes

Specific Gravity (H$_2$O = 1): 1.6

Vapor Pressure (mm Hg): NA

Percent Volatile by Volume (%): Negligible

Vapor Density (air = 1): NA

Freezing Point: NA

Evaporation Rate: NA

pH: NA

Solubility in Water: Negligible

Appearance/Color/Odor: Off-white color.

--- FIRE AND EXPLOSION HAZARDS ---

Flash Point: 120°C (248°F) SETA-CC

Flammability Limits: Unknown

Autoignition: NA

Extinguishing Media: Water spray, dry chemical, carbon dioxide.

Special Fire Fighting Procedures: Use water to keep fire-exposed containers
cool. Use self-contained breathing apparatus.

Unusual Fire and Explosion Hazards: Exposure of containers to fire results in rapid product decomposition, container pressure buildup and failure, followed by vigorous burning with flare effect. Cleanup should not be attempted until all the product has cooled completely. May form flammable dust-air mixtures.

Hazardous Products of Combustion and Decomposition: Acetophenone, cumyl alcohol, alpha-methyl styrene, methane, and ethane. Under acidic decomposition conditions, phenols may also be formed. Combustion products often depend on other products present in the fire and on fire conditions. In most cases, these will be carbon monoxide and carbon dioxide. Usually, carbon dioxide will be the principal combustion product. In some cases, irritating partial combustion products will be formed.

Hazardous Polymerization: Will not occur.

--- HEALTH HAZARD DATA ---

Effects of Overexposure: May cause mild, temporary eye, skin, and respiratory irritation. Decomposition products may cause eye, skin, and respiratory irritation and may be skin sensitizers. Decomposition products and fumes from vulcanizing and crosslinking operations can cause eye, skin, and respiratory irritation and may be skin sensitizers.

Additional Information on Effects: Slight primary irritation, but no evidence of sensitization was seen in a human patch test with DI-CUP T, which is dicumyl peroxide slightly less refined than DI-CUP R. Increased numbers of blood vessels in the nasal septum have been noted in workers exposed to dicumyl peroxide as compared to other groups examined. Overexposure to cumene will cause irritation of the eyes, skin, and mucous membranes. Headache, dizziness, slight incoordination, narcosis, and coma may also be seen. It causes irritation when inhaled at 200 ppm.

Emergency Response and First Aid:
Skin: Wash thoroughly with soap and running water. Remove contaminated clothing and wash before reuse.
Eyes: In case of contact, immediately flush with plenty of low-pressure water for at least 15 min. Remove any contact lenses to ensure thorough flushing. Call a physician.
Inhalation: Remove victim to fresh air and, if indicated, give artificial respiration. If breathing is difficult, give oxygen. Call a physician. Treat immediately.

Ingestion: If conscious, the person should immediately drink large quantities of liquid to dilute this product. *Never* give liquids to an unconscious person. Call a physician.

Cancer Information: Not listed as a carcinogen by NTP (National Toxicology Program); not regulated as a carcinogen by OSHA; not evaluated by IARC (International Agency for Research on Cancer).

--- TOXICITY DATA ---

Animal Effects: Rat oral LD50 (studies conducted with active ingredient, DI-CUP T oil, 4100 mg/kg administered as 20% solution in corn). The oral LD50 of cumene was 1400 mg/kg.

Rabbit eye irritation: Mild conjunctivitis but no corneal effects were seen in testing of a 50% solution of DI-CUP T in corn oil. Eyes returned to normal during the next 48 hr. Cumene was the irritant.

Guinea pig sensitization: Not a sensitizer. No primary irritation was obtained from repeated intradermal injection of 0.1% solution of DI-CUP T.

Rat and rabbit: Acute inhalation study: No noticeable effects in animals exposed at 21 to 224 mg dust/m^3 for 6 hr. No observable signs of toxicity during the subsequent 14-day postexposure period.

Other: Dicumyl peroxide has shown negative in the Ames test. Dicumyl peroxide has been reported to cause embryotoxicity (death and/or malformations) in 3-day-old chick embryos with an ED50 of 1.2 micromole(s) (weight) per egg.

--- REACTIVITY DATA ---

Stability and Conditions to Avoid: Stable at room temperatures. Avoid temperatures above 55°C (131°F) or contact with materials listed below. Higher temperatures or contact with listed materials promote exothermic decomposition. Static charges generated by emptying package in or near flammable vapors may cause flash fire. Dust may form flammable dust-air mixture.

Materials to Avoid (Incompatibility): Acids and acidic-type materials such as Friedel-Crafts catalysts, strong oxidizing catalysts, copper, copper alloys, lead, and iron. Glass and stainless steel are preferred structural materials for handling this product because corroded metals and acid clays may be catalytically active in causing decomposition of the peroxide.

Hazard Decomposition Products: Acetophenone, cumyl alcohol, alpha-methyl styrene, methane, and ethane.   Under acidic decomposition conditions, phenols may also be formed.

Hazardous Polymerization: Will not occur.

--- SPILL AND EMERGENCY RESPONSE ---

For Spills: If material is not contaminated, scoop it into clean containers for use. If it is contaminated, scoop it into containers for disposal.

Waste Disposal Method: An acceptable method of disposal would be to incinerate after diluting to 4% concentration in No. 2 fuel oil. Use an RCRA-approved incinerator. *Do not* attempt to burn product in containers. *Do not* mix concentrations or solutions of product with other chemical wastes. *Do not* put solutions containing product into sewer systems. Refer to Regulations Section for specific federal environmental and regulatory data regarding use or disposal of this product.

Empty Containers: The container for this product can present explosion or fire hazards, even when emptied.   To avoid the risk of injury, do not cut, puncture, or weld on or near this container.

--- SPECIAL PROTECTION AND PRECAUTIONS ---

Respiratory Protection: Appropriate respirator selected and used in accordance with OSHA Subpart I (29 CFR 1910.134) and manufacturer's recommendations is required if airborne dust is not adequately controlled or if excessive fumes are generated from product decomposition. Self-contained breathing apparatus should be worn in high vapor and dust concentrations.

Type of Ventilation Required: General mechanical. Avoid breathing dusts and vapors. Adequate ventilation should be provided to keep mist and dust concentrations below acceptable exposure limits.   Discharge from the ventilation system should comply with applicable air-pollution regulations.

Eye Protection Requirements: Chemical goggles. Eyewash fountains and safety showers should be easily accessible.

Protective Gloves and Other: Impervious gloves and clothing should be worn.

Storage and Handling Precautions: Store in cool, dry location away from all sources of heat, spark, and open flames. Avoid ignition sources such as sparks and flames. Ground all equipment. In addition, when emptying bags where flammable vapors may be present, blanket vessel with inert gas, ground operator, and pour material slowly into conductive, grounded chute.

Store below 35°C (95°F) in original containers in a ventilated area out of direct sunlight and away from heat sources. Storage at higher temperatures risks melting of the products organic peroxide and the information of lumps and cakes.

This material may react with acids and acid-like materials, strong oxidizing and reducing agents, oxidation catalysts, copper, copper alloys, lead, and iron and should not be stored near such materials.

Other Precautions: Use all standard practices for good personal hygiene. Completely isolate and thoroughly clean all equipment, piping, or vessels before beginning maintenance or repairs. Keep area clean.

--- REGULATORY INFORMATION ---

SARA TITLE III (see footnotes):

Component No.	SEC. 304 EHS RQ (lb)	SEC.302 EHS TPQ (lb)	SEC.311/312 Hazard Category	SEC.313 Toxic Chemical (Yes, No)
P	NA	NA	HC-1, HC-3, HC-4	NA
1	NA	NA	HC-1	NO
2	NA	NA	HC-1	NO
3	5000	NA	HC-1	YES

CERCLA (40 CFR 302.4 Hazardous Substance and Reportable Quantities): This product may contain hazardous substances listed under 40 CFR 302.4. However, they will be present only in trace or residual quantities. Under reasonably anticipated conditions of shipment or handling, it is highly unlikely that an accidental release would exceed the reportable quantity of the listed chemical.

RCRA Information: This product is not listed in federal hazardous waste regulation 40 CFR 261.33, paragraph (e) or (f), i.e., chemical products that are considered hazardous if they become wastes. It does not exhibit any of the hazardous characteristics listed in 40 CFR 261, Subpart C. State or local hazardous waste regulations may apply if they are different from the federal regulation.

Other: This product may release phenol upon decomposition, which is listed as a "Toxic Pollutant" under Section 307 of the Clean Water Act, and specific discharge limitations on wastewaters containing it may apply. Refer to the Effluent Guidelines for your industry (40 CFR 401 through 469).

Footnotes to Regulatory Data:
SEC. 302 - Threshold Planning Quantity, Extremely Hazardous Substance. (EHS) (40 CFR 355 Emergency Planning and Notification Regulations).

NA - This chemical is not an EHS. Therefore, there is no Threshold Planning Quantity (TPQ).

SEC. 304 - Reportable Quantity for Releases of an EHS (40 CFR 355, Appendix A).

SEC. 311/312 - 40 CFR 370 Hazardous Chemical Reporting Requirements. The "hazard categories" are:

HC-1 Immediate (acute) health hazard
HC-2 Delayed (chronic) health hazard
HC-3 Fire hazard
HC-4 Sudden release of pressure hazard
HC-5 Reactive hazard
NHH Not a health hazard
NPH Not a physical hazard

SEC. 313 - 40 CFR 372 Toxic Chemical Release Reporting Requirements: NO: This component is *not* subject to the reporting requirements of Section 313 of Title III of the Superfund Amendments and Reauthorization Act of 1986 and 40 CFR Part 372 Toxic Chemical Reporting requirements.

YES: This component is subject to the reporting requirements of Section 313 of Title III of the Superfund Amendments and Reauthorization Act of 1986 and 40 CFR Part 372 Toxic Chemical Reporting requirements. Percent composition (or estimated range) is listed above.

NA: This product is a mixture. As such, it is not listed as a Toxic Chemical under 40 CFR 372, Section 313 reporting requirements. Reportable constituents are listed individually where they exceed threshold concentration limits.

▼   ▼   ▼   ▼   ▼   ▼   ▼   ▼   ▼   ▼   ▼   ▼   ▼

--- IDENTIFIERS ---

Chemical Name and Synonyms: ORGANIC PEROXIDE MIXTURE

Trade Name and Synonyms: DI-CUP 40 kE Organic Peroxide

Chemical Family: Organic Peroxide, supported on a carrier silane modified clay.

Manufacturer: Hercules, Inc., Wilmington, Delaware

UN ID Number: NA

--- INGREDIENTS AND COMPOSITION DATA ---

Material	Cas No.	%	TLV
Dicumyl peroxide [(peroxide, bis (alpha, alpha-dimethylbenzyl)]	80-43-3	38.8	Not established*
Silane modified clay	64402-68-4/ 1067-53-4	60	10 mg/m^3
Cumene	98-82-8	1.2	50 ppm (skin)

*This material is stated by manufacturer as not being expected to cause physiological impairment at low concentration. Until a specific TLV is adopted by the ACGIH or an OSHA standard is issued, the manufacturer recommends that the product be treated as a NUISANCE DUST OR PARTICULATE in accordance with the recommendations of the ACGIH.

Other Data: Recommended OSHA PEL is:

Dicumyl peroxide	PEL not established
Silane modified clay	PEL is 15 mg/m^3
Cumene	PEL is 50 ppm (skin)

Additional Data: 10 ppm [Gage, J.C., *British Journal of Industrial Medicine* 27:1-18 (1970)]

--- PHYSICAL DATA ---

Boiling Point (°F): NA, Decomposes

Specific Gravity (H$_2$O = 1): 1.6

Vapor Pressure (mm Hg): NA

Percent Volatile by Weight (%): Negligible

Vapor Density (air = 1): NA

Freezing Point: NA

Evaporation Rate: NA

pH: NA

Solubility in Water: Negligible

Melting Point: 30 to 35°C (86 to 95°F) (peroxide only)

Appearance/Color/Odor: Off-white color.

--- FIRE AND EXPLOSION HAZARDS ---

Flash Point: 127°C (260°F) SETA-CC

Flammability Limits: Unknown

Autoignition Temperature: NA

Extinguishing Media: Water spray, dry chemical, carbon dioxide.

Special Fire Fighting Procedures: Use water to keep fire-exposed containers cool. Use self-contained breathing apparatus.

Unusual Fire and Explosion Hazards: Exposure of containers to fire results in rapid product decomposition, container pressure buildup and failure, followed by vigorous burning with flare effect. Cleanup should not be attempted until all the product has cooled completely. May form flammable dust-air mixtures.

Hazardous Products of Combustion and Decomposition: Acetophenone, cumyl alcohol, alpha-methyl styrene, methane, and ethane. Under acidic decomposition conditions, phenols may also be formed. Combustion products often depend on other products present in the fire and on fire conditions. In most cases, these will be carbon monoxide and carbon dioxide. Usually, carbon dioxide will be the principal combustion product. In some cases, irritating partial combustion products will be formed.

Hazardous Polymerization: Will not occur.

--- HEALTH HAZARD DATA ---

Effects of Overexposure: Overexposure can cause mild eye and skin irritation. Inhalation of particulate clay dusts of silane modified clay have caused a low prevalence of pulmonary fibrosis in miners. Decomposition products and fumes from vulcanizing and crosslinking operations can cause eye, skin, and respiratory irritation and may be skin sensitizers. Overexposure to cumene will cause irritation to eyes, skin, and mucous membranes. Headache, dizziness, slight incoordination, narcosis, and coma may also occur. Dermatitis may result from prolonged or repeated contact. It causes irritation when inhaled at 200 ppm.

Emergency Response and First Aid:
Skin: Wash thoroughly with soap and running water. Remove contaminated clothing and wash before reuse.

Eyes: In case of contact, immediately flush with plenty of low-pressure water for at least 15 min. Remove any contact lenses to ensure thorough flushing. Call a physician.

Inhalation: Remove victim to fresh air and, if indicated, give artificial respiration. If breathing is difficult, give oxygen. Call a physician. Treat immediately.

Ingestion: If conscious, the person should immediately drink large quantities of liquid to dilute this product. *Never* give liquids to an unconscious person. Call a physician.

Cancer Information: Not listed as a carcinogen by NTP (National Toxicology Program); not regulated as a carcinogen by OSHA; not evaluated by IARC (International Agency for Research on Cancer).

--- TOXICITY DATA ---

Animal Effects: Rat oral LD50 (studies conducted with active ingredient, DI-CUP T oil, 4100 mg/kg administered as 20% solution in corn oil). The oral LD50 of cumene was 1400 mg/kg.

Rabbit eye irritation: Mild conjunctivitis but no corneal effects were seen in testing of a 50% solution of DI-CUP T in corn oil. Eyes returned to normal during the next 48 hr. Cumene was the irritant.

Guinea pig sensitization: Not a sensitizer. No primary irritation was obtained from repeated intradermal injection of 0.1% solution of DI-CUP T.

Rat and rabbit: Acute inhalation study: No noticeable effects in animals exposed at 21 to 224 mg dust/m^3 for 6 hr. No observable signs of toxicity during the subsequent 14-day postexposure period.

Other: Dicumyl peroxide has shown negative in the Ames test. Dicumyl peroxide has been reported to cause embryotoxicity (death and/or malformations) in 3-day-old chick embryos with an ED50 of 1.2 micromole(s) (weight) per egg.

--- REACTIVITY DATA ---

Stability and Conditions to Avoid: Stable at room temperatures. Avoid temperatures above 55°C (131°F) or contact with materials listed below. Higher temperatures or contact with listed materials promote exothermic decomposition. Static charges generated by emptying package in or near flammable vapors may cause flash fire. Dust may form flammable dust-air mixture.

Materials to Avoid (Incompatibility): Acids and acidic-type materials such as Friedel-Crafts catalysts, strong oxidizing catalysts, copper, copper alloys, lead, and iron. Glass and stainless steel are preferred structural materials for handling this product because corroded metals and acid clays may be catalytically active in causing decomposition of the peroxide.

Hazard Decomposition Products: Acetophenone, cumyl alcohol, alpha-methyl styrene, methane, and ethane. Under acidic decomposition conditions, phenols may also be formed.

Hazardous Polymerization: Will not occur.

--- SPILL AND EMERGENCY RESPONSE ---

For Spills: If material is not contaminated, scoop it into clean containers for use. If it is contaminated, scoop it into containers for disposal.

Waste Disposal Method: An acceptable method of disposal would be to incinerate after diluting to 5% concentration in No. 2 fuel oil. Use an RCRA-approved incinerator. *Do not* attempt to burn product in containers. *Do not* mix concentrations or solutions of product with other chemical wastes. *Do not* put solutions containing product into sewer systems. Refer to Regulations Section for specific federal environmental and regulatory data regarding use or disposal of this product.

Empty Containers: The container for this product can present explosion or fire hazards, even when emptied. To avoid the risk of injury, do not cut, puncture, or weld on or near this container.

--- SPECIAL PROTECTION AND PRECAUTIONS ---

Respiratory Protection: Appropriate respirator selected and used in accordance with OSHA Subpart I (29 CFR 1910.134) and manufacturer's recommendations is required if airborne dust is not adequately controlled or if excessive fumes are generated from product decomposition. Self-contained breathing apparatus should be worn in high vapor and dust concentrations.

Type of Ventilation Required: General mechanical. Avoid breathing dusts and vapors. Adequate ventilation should be provided to keep mist and dust concentrations below acceptable exposure limits. Discharge from the ventilation system should comply with applicable air-pollution regulations.

Eye Protection Requirement: Chemical goggles. Eyewash fountains and safety showers should be easily accessible.

Protective Gloves and Other: Impervious gloves and clothing should be worn.

Storage and Handling Precautions: Store in cool, dry location away from all sources of heat, spark, and open flames. Avoid ignition sources such as sparks and flames. Ground all equipment. In addition, when emptying bags where flammable vapors may be present, blanket vessel with inert gas, ground operator, and pour material slowly into conductive, grounded chute.

Store below 35°C (95°F) in original containers in a ventilated area out of direct sunlight and away from heat sources. Storage at higher temperatures risks melting of the products organic peroxide and the information of lumps and cakes.

This material may react with acids and acid-like materials, strong oxidizing and reducing agents, oxidation catalysts, copper, copper alloys, lead, and iron and should not be stored near such materials.

Other Precautions: Use all standard practices for good personal hygiene. Completely isolate and thoroughly clean all equipment, piping, or vessels before beginning maintenance or repairs. Keep area clean. Note that this product will burn.

--- REGULATORY INFORMATION ---

SARA TITLE III (see footnotes):

Component No.	SEC. 304 EHS RQ (lb)	SEC.302 EHS TPQ (lb)	SEC.311/312 Hazard Category	SEC.313 Toxic Chemical (Yes, No)
P	NA	NA	HC-1, HC-3, HC-4	NA
1	NA	NA	HC-1	NO
2	NA	NA	HC-1	NO

Component No.	SEC. 304 EHS RQ (lb)	SEC.302 EHS TPQ (lb)	SEC.311/312 Hazard Category	SEC.313 Toxic Chemical (Yes, No)
3	5000	NA	HC-1	YES

CERCLA (40 CFR 302.4 Hazardous Substance and Reportable Quantities):
This product may contain hazardous substances listed under 40 CFR 302.4.
However, they will be present only in trace or residual quantities. Under
reasonably anticipated conditions of shipment or handling, it is highly unlikely
that an accidental release would exceed the reportable quantity of the listed
chemical.

RCRA Information: This product is not listed in federal hazardous waste
regulation 40 CFR 261.33, paragraph (e) or (f), i.e., chemical products that
are considered hazardous if they become wastes. It does not exhibit any of
the hazardous characteristics listed in 40 CFR 261, Subpart C. State or local
hazardous waste regulations may apply if they are different from the federal
regulation.

Other: This product may release phenol upon decomposition, which is listed
as a "Toxic Pollutant" under Section 307 of the Clean Water Act, and specific
discharge limitations on wastewaters containing it may apply. Refer to the
Effluent Guidelines for your industry (40 CFR 401 through 469).

Footnotes to Regulatory Data:

SEC. 302 - Threshold Planning Quantity, Extremely Hazardous Substance.
(EHS) (40 CFR 355 Emergency Planning and Notification Regulations).

NA - This chemical is not an EHS. Therefore, there is no Threshold
Planning Quantity (TPQ).

SEC. 304 - Reportable Quantity for Releases of an EHS (40 CFR 355,
Appendix A).

SEC. 311/312 - 40 CFR 370 Hazardous Chemical Reporting Requirements.
The "hazard categories" are:

HC-1 Immediate (acute) health hazard
HC-2 Delayed (chronic) health hazard

HC-3 Fire hazard
HC-4 Sudden release of pressure hazard
HC-5 Reactive hazard
NHH Not a health hazard
NPH Not a physical hazard

SEC. 313 - 40 CFR 372 Toxic Chemical Release Reporting Requirements:
NO: This component is *not* subject to the reporting requirements of Section 313 of Title III of the Superfund Amendments and Reauthorization Act of 1986 and 40 CFR Part 372 Toxic Chemical Reporting requirements.

YES: This component is subject to the reporting requirements of Section 313 of Title III of the Superfund Amendments and Reauthorization Act of 1986 and 40 CFR Part 372 Toxic Chemical Reporting requirements. Percent composition (or estimated range) is listed above.

NA: This product is a mixture. As such, it is not listed as a Toxic Chemical under 40 CFR 372, Section 313 reporting requirements. Reportable constituents are listed individually where they exceed threshold concentration limits.

▼   ▼   ▼   ▼   ▼   ▼   ▼   ▼   ▼   ▼   ▼   ▼   ▼   ▼

--- IDENTIFIERS ---

Name: DICUMYL PEROXIDE
Synonyms: Active Dicumyl Peroxide; Bis(alpha,alpha-Dimethylbenzyl)-Peroxide; Cumene Peroxide; Cumyl Peroxide; Dicumyl Peroxide; Dicumyl Peroxide (DOT); Di-alpha-Cumyl Peroxide; DI-CUP; Diisopropylbenzene Peroxide; Isopropylbenzene Peroxide
CAS: 80-43-3; **RTECS:** SD8150000
Formula: $C_{18}H_{22}O_2$; **Mol Wt:** 270.40
WLN: 1X1
Chemical Class: Organic peroxide

See other identifiers listed below under Regulations.

--- PROPERTIES ---

Boiling Point: NA
Melting Point: 312.16 to 314.16 K; 39 to 41°C; 102.2 to 105.8°F
Flash Point: >383.16 K; >10°C; >230°F

Autoignition: NA
UEL: NA
LEL: NA
Vapor Density: No data
Specific Gravity: No data

Reactivity with Water: No data on water reactivity
Reactivity with Common Materials: No data
Stability during Transport: No data
Neutralizing Agents: No data
Polymerization Possibilities: No data

Toxic Fire Gases: Acrid smoke and irritating fumes
Odor Detected at (ppm): Unknown
Odor Description: No data
100% Odor Detection: No data

--- REGULATIONS ---

DOT hazard class: 5.2
DOT guide: 48
Identification number: UN3110
DOT shipping name: Organic peroxide type F, solid
Packing group: II
Label(s) required: ORGANIC PEROXIDE
Packaging exceptions: 173.152
Nonbulk packaging: 173.225
Bulk packaging: 173.None
Quantity limitations:
  Passenger air/rail: 10 kg
  Cargo aircraft only: 25 kg
  Vessel stowage: D
  Other stowage provisions: 12, 40

Clean Water Act Sect. 307: No
Clean Water Act Sect. 311: No
Clean Air Act: Not listed
EPA Waste Number: None
CERCLA Ref: Not listed
RQ Designation: Not listed
SARA TPQ Value: Not listed

SARA Sect. 312 Categories:
Reactive hazard: organic peroxide.

U.S. Postal Service Mailability:
Hazard class: Organic peroxide
Mailability: Domestic service and air transportation shipper's declaration
Max per parcel: 1 lb

NFPA Codes:
Health Hazard (blue): Unspecified
Flammability (red): Unspecified
Reactivity (yellow): Unspecified
Special: Unspecified

--- TOXICITY DATA ---

Short-term Toxicity: Unknown

Long-term Toxicity: Unknown

Conc IDLH: Unknown

NIOSH REL: Not given

ACGIH TLV: Not listed
ACGIH STEL: Not listed

OSHA PEL: Not in Table Z-1-A

MAK Information: Not listed

Carcinogen: N; **Status:** See below

Carcinogen Lists: *IARC*: Not listed; *MAK*: Not listed; *NIOSH*: Not listed; *NTP:* Not listed; *ACGIH:* Not listed; *OSHA*: Not listed.

LD50 Value: orl-rat LD50:4100 mg/kg

Other Species Toxicity Data: (Source: NIOSH RTECS 1992)
orl-rat LD50:4100 mg/kg
unr-rat LD50:3500 mg/kg

Reproductive Toxicity (1992 RTECS): This chemical has no known mammalian reproductive toxicity.

--- PROTECTION AND FIRST AID ---

First Aid Source: DOT Emergency Response Guide 1993.
Move victim to fresh air; call emergency medical care. In case of contact with material, immediately flush eyes with running water for at least 15 min. Wash skin with soap and water. Remove and isolate contaminated clothing and shoes at the site. Keep victim quiet and maintain normal body temperature.

--- INITIAL INCIDENT RESPONSE ---

U.S. Department of Transportation Guide to Hazardous Materials Transport Information - Publication DOT 5800.5 (1990).
DOT Shipping Name: Organic peroxide type F, solid
DOT ID Number: UN3110

ERG90   GUIDE 48
* POTENTIAL HAZARDS *

*Health Hazards
Contact may cause burns to skin and eyes.
Fire may produce irritating or poisonous gases.
Runoff from fire control or dilution water may cause pollution.

*Fire or Explosion
May be ignited by heat, sparks, and flames.
Container may explode in heat of fire.
May explode from heat or contamination.
Runoff to sewer may create fire or explosion hazard.

* EMERGENCY ACTION *

Keep unnecessary people away; isolate hazard area and deny entry.
Stay upwind; keep out of low areas.
Positive pressure self-contained breathing apparatus (SCBA) and structural firefighters' protective clothing will provide limited protection.
CALL CHEMTREC AT 1-800-424-9300 FOR EMERGENCY ASSISTANCE. If water pollution occurs, notify the appropriate authorities.

*Fire
Small fires: Dry chemical, CO_2, water spray, or regular foam.
Large fires: Flood fire area with water.
Apply cooling water to sides of containers that are exposed to flames until well after fire is out. Stay away from ends of tanks.
For massive fire in cargo area, use unmanned hose holder or monitor nozzles; if this is impossible, withdraw from area and let fire burn.

*Spill or Leak
Shut off ignition sources; no flares, smoking, or flames in hazard area.
Do not touch or walk through spilled material; stop leak if you can do it without risk.
Small spills: Take up with inert, damp noncombustible material; move containers from spill area.
Large spills: Wet down with water and dike for later disposal.

*First Aid
Move victim to fresh air; call emergency medical care.
In case of contact with material, immediately flush eyes with running water for at least 15 min. Wash skin with soap and water.
Remove and isolate contaminated clothing and shoes at the site.
Keep victim quiet and maintain normal body temperature.

▼ ▼ ▼ ▼ ▼ ▼ ▼ ▼ ▼ ▼ ▼ ▼ ▼ ▼

--- IDENTIFIERS ---

Name: 2,5-DIMETHYL-2,5-DI(TERT-BUTYLPEROXY)HEXYNE-3
Synonyms: 2,5-Dimethyl-2,5-Di(*t*-Butylperoxy)Hexyne-3; 3-Hexyne, 2,5-Dimethyl-2,5-Di (*T*-Butylperoxy)-
CAS: 1068-27-5; **RTECS:** MQ750000
Formula: $C_{16}H_{30}O_4$; **Mol Wt:** 286.46
WLN: 1X1
Chemical Class: Organic peroxide

See other identifiers listed below under Regulations.

--- PROPERTIES ---

Boiling Point: 340.16 K at 2 mm; 67°C at 2 mm; 152.6°F at 2 mm
Melting Point: NA
Flash Point: 358.16 K; 85°C; 185°F

Autoignition: NA
UEL: NA
LEL: NA
Vapor Density: No data
Specific Gravity: No data
Density: 0.881

Reactivity with Water: No data on water reactivity
Reactivity with Common Materials: No data
Stability during Transport: No data
Neutralizing Agents: No data
Polymerization Possibilities: No data

Toxic Fire Gases: None reported other than possible unburned vapors
Odor Detected at (ppm): Unknown
Odor Description: No data
100% Odor Detection: No data

--- REGULATIONS ---

DOT hazard class: 5.2
DOT guide: 49
Identification number: UN3103
DOT shipping name: ORGANIC PEROXIDE TYPE C, LIQUID
Packing group: II
Label(s) required: ORGANIC PEROXIDE
Packaging exceptions: 173.152
Nonbulk packaging: 173.225
Bulk packaging: 173.None
Quantity limitations:
  Passenger air/rail: 5 L
  Cargo aircraft only: 10 L
  Vessel stowage: D
  Other stowage provisions: 12, 40

STCC Number: 4919264

Clean Water Act Sect. 307: No
Clean Water Act Sect. 311: No
Clean Air Act: Not listed
EPA Waste Number: None
CERCLA Ref: Not listed

RQ Designation: Not listed
SARA TPQ Value: Not listed
SARA Sect. 312 Categories:
Reactive hazard: organic peroxide.

U.S. Postal Service Mailability: Not given

NFPA Codes:
Health Hazard (blue): Unspecified
Flammability (red): Unspecified
Reactivity (yellow): Unspecified
Special: Unspecified

--- TOXICITY DATA ---

Short-term Toxicity: Unknown

Long-term Toxicity: Unknown

Conc IDLH: Unknown

NIOSH REL: Not given

ACGIH TLV: Not listed
ACGIH STEL: Not listed

OSHA PEL: Not in Table Z-1-A

MAK Information: Not listed

Carcinogen: N; **Status:** See below

Carcinogen Lists: *IARC*: Not listed; *MAK*: Not listed; *NIOSH*: Not listed; *NTP*: Not listed; *ACGIH*: Not listed; *OSHA*: Not listed.

LD50 Value: No LD50 in RTECS 1992

Other Species Toxicity Data: (Source: NIOSH RTECS 1992)
ipr-mus LD50:1850 mg/kg

Reproductive Toxicity (1992 RTECS): This chemical has no known mammalian reproductive toxicity.

--- PROTECTION AND FIRST AID ---

First Aid Source: NIOSH.
Inhalation: None given
Ingestion: None given
Skin: None given
Eyes: None given

First Aid Source: DOT Emergency Response Guide 1993.
Move victim to fresh air; call emergency medical care. In case of contact with material, immediately flush eyes with running water for at least 15 min. Wash skin with soap and water. Remove and isolate contaminated clothing and shoes at the site. Keep victim quiet and maintain normal body temperature.

--- INITIAL INCIDENT RESPONSE ---

U.S. Department of Transportation Guide to Hazardous Materials Transport Information - Publication DOT 5800.5 (1990).
DOT Shipping Name: ORGANIC PEROXIDE TYPE C, LIQUID
DOT ID Number: UN3103

ERG90   GUIDE 49
* POTENTIAL HAZARDS *

*Health Hazards
Contact may cause burns to skin and eyes.
Fire may produce irritating or poisonous gases.
Runoff from fire control or dilution water may cause pollution.

*Fire or Explosion
May be ignited by heat, sparks, and flames.
May burn rapidly with flare-burning effect.
Container may explode in heat of fire.
May explode from friction, heat, or contamination.
Runoff to sewer may create fire or explosion hazard.

* EMERGENCY ACTION *

Keep unnecessary people away; isolate hazard area and deny entry.
Stay upwind; keep out of low areas.

Positive pressure self-contained breathing apparatus (SCBA) and structural firefighters' protective clothing will provide limited protection.
Isolate for 1/2 mile in all directions if tank, rail car, or tank truck is involved in fire.
CALL CHEMTREC AT 1-800-424-9300 FOR EMERGENCY ASSIS-TANCE. If water pollution occurs, notify the appropriate authorities.

*Fire
Small fires: Dry chemical, CO_2, water spray, or regular foam.
Large fires: Flood fire area with water from a distance.
Do not move cargo or vehicle if cargo has been exposed to heat.
If fire can be controlled, cool container with water from unmanned hose holder or monitor nozzles until well after fire is out.
If this is impossible, withdraw from area and let fire burn.

*Spill or Leak
Shut off ignition sources; no flares, smoking, or flames in hazard area.
Do not touch or walk through spilled material; stop leak if you can do it without risk.
Small spills: Take up with inert, damp noncombustible material; move containers from spill area.
Large spills: Wet down with water and dike for later disposal.

*First Aid
Move victim to fresh air; call emergency medical care.
In case of contact with material, immediately flush eyes with running water for at least 15 min. Wash skin with soap and water.
Remove and isolate contaminated clothing and shoes at the site.
Keep victim quiet and maintain normal body temperature.

▼   ▼   ▼   ▼   ▼   ▼   ▼   ▼   ▼   ▼   ▼   ▼   ▼   ▼

--- IDENTIFIERS ---

Name: 2,5-DIMETHYL-2,5-DI(TERT-BUTYLPEROXY)HEXANE
CAS: 78-63-7; RTECS: MO1835000
Formula: $C_{16}H_{34}O_4$; Mol Wt: 290.50
WLN: 1X1
Chemical Class: Organic peroxide

See other identifiers listed below under Regulations.

--- PROPERTIES ---

Physical Description: Colorless to light-yellow liquid
Boiling Point: 523.15 K; 250°C; 482°F
Melting Point: 281.15 K; 8°C; 46.4°F
Flash Point: NA
Autoignition: NA
UEL: NA
LEL: NA
Vapor Density: No data
Specific Gravity: No data
Density: 0.85
Water Solubility: INSOL

Reactivity with Water: No data on water reactivity
Reactivity with Common Materials: No data
Stability during Transport: No data
Neutralizing Agents: No data
Polymerization Possibilities: No data

Toxic Fire Gases: None reported other than possible unburned vapors
Odor Detected at (ppm): Unknown
Odor Description: No data
100% Odor Detection: No data

--- REGULATIONS ---

DOT hazard class: 5.2
DOT guide: 48
Identification number: UN3105
DOT shipping name: ORGANIC PEROXIDE TYPE D, LIQUID
Packing group: II
Label(s) required: ORGANIC PEROXIDE
Packaging exceptions: 173.152
Nonbulk packaging: 173.225
Bulk packaging: 173.None
Quantity limitations:
  Passenger air/rail: 5 L
  Cargo aircraft only: 10 L
  Vessel stowage: D
  Other stowage provisions: 12, 40

STCC Number: 4919263

Clean Water Act Sect. 307: No
Clean Water Act Sect. 311: No
Clean Air Act: Not listed
EPA Waste Number: None
CERCLA Ref: Not listed
RQ Designation: Not listed
SARA TPQ Value: Not listed
SARA Sect. 312 Categories:
  Reactive hazard: organic peroxide.

U.S. Postal Service Mailability: Not given

NFPA Codes:
  Health Hazard (blue): Unspecified
  Flammability (red): Unspecified
  Reactivity (yellow): Unspecified
  Special: Unspecified

--- TOXICITY DATA ---

Short-term Toxicity: Unknown

Long-term Toxicity: Unknown

Conc IDLH: Unknown

NIOSH REL: Not given

ACGIH TLV: Not listed
ACGIH STEL: Not listed

OSHA PEL: Not in Table Z-1-A

MAK Information: Not listed

Carcinogen: N; **Status:** See below

Carcinogen Lists: *IARC*: Not listed; *MAK*: Not listed; *NIOSH*: Not listed; *NTP*: Not listed; *ACGIH*: Not listed; *OSHA*: Not listed.

LD50 Value: No LD50 in RTECS 1992

Other Species Toxicity Data: (Source: NIOSH RTECS 1992)
ipr-mus LDLo:1700 mg/kg

Reproductive Toxicity (1992 RTECS): This chemical has no known mammalian reproductive toxicity.

--- PROTECTION AND FIRST AID ---

First Aid Source: NIOSH.
Inhalation: None given
Ingestion: None given
Skin: None given
Eyes: None given

First Aid Source: DOT Emergency Response Guide 1993.
Move victim to fresh air; call emergency medical care. In case of contact with material, immediately flush eyes with running water for at least 15 min. Wash skin with soap and water. Remove and isolate contaminated clothing and shoes at the site. Keep victim quiet and maintain normal body temperature.

--- INITIAL INCIDENT RESPONSE ---

Fire Extinguishment: Fight fire with water, spray, dry chemical (*Note:* CHRIS 91).

U.S. Department of Transportation Guide to Hazardous Materials Transport Information - Publication DOT 5800.5 (1990).
DOT Shipping Name: ORGANIC PEROXIDE TYPE D, LIQUID
DOT ID Number: UN3105

ERG90   GUIDE 48
* POTENTIAL HAZARDS *

*Health Hazards
Contact may cause burns to skin and eyes.
Fire may produce irritating or poisonous gases.
Runoff from fire control or dilution water may cause pollution.

*Fire or Explosion
May be ignited by heat, sparks, and flames.
Container may explode in heat of fire.
May explode from heat or contamination.
Runoff to sewer may create fire or explosion hazard.

* EMERGENCY ACTION *

Keep unnecessary people away; isolate hazard area and deny entry.
Stay upwind; keep out of low areas.
Positive pressure self-contained breathing apparatus (SCBA) and structural
   firefighters' protective clothing will provide limited protection.
**CALL CHEMTREC AT 1-800-424-9300 FOR EMERGENCY ASSIS-
TANCE.** If water pollution occurs, notify the appropriate authorities.

*Fire
Small fires: Dry chemical, CO_2, water spray, or regular foam.
Large fires: Flood fire area with water.
Apply cooling water to sides of containers that are exposed to flames until
   well after fire is out. Stay away from ends of tanks.
For massive fire in cargo area, use unmanned hose holder or monitor nozzles;
   if this is impossible, withdraw from area and let fire burn.

*Spill or Leak
Shut off ignition sources; no flares, smoking, or flames in hazard area.
Do not touch or walk through spilled material; stop leak if you can do it
   without risk.
Small spills: Take up with inert, damp noncombustible material; move
   containers from spill area.
Large spills: Wet down with water and dike for later disposal.

*First Aid
Move victim to fresh air; call emergency medical care.
In case of contact with material, immediately flush eyes with running water
   for at least 15 min. Wash skin with soap and water.
Remove and isolate contaminated clothing and shoes at the site.
Keep victim quiet and maintain normal body temperature.

▼   ▼   ▼   ▼   ▼   ▼   ▼   ▼   ▼   ▼   ▼   ▼   ▼   ▼

--- IDENTIFIERS ---

Chemical Name and Synonyms: TERT-BUTYL PEROXYMALEIC ACID, $C_8H_{12}O_5$

Trade Name and Synonyms: Luperox PMA

Chemical Family: Organic peroxide

Manufacturer: Lucidol Division, Pennwalt Corp., Buffalo, New York

UN ID Number: NA

--- INGREDIENTS AND COMPOSITION DATA ---

Material	Cas No.	%	TLV
t-Butyl peroxymaleic acid	NA	98	16 mg/kg (mouse)

--- PHYSICAL DATA ---

Boiling Point (°F): NA

Specific Gravity (H_2O = 1): NA

Vapor Pressure (mm Hg): NA

Percent Volatile by Volume (%): NA

Vapor Density (air = 1): NA

Freezing Point: NA

Evaporation Rate: NA

pH: NA

Solubility in Water: NA

Melting Point: 114° to 116°C (with decomposition)

Appearance/Color/Odor: Fine, white powder.

--- FIRE AND EXPLOSION HAZARDS ---

Flash Point: NA

Flammability Limits: NA

Autoignition Temperature: NA

Extinguishing Media: Foam or water spray, dry chemical, carbon dioxide.

Special Fire Fighting Procedures: Use water to keep fire-exposed containers cool. Use self-contained breathing apparatus. If fire is large, evacuate area and fight fire from a safe distance. Cool surrounding area with water.

If confined during exposure to a fire, can decompose with force. Material is subject to decomposition under friction or heavy shock.

Unusual Fire and Explosion Hazards: Exposure of containers to fire results in rapid product decomposition, container pressure buildup, and failure, followed by vigorous burning with flare effect. Cleanup should not be attempted until all the product has cooled completely. May form flammable dust-air mixtures.

Hazardous Products of Combustion and Decomposition: No information provided by manufacturer.

Hazardous Polymerization: Will not occur.

--- HEALTH HAZARD DATA ---

Effects of Overexposure: Unknown

Emergency Response and First Aid:
Skin: Wash thoroughly with soap and running water. Remove contaminated clothing and wash before reuse.
Eyes: In case of contact, immediately flush with plenty of low-pressure water for at least 15 min. Remove any contact lenses to ensure thorough flushing. Call a physician.
Inhalation: Remove to fresh air and, if indicated, give artificial respiration. If breathing is difficult, give oxygen. Call a physician. Treat immediately.

Ingestion: If conscious, the person should immediately drink large quantities of liquid to dilute this product. *Never* give liquids to an unconscious person. Call a physician.

Cancer Information: Not listed as a carcinogen by NTP (National Toxicology Program); not regulated as a carcinogen by OSHA; not evaluated by IARC (International Agency for Research on Cancer).

--- TOXICITY DATA ---

Animal Effects: LD50 for mouse reported at approximately 16 mg/kg.

--- REACTIVITY DATA ---

Stability and Conditions to Avoid: Extended exposure to 120°F or higher temperatures may lead to rapid decomposition. Do not subject dry peroxide to grinding or other sources of friction.

Materials to Avoid (Incompatibility): Strong oxidizing and reducing agents, mineral acids, and accelerators.

Hazard Decomposition Products: Acetophenone, cumyl alcohol, alpha-methyl styrene, methane, and ethane. Under acidic decomposition conditions, phenols may also be formed.

Hazardous Polymerization: Will not occur.

--- SPILL AND EMERGENCY RESPONSE ---

For Spills: Spilled material may be wetted with water and/or blended with a noncombustible absorbent such as vermiculite. Sweep up with nonsparking utensils.

Waste Disposal Method: Place absorbed material in shallow trench and ignite from safe distance with a long torch (6 ft minimum). Any disposal must comply with local, state, and federal regulations.

Empty Containers: The container for this product can present explosion or fire hazards, even when emptied. To avoid the risk of injury, do not cut, puncture, or weld on or near this container.

--- SPECIAL PROTECTION AND PRECAUTIONS ---

Respiratory Protection: Appropriate respirator selected and used in accordance with OSHA Subpart I (29 CFR 1910.134) and manufacturer's recommendations is required if airborne dust is not adequately controlled or if excessive fumes are generated from product decomposition. Self-contained breathing apparatus should be worn in high vapor and dust concentrations.

Type of Ventilation Required: General mechanical. Avoid breathing dusts and vapors. Adequate ventilation should be provided to keep mist and dust concentrations below acceptable exposure limits. Discharge from the ventilation system should comply with applicable air-pollution regulations.

Eye Protection Requirement: Chemical goggles. Eyewash fountains and safety showers should be easily accessible.

Protective Gloves and Other: Impervious gloves and clothing should be worn.

Storage and Handling Precautions: Store in cool, dry location away from all sources of heat, spark, and open flames. Avoid ignition sources such as sparks and flames. Ground all equipment. In addition, when emptying bags where flammable vapors may be present, blanket vessel with inert gas, ground operator, and pour material slowly into conductive, grounded chute.

Store below 86°C in original containers in a ventilated area out of direct sunlight and away from heat sources. Storage at higher temperatures risks melting of the products organic peroxide and the information of lumps and cakes. Also store below 86°F to maintain active oxygen content.

This material may react with acids and acid-like materials, strong oxidizing and reducing agents, oxidation catalysts, copper, copper alloys, lead, and iron and should not be stored near such materials.

Other Precautions: Use all standard practices for good personal hygiene. Completely isolate and thoroughly clean all equipment, piping, or vessels before beginning maintenance or repairs. Keep area clean. Note that this product will burn.

--- REGULATORY INFORMATION ---

Refer to other organic peroxides reported in this section.

▼   ▼   ▼   ▼   ▼   ▼   ▼   ▼   ▼   ▼   ▼   ▼   ▼   ▼

--- IDENTIFIERS ---

Name: SULFUR
Synonyms: Bensulfoid; Brimstone; Collokit; Colloidal-S; Colloidal Sulfur;
Colsul; Corosul D; Corosul S; Cosan; Cosan 80; Crystex; Flour Sulfur;
Flowers of Sulfur; Ground Vocle Sulfur; Hexasul; Kolofog; Kolospray;
Kumulus; Magnetic 70; Magnetic 90; Magnetic 95; Microflotox; Precipitated
Sulfur; Sofril; Sperlox-S; Spersul; Spersul Thiovit; Sublimed Sulfur; Sulfidal;
Sulforon; Sulfur, Solid; Sulkol; Super Cosan; Sulphur; Sulfur (DOT); Sulsol;
Technetium Tc 99M Sulfur Colloid; Tesuloid; Thiolux; Thiovit
CAS: 7704-34-9; **RTECS:** WS4250000
Formula: S; **Mol Wt:** 32.06
WLN: S
Chemical Class: Nonmetal

See other identifiers listed below under Regulations.

--- PROPERTIES ---

Boiling Point: 717.75 K; 444.6°C; 832.2°F
Melting Point: 392.15 K; 119°C; 246.2°F
Flash Point: NA
Autoignition: NA
UEL: NA
LEL: NA
Ionization Potential (eV): 16.80
Vapor Density: No data
Specific Gravity: No data
Density: 2.070

Reactivity with Water: No data on water reactivity
Reactivity with Common Materials: No data
Stability during Transport: No data
Neutralizing Agents: No data
Polymerization Possibilities: No data

Toxic Fire Gases: None reported other than possible unburned vapors
Odor Detected at (ppm): Unknown
Odor Description: No data
100% Odor Detection: No data

--- REGULATIONS ---

DOT hazard class: 9 CLASS 9
DOT guide: 32
Identification number: NA1350
DOT shipping name: Sulfur
Packing group: III
Label(s) required: CLASS 9
Special Provisions: A1
Packaging exceptions: 173.151
Nonbulk packaging: 173.213
Bulk packaging: 173.240
Quantity limitations:
Passenger air/rail: 25 kg
Cargo aircraft only: 100 kg
Vessel stowage: A
Other stowage provisions: 19, 74

STCC Number: Not listed

Clean Water Act Sect. 307: No
Clean Water Act Sect. 311: No
Clean Air Act: Not listed
EPA Waste Number: None
CERCLA Ref: Not listed
RQ Designation: Not listed
SARA TPQ Value: Not listed
SARA Sect. 312 Categories:
Fire hazard: flammable.

U.S. Postal Service Mailability:
Hazard class: ORM-C
Mailability: Domestic service and air transportation shipper's declaration
Max per parcel: 25 lb

NFPA Codes:
Health Hazard (blue): (2) Hazardous to health. Area may be entered with self-contained breathing apparatus.
Flammability (red): (1) This material must be preheated before ignition can occur.
Reactivity (yellow): (0) Stable even under fire conditions.
Special: Unspecified.

--- TOXICITY DATA ---

Short-term Toxicity: Unknown

Long-term Toxicity: Unknown

Symptoms: Human eye irritant at 6 ppm; chronic inhalation can cause irritation of mucous membrane; conjunctivitis, inflammation of skin, irritation of respiratory tract (Source: SAX; THIC).

Conc IDLH: Unknown

NIOSH REL: Not given

ACGIH TLV: Not listed
ACGIH STEL: Not listed

OSHA PEL: Not in Table Z-1-A

MAK Information: Not listed

Carcinogen: N; **Status:** See below

Carcinogen Lists: *IARC*: Not listed; *MAK*: Not listed; *NIOSH*: Not listed; *NTP*: Not listed; *ACGIH*: Not listed; *OSHA*: Not listed.

Human Toxicity Data: (Source: NIOSH RTECS)

LD50 Value: No LD50 in RTECS 1992

Other Species Toxicity Data: (Source: NIOSH RTECS 1992)
ivn-rat LDLo:8 mg/kg
ivn-dog LDLo:10 mg/kg
orl-rbt LDLo:175 mg/kg
ivn-rbt LDLo:5 mg/kg
ipr-gpg LDLo:55 mg/kg

Irritation Data: (Source: NIOSH RTECS 1992)
eye-hmn 8 ppm

Reproductive Toxicity (1992 RTECS): This chemical has no known mammalian reproductive toxicity.

--- PROTECTION AND FIRST AID ---

First Aid Source: DOT Emergency Response Guide 1993.
Move victim to fresh air; call emergency medical care. In case of contact
with material, immediately flush eyes with running water for at least 15 min.
Removal of solidified molten material from skin requires medical assistance.
Remove and isolate contaminated clothing and shoes at the site.

--- INITIAL INCIDENT RESPONSE ---

U.S. Department of Transportation Guide to Hazardous Materials Transport
Information - Publication DOT 5800.5 (1990).
DOT Shipping Name: Sulfur
DOT ID Number: NA1350

ERG90   GUIDE 32
* POTENTIAL HAZARDS *

*Health Hazards
Contact may cause burns to skin and eyes.
Fire may produce irritating or poisonous gases.
Runoff from fire control or dilution water may cause pollution.

*Fire or Explosion
Flammable/combustible material; may be ignited by heat, sparks, or flames.
May burn rapidly with flare-burning effect.

* EMERGENCY ACTION *

Keep unnecessary people away; isolate hazard area and deny entry.
Stay upwind; keep out of low areas.
Positive pressure self-contained breathing apparatus (SCBA) and structural
    firefighters' protective clothing will provide limited protection.
**CALL CHEMTREC AT 1-800-424-9300 FOR EMERGENCY ASSIS-
TANCE.** If water pollution occurs, notify the appropriate authorities.

*Fire
Small fires: Dry chemical, sand, earth, water spray, or regular foam.
Large fires: Water spray, fog, or regular foam.
Move container from fire area if you can do it without risk.
Apply cooling water to sides of containers that are exposed to flames until
    well after fire is out. Stay away from ends of tanks.

For massive fire in cargo area, use unmanned hose holder or monitor nozzles; if this is impossible, withdraw from area and let fire burn.
Magnesium fires: Use dry sand, Met-L-X R powder, or G-1 graphite powder.

*Spill or Leak
Shut off ignition sources; no flares, smoking, or flames in hazard area.
Do not touch or walk through spilled material; stop leak if you can do it without risk.
Small dry spills: With clean shovel, place material into clean, dry container and cover loosely; move containers from spill area.
Large spills: Wet down with water and dike for later disposal.

*First Aid
Move victim to fresh air; call emergency medical care.
In case of contact with material, immediately flush eyes with running water for at least 15 min.
Removal of solidified molten material from skin requires medical assistance.
Remove and isolate contaminated clothing and shoes at the site.

▼　▼　▼　▼　▼　▼　▼　▼　▼　▼　▼　▼　▼　▼

--- IDENTIFIERS ---

Chemical Name and Synonyms: *N*-OXYDIETHYLENEBENZOTHIA-ZOLE-2-SULFENAMIDE, $C_2H_4SNCSNC_4H_8O$

Trade Name and Synonyms: Amax

Chemical Family: Sulfenamide

Manufacturer: R. T. Vanderbilt Co., Inc., Norwalk, Connecticut

UN ID Number: NA

--- INGREDIENTS AND COMPOSITION DATA ---

Material	Cas No.	%	TLV
N-Oxydiethylenebenzo-thiazole-2-Sulfenamise	NA	98 min.	Acute oral, LD50, 4000 mg/kg mice

--- PHYSICAL DATA ---

Boiling Point (°F): NA

Specific Gravity (H₂O = 1): 1.37

Vapor Pressure (mm Hg): NA

Percent Volatile by Volume (%): NA

Vapor Density (air = 1): NA

Freezing Point: NA

Evaporation Rate: NA

pH: NA

Solubility in Water: Negligible

Melting Point: NA

Appearance/Color/Odor: Tan flakes.

--- FIRE AND EXPLOSION HAZARDS ---

Flash Point: NA

Flammability Limits: NA

Autoignition Temperature: NA

Special Fire Fighting Procedures: Use water to keep fire-exposed containers cool. Use self-contained breathing apparatus. If fire is large, evacuate area and fight fire from a safe distance. Cool surrounding area with water.

If confined during exposure to a fire, can decompose with force. Material is subject to decomposition under friction or heavy shock.

Unusual Fire and Explosion Hazards: Exposure of containers to fire results in rapid product decomposition, container pressure buildup and failure, followed by vigorous burning with flare effect. Cleanup should not be attempted until all the product has cooled completely. May form flammable dust-air mixtures.

Hazardous Products of Combustion and Decomposition: Toxic fumes of sulfur and nitrogen oxides at decomposition temperatures.

Hazardous Polymerization: Will not occur.

--- HEALTH HAZARD DATA ---

Effects of Overexposure: Unknown

Emergency Response and First Aid:
Skin: Wash thoroughly with soap and running water. Remove contaminated clothing and wash before reuse.
Eyes: In case of contact, immediately flush with plenty of low-pressure water for at least 15 min. Remove any contact lenses to ensure thorough flushing. Call a physician.
Inhalation: Remove victim to fresh air and, if indicated, give artificial respiration. If breathing is difficult, give oxygen. Call a physician. Treat immediately.
Ingestion: If conscious, the person should immediately drink large quantities of liquid to dilute this product. *Never* give liquids to an unconscious person. Call a physician.

Cancer Information: Not listed as a carcinogen by NTP (National Toxicology Program); not regulated as a carcinogen by OSHA; not evaluated by IARC (International Agency for Research on Cancer).

--- TOXICITY DATA ---

Animal Effects: LD50 for mouse reported at approximately 4000 mg/kg.

--- REACTIVITY DATA ---

Stability and Conditions to Avoid: None reported by manufacturer.

Materials to Avoid (Incompatibility): Avoid contact with acids or acidic conditions.

Hazard Decomposition Products: Unknown

Hazardous Polymerization: Will not occur.

--- SPILL AND EMERGENCY RESPONSE ---

For Spills: Sweep spillage; wash residuals with soap and water and transfer to a closed container.

Waste Disposal Method: Same as for organic chemicals.

Empty Containers: The container for this product can present explosion or fire hazards, even when emptied. To avoid the risk of injury, do not cut, puncture, or weld on or near this container.

--- SPECIAL PROTECTION AND PRECAUTIONS ---

Respiratory Protection: Appropriate respirator selected and used in accordance with OSHA Subpart I (29 CFR 1910.134) and manufacturer's recommendations is required if airborne dust is not adequately controlled or if excessive fumes are generated from product decomposition. Self-contained breathing apparatus should be worn in high vapor and dust concentrations. Under normal use, manufacturer recommends a dust mask.

Type of Ventilation Required: General mechanical. Avoid breathing dusts and vapors. Adequate ventilation should be provided to keep mist and dust concentrations below acceptable exposure limits. Discharge from the ventilation system should comply with applicable air-pollution regulations.

Eye Protection Requirements: Chemical goggles. Eyewash fountains and safety showers should be easily accessible.

Protective Gloves and Other: Impervious gloves and clothing should be worn. Manufacturer recommends rubber gloves be used.

Storage and Handling Precautions: Store in cool, dry location away from all sources of heat, spark, and open flames. Avoid ignition sources such as sparks and flames.

This material may react with acids and acid-like materials and strong oxidizing and reducing agents and should not be stored near such materials.

Other Precautions: Use all standard practices for good personal hygiene. Completely isolate and thoroughly clean all equipment, piping, or vessels before beginning maintenance or repairs. Keep area clean. Note that this product will burn.

--- REGULATORY INFORMATION ---

No information found.

▼ ▼ ▼ ▼ ▼ ▼ ▼ ▼ ▼ ▼ ▼ ▼ ▼ ▼

--- IDENTIFIERS ---

Name: VALERIC ACID, 4,4-BIS(tert-BUTYLPEROXY)-, BUTYL ESTER
Synonyms: 4,4-Bis(tert-Butylperoxy)Valeric Acid Butyl Ester; *n*-Butyl-4,4-Di(tert-Butylperoxy)Valerate, technical pure (DOT); Pentanoic Acid, 4,4-Bis((1,1-Dimethylethyl)Dioxy)-, Butyl Ester; Trigonox 17/40; UN 2140 (DOT)
CAS: 995-33-5; **RTECS:** YV6191200
Formula: $C_{17}H_{34}O_6$; **Mol Wt:** 334.51
Chemical Class: Ester

See other identifiers listed below under Regulations.

--- PROPERTIES ---

Boiling Point: NA
Melting Point: NA
Flash Point: NA
Autoignition: NA
UEL: NA
LEL: NA
Vapor Density: No data
Specific Gravity: No data

Reactivity with Water: No data on water reactivity
Reactivity with Common Materials: No data
Stability during Transport: No data
Neutralizing Agents: No data
Polymerization Possibilities: No data

Toxic Fire Gases: None reported other than possible unburned vapors
Odor Detected at (ppm): Unknown
Odor Description: No data
100% Odor Detection: No data

--- REGULATIONS ---

DOT hazard class: 5.2
DOT guide: 48
Identification number: UN3103
DOT shipping name: ORGANIC PEROXIDE TYPE C, LIQUID
Packing group: II
Label(s) required: ORGANIC PEROXIDE
Packaging exceptions: 173.152
Nonbulk packaging: 173.225
Bulk packaging: 173.None
Quantity limitations:
  Passenger air/rail: 5 L
  Cargo aircraft only: 10 L
  Vessel stowage: D
  Other stowage provisions: 12, 40

STCC Number: Not listed

Clean Water Act Sect. 307: No
Clean Water Act Sect. 311: No
Clean Air Act: Not listed
EPA Waste Number: None
CERCLA Ref: Not listed
RQ Designation: Not listed
SARA TPQ Value: Not listed
SARA Sect. 312 Categories: No category

U.S. Postal Service Mailability: Not given

--- TOXICITY DATA ---

Short-term Toxicity: Unknown

Long-term Toxicity: Unknown

Conc IDLH: Unknown

NIOSH REL: Not given

ACGIH TLV: Not listed
ACGIH STEL: Not listed

OSHA PEL: Not in Table Z-1-A

MAK Information: Not listed

Carcinogen: N; **Status:** See below

Carcinogen Lists: *IARC*: Not listed; *MAK*: Not listed; *NIOSH*: Not listed; *NTP*: Not listed; *ACGIH*: Not listed; *OSHA*: Not listed.

LD50 Value: No LD50 in RTECS 1992

Reproductive Toxicity (1992 RTECS): This chemical has no known mammalian reproductive toxicity.

--- PROTECTION AND FIRST AID ---

First Aid Source: DOT Emergency Response Guide 1993.
Move victim to fresh air; call emergency medical care. In case of contact with material, immediately flush eyes with running water for at least 15 min. Wash skin with soap and water. Remove and isolate contaminated clothing and shoes at the site. Keep victim quiet and maintain normal body temperature.

--- INITIAL INCIDENT RESPONSE ---

U.S. Department of Transportation Guide to Hazardous Materials Transport Information - Publication DOT 5800.5 (1990).
DOT Shipping Name: ORGANIC PEROXIDE TYPE C, LIQUID
DOT ID Number: UN3103

ERG90   GUIDE 48
                            * POTENTIAL HAZARDS *

*Health Hazards
Contact may cause burns to skin and eyes.
Fire may produce irritating or poisonous gases.
Runoff from fire control or dilution water may cause pollution.

*Fire or Explosion
May be ignited by heat, sparks, and flames.
Container may explode in heat of fire.
May explode from heat or contamination.
Runoff to sewer may create fire or explosion hazard.

* EMERGENCY ACTION *

Keep unnecessary people away; isolate hazard area and deny entry.
Stay upwind; keep out of low areas.
Positive-pressure self-contained breathing apparatus (SCBA) and structural
firefighters' protective clothing will provide limited protection.
**CALL CHEMTREC AT 1-800-424-9300 FOR EMERGENCY ASSIS-
TANCE.** If water pollution occurs, notify the appropriate authorities.

*Fire
Small fires: Dry chemical, CO_2, water spray, or regular foam.
Large fires: Flood fire area with water.
Apply cooling water to sides of containers that are exposed to flames until
well after fire is out. Stay away from ends of tanks.
For massive fire in cargo area, use unmanned hose holder or monitor nozzles;
if this is impossible, withdraw from area and let fire burn.

*Spill or Leak
Shut off ignition sources; no flares, smoking, or flames in hazard area.
Do not touch or walk through spilled material; stop leak if you can do it
without risk.
Small spills: Take up with inert, damp noncombustible material; move
containers from spill area.
Large spills: Wet down with water and dike for later disposal.

*First Aid
Move victim to fresh air; call emergency medical care.
In case of contact with material, immediately flush eyes with running water
for at least 15 min. Wash skin with soap and water.
Remove and isolate contaminated clothing and shoes at the site.
Keep victim quiet and maintain normal body temperature.

▼   ▼   ▼   ▼   ▼   ▼   ▼   ▼   ▼   ▼   ▼   ▼   ▼   ▼

--- IDENTIFIERS ---

Chemical Name and Synonyms: MIXTURE OF INSOLUBLE SULFUR
AND RUBBER PROCESS OIL

Trade Name and Synonyms: Crystex

Chemical Family: Polymeric Sulfur

Manufacturer: Akzo Chemicals, Inc., Chicago, Illinois

UN ID Number: UN1350

--- INGREDIENTS AND COMPOSITION DATA ---

Material	Cas No.	%	TLV
Polymeric sulfur	9035-99-8	72	Not established
Sulfur	7704-34-9	8	
Naphthenic oils	64741-96-4/ 64742-52-5	20	

--- PHYSICAL DATA ---

Boiling Point (°F): 560°F (293°C) oil fraction;
          832°F (444°C) sulfur fraction

Specific Gravity (H$_2$O = 1): 1.57 at 68°F/68°F (20°C/20°C)

Vapor Pressure (mm Hg): NA

Percent Volatile by Volume (%): NA

Vapor Density (air = 1): NA

Freezing Point: NA

Evaporation Rate: NA

pH: NA

Solubility in Water: Not in water.   Oil fraction soluble in most organic solvents; sulfur fraction slightly soluble in organic solvents.   Sulfur fraction is soluble up to 10% in carbon disulfide.

Melting Point: NA

Appearance/Color/Odor: Yellow powder.

--- FIRE AND EXPLOSION HAZARDS ---

Flash Point: NA

Flammability Limits: NA

Autoignition: NA

Extinguishing Media: Foam or water spray, dry chemical, carbon dioxide.

Special Fire Fighting Procedures: As in any fire, prevent human exposure to fire, smoke, fumes, or products of combustion.   Evacuate nonessential personnel from the fire area.   Firefighters should wear full-face, self-contained breathing apparatus and impervious protective clothing such as gloves, hoods, suits, and rubber boots.

Use water fog, dry chemicals, or carbon dioxide to extinguish fires involving this material.   Solid streams of water should not be used because of the possibility of dispersing dust clouds of sulfur in the air, possibly causing an explosion.   Cool the surrounding area with water fog to prevent ignition.   Fires will rekindle before the mass is cooled down below 428°F (220°C).

Unusual Fire and Explosion Hazards: Sulfur dust is considered flammable and a fire hazard.   Sulfur dust that is suspended in air ignites easily and results in an explosion in confined areas.   Ignition can be caused by heat sources, friction, and static electricity developed by the movement of sulfur dust in the air.   Supports combustion and decomposes under fire conditions to give off toxic sulfur dioxide.   *Do not* use welding or cutting torch on or near any container of this material, even empty, because an explosion could occur.   *Do not* use, pour, spill, or store near heat or open flame.

Hazardous Products of Combustion and Decomposition: Toxic sulfur dioxide.

Hazardous Polymerization: Will not occur.

--- HEALTH HAZARD DATA ---

Effects of Overexposure: Principal routes of exposure are skin contact and inhalation.

Exposure to large amounts of the dust may cause eye and respiratory irritation accompanied by coughing, sneezing, and tears. If ignited, the material generates toxic fumes of sulfur dioxide, large quantities of which are capable of producing irritation of the skin, eyes, and mucous membranes. Serious cases may result in pulmonary edema and may be fatal.

Ingestion of the product in large quantities may cause mild digestive disorders, including diarrhea.

There are no data available that address medical conditions that are generally recognized as being aggravated by exposure to this product. Refer to the section on Toxicology Data.

Emergency Response and First Aid:
Skin: Flush all affected areas with plenty of water for several minutes. Remove and clean any contaminated clothing and shoes. Seek medical attention if skin irritation occurs.
Eyes: Flush the eyes with plenty of running water for at least 15 min. Hold the eyelids apart during the flushing to ensure rinsing of the surface of the eye and lids with water. Seek medical attention if eye irritation occurs.
Inhalation: Remove to fresh air and, if indicated, give artificial respiration. If breathing is difficult, give oxygen. Call a physician. Treat immediately.
Ingestion: If swallowed, immediately give several glasses of water and induce vomiting by gagging the victim with a finger placed on the back of the victim's tongue. Give fluids until vomitus is clear. If victim is unconscious or convulsing, do not induce vomiting or give anything by mouth.

--- TOXICITY DATA ---

Animal Effects:
Ingestion: The acute oral LD50 is approximately 4640 mg/kg in male rats.
Skin contact: Nonirritant to rabbit skin following a 4-hr exposure.

Eye contact: Mild irritant to rabbit eyes.

--- REACTIVITY DATA ---

Stability and Conditions to Avoid: Heat, open flames.

Materials to Avoid (Incompatibility): Material is corrosive to copper and copper alloys. In the presence of water, sulfur will attack steel. Carbon steel is the most common material of construction for process equipment for the dry product, although stainless steels are sometimes used. Material is incompatible with chlorates, nitrates, oxidizing agents, amines, strong bases.

Hazard Decomposition Products: Sulfur dioxide

Hazardous Polymerization: Will not occur.

--- SPILL AND EMERGENCY RESPONSE ---

For Spills: Sweep spillage; wash residuals with soap and water and transfer to a closed container.

Waste Disposal Method: Same as for organic chemicals.

Empty Containers: The container for this product can present explosion or fire hazards, even when emptied. To avoid the risk of injury, do not cut, puncture, or weld on or near this container.

--- SPECIAL PROTECTION AND PRECAUTIONS ---

Respiratory Protection: Appropriate respirator selected and used in accordance with OSHA Subpart I (29 CFR 1910.134) and manufacturer's recommendations is required if airborne dust is not adequately controlled or if excessive fumes are generated from product decomposition. Self-contained breathing apparatus should be worn in high vapor and dust concentrations. Under normal use, manufacturer recommends a dust mask.

Type of Ventilation Required: General mechanical. Avoid breathing dusts and vapors. Adequate ventilation should be provided to keep mist and dust concentrations below acceptable exposure limits. Discharge from the ventilation system should comply with applicable air-pollution regulations.

Eye Protection Requirements: Chemical goggles. Eyewash fountains and safety showers should be easily accessible.

Protective Gloves and Other: Impervious gloves and clothing should be worn. Manufacturer recommends rubber gloves be used.

Storage and Handling Precautions: Containers should be stored in a cool, dry, well-ventilated area. Store away from flammable materials, sources of heat, flame, sparks, and foodstuffs. Exercise due caution to prevent damage to or leakage from the container. Avoid any conditions that might tend to create a dust explosion. Maintain good housekeeping practices to minimize dust buildup and dispersion. Separate from chlorates, nitrates, and other oxidizing agents. Store away from basic amines and other strong bases.

This material may react with acids and acid-like materials and strong oxidizing and reducing agents and should not be stored near such materials.

Other Precautions: Use all standard practices for good personal hygiene. Completely isolate and thoroughly clean all equipment, piping, or vessels before beginning maintenance or repairs. Keep area clean. Note that this product will burn.

--- REGULATORY INFORMATION ---

Refer to DOT shipping regulations.

▼   ▼   ▼   ▼   ▼   ▼   ▼   ▼   ▼   ▼   ▼   ▼   ▼   ▼

--- IDENTIFIERS ---

Name: BIS(DIMETHYLTHIOCARBAMYL) DISULFIDE
Synonyms: Thiram; Bis((Dimethylamino)Carbonothioyl) Disulphide; Disolfuro Di Tetrametiltiourame (Italian); Disulfure De Tetramethylthiourame; Alpha,Alpha'-Dithiobis(Dimethylthio) Formamide; N,N'-(Dithiodicarborothioyl)Bis(N-Methyl-Methanamine); Methyl Thiuramdisulfide; Tetramethyl Thiuram Disulfide; Tetramethyl Diurane Sulfite; Tetramethylene Thiuram Disulfide; Tetramethylthiocarbamoyl Disulfide; Tetramethylthioperoxydicarbonic Diamide; Tetramethyl Thioram Disulfide (Dutch); Tetramethylthiram Disulfid (German); Tetramethylthiuram Bisulfide; Tetramethylthiuram Disulfide; N,N,N',N'-Tetramethyl Thiuram Disulfide; Tetramethyl Thiurane Disulfide; Tetramethylthiuram Disulfide; Thiram; Tiuram (Polish); USAF B-30; USAF EK-2089; USAF P-5; TMTD; ENT 987; SQ 1489; NSC 1771;

Thiurad; Thiosan; Thylate; Tiuramyl; Thiuramyl; Duralyn; Fernasan; Nowergan; Pezifilm; Tersan; Tuads; Tulisan; Arasan; HCDB
CAS: 137-26-8; **RTECS:** JO1400000
Formula: $C_6H_{12}N_2S_4$; **Mol Wt:** 240.44
WLN: 1N1
Chemical Class: Organic sulfide

See other identifiers listed below under Regulations.

--- PROPERTIES ---

Physical Description: Colorless to cream solid (some commercial products dyed blue)
Boiling Point: Decomposes
Melting Point: 413.15 K; 140°C; 284°F
Flash Point: 362 K; 88.8°C; 191.9°F
Autoignition: NA
Vapor Pressure: About 0 mm
UEL: NA
LEL: NA
Vapor Density: No data
Specific Gravity: 1.43 at 20°C
Density: 1.43 g/cc or 13.299 lb/gal
Water Solubility: INSOL
Incompatibilities: Strong oxidizers, strong acids, oxidizable materials

Reactivity with Water: No data on water reactivity
Reactivity with Common Materials: No data
Stability during Transport: No data
Neutralizing Agents: No data
Polymerization Possibilities: No data

Toxic Fire Gases: Oxides of nitrogen, oxides of sulfur
Odor Detected at (ppm): Data not available
Odor Description: Data not available (Source: CHRIS)
100% Odor Detection: No data

--- REGULATIONS ---

DOT hazard class: 9 CLASS 9
DOT guide: 31
Identification number: UN3077

DOT shipping name: ENVIRONMENTALLY HAZARDOUS SUBSTANCES, SOLID, N.O.S.
Packing group: III
Label(s) required: CLASS 9
Special Provisions: 8, B54
Packaging exceptions: 173.155
Nonbulk packaging: 173.213
Bulk packaging: 173.240
Quantity limitations:
Passenger air/rail: None
Cargo aircraft only: None
Vessel stowage: A

STCC Number: 4921632, 4921631, 4921634, 4921633

Clean Water Act Sect. 307: No
Clean Water Act Sect. 311: No
Clean Air Act: Not listed
EPA Waste Number: U244
CERCLA Ref: Y
RQ Designation: A; 10 lb (4.54 kg) CERCLA
SARA TPQ Value: Not listed
SARA Sect. 312 Categories:
Acute toxicity: toxic. LD50 > 50 and < = 500 mg/kg (oral rat).
Acute toxicity: irritant.
Acute toxicity: adverse effect to target organs.
Chronic toxicity: adverse effect to target organs after long periods of exposure.
Chronic toxicity: reproductive toxin.

U.S. Postal Service Mailability:
Hazard class: ORM-A
Mailability: Domestic service and air transportation; shipper's declaration
Max per parcel: 70 lb; 5 lb (air)

NFPA Codes:
Health Hazard (blue): Unspecified
Flammability (red): Unspecified
Reactivity (yellow): Unspecified
Special: Unspecified

--- TOXICITY DATA ---

Short-term Toxicity: *Inhalation:* No information on human exposure is available. Animal studies indicate that irritation of the nose and throat may occur at levels above 5 mg/m^3. *Eyes:* May cause irritation, tearing, and sensitivity to light. *Skin:* Exposure to spray containing 45% thiram resulted in irritation and skin sensitization. *Ingestion:* No information available on human exposure. In animal studies, 38 ppm in food caused nausea, vomiting, diarrhea, hyperexcitability, weakness, and loss of muscle control. Death may occur from ingestion of approximately one teaspoonful. (NYDH)

Long-term Toxicity: Occupational exposures to 0.03 mg/m^3 over a 5-yr period has caused mild irritation of the nose and throat. Prolonged contact has caused eye irritation, tearing, increased sensitivity to light, reduced night vision, and blurred vision. Thiram has caused birth defects in laboratory animals. Whether it has this effect in humans is not known. (NYDH)

Target Organs: Skin, mucous membranes, kidneys, respiratory system, lungs.

Symptoms: Inhalation of dust may cause respiratory irritation. Liquid irritates eyes and skin and may cause allergic eczema in sensitive individuals. Ingestion causes nausea, vomiting, and diarrhea, all of which may be persistent; paralysis may develop. (Source: CHRIS)

Conc IDLH: 1500 mg/m^3

ACGIH TLV: TLV = 1 mg/m^3
ACGIH STEL: Not listed

OSHA PEL: Transitional Limits: PEL = 5 mg/m^3; Final Rule Limits: TWA = 5 mg/m^3

MAK Information: 5 calculated as total dust mg/m^3; substance with systemic effects, onset of effect less than or equal to 2 hr: Peak = 5xMAK for 30 min, 2 times per shift of 8 hr.

Carcinogen: N; **Status:** See below
References: Animal indefinite IARC** 12, 225, 76; Human indefinite IARC** 12, 225, 76

Carcinogen Lists: *IARC*: Not classified as to human carcinogenicity or probably not carcinogenic to humans; *MAK*: Not listed; *NIOSH*: Not listed; *NTP*: Not listed; *ACGIH*: Not listed; *OSHA*: Not listed.

Human Toxicity Data: (Source: NIOSH RTECS)
ihl-hmn TCLo:30 mg/m^3/5Y-I VRDEA5 (10), 136, 71
  Sense Organs
    Eye
      Conjunctive irritation
  Cardiac
    Other changes
  Lungs, Thorax, or Respiration
    Structural of function changes in trachea or bronchi

LD50 Value: orl-rat LD50:560 mg/kg

Other Species Toxicity Data: (Source: NIOSH RTECS 1992)
orl-rat LD50:560 mg/kg
ihl-rat LC50:500 mg/m^3/4H
skn-rat LD50: >5 g/kg
ipr-rat LD50:138 mg/kg
scu-rat LD50:646 mg/kg
unr-rat LD50:740 mg/kg
orl-mus LD50:1350 mg/kg
ipr-mus LD50:70 mg/kg
scu-mus LD50:1109 mg/kg
unr-mus LD50:1150 mg/kg
orl-cat LDLo:230 mg/kg
orl-rbt LD50:210 mg/kg
skn-rbt LDLo:1 g/kg
unr-rbt LD50:210 mg/kg
unr-ckn LD50:840 mg/kg
unr-dom LD50:225 mg/kg
unr-mam LD50:400 mg/kg
orl-bwd LD50:300 mg/kg

Reproductive Toxicity (1992 RTECS): This chemical is a mammalian reproductive toxin.

Reproductive Toxicity Data (1992 RTECS):
orl-rat TDLo:1200 mg/kg (7-12D preg) TXAPA9, 35, 83, 76
  Effects on Fertility
    Preimplantation mortality
    Postimplantation mortality
    Litter size (number of fetuses per litter; measured before birth)

orl-rat TDLo:300 mg/kg (15D preg) GISAAA 43(6), 37, 78
  Effects on Embryo or Fetus
    Fetotoxicity (except death, e.g., stunted fetus)
    Fetal death
  Specific Developmental Abnormalities
    Other developmental abnormalities

orl-rat TDLo:1190 mg/kg (16-22D preg/21D post) TXAPA9 35, 83, 76
  Effects on Newborn
    Growth statistics (e.g., reduced weight gain)

orl-rat TDLo:550 mg/kg (1-22D preg) AEEDDS 2, 215, 76
  Effects on Newborn
    Behavioral

orl-rat TDLo:420 mg/kg (1-20D preg) GISAAA 51(6), 23, 86
  Specific Developmental Abnormalities
    Cardiovascular (circulatory) system

par-rat TDLo:400 mg/kg (4-11D preg) BEXBAN 93, 107, 82
  Effects on Embryo or Fetus
    Fetotoxicity (except death, e.g., stunted fetus)
    Fetal death

par-rat TDLo:800 mg/kg (2D male/2D pre) BEXBAN 93, 107, 82
  Effects on Embryo or Fetus
    Fetotoxicity (except death, e.g., stunted fetus)
    Fetal death

orl-mus TDLo:100 mg/kg (6-15D preg) ATXKA8 30, 251, 73
  Specific Developmental Abnormalities
    Craniofacial (including nose and tongue)
    Musculoskeletal system

orl-mus TDLo:300 mg/kg (6-15D preg) ATXKA8 30, 251, 73
  Effects on Fertility
    Postimplantation mortality

orl-mus TDLo:80 mg/kg (3D male) FCTOD7 25, 709, 87
  Paternal Effects
    Spermatogenesis

scu-mus TDLo:90 mg/kg (6014D preg) NTIS** PB223-160
  Effects on Fertility
    Preimplantation mortality
    Litter size (number of fetuses per litter; measured before birth)
  Effects on Embryo or Fetus
    Extra embryonic features (e.g., placenta, umbilical cord)

scu-mus TDLo:1035 mg/kg (6-14D preg) NTIS** PB223-160
  Effects on Embryo or Fetus
    Fetotoxicity (except death, e.g., stunted fetus)

orl-ham TDLo:125 mg/kg (7D preg) TXAPA9 15, 152, 69
  Specific Developmental Abnormalities
    Musculoskeletal system

orl-ham TDLo:250 mg/kg (8D preg) TXAPA9 15, 152, 69
  Specific Developmental Abnormalities
    Craniofacial (including nose and tongue)

orl-ham TDLo:250 mg/kg (7D preg) TXAPA9 15, 152, 69
  Specific Developmental Abnormalities
    Body wall

orl-ham TDLo:300 mg/kg (7D preg) TXAPA9 15, 152, 69
  Specific Developmental Abnormalities
    Central nervous system

--- PROTECTION AND FIRST AID ---

NIOSH Pocket Guide to Chemical Hazards:
  **Wear Appropriate Equipment to Prevent: Reasonable probability of skin
  contact.

  **Wear Eye Protection to Prevent: Reasonable probability of eye contact.

**Exposed Personnel Should Wash: Promptly when skin becomes contaminated.

**Exposed Clothing Should be Changed Daily: If there is any reasonable possibility that the clothing may be contaminated.

**Remove Clothing: Promptly remove nonimpervious clothing that becomes contaminated.

**Reference: NIOSH

Recommended Respiration Protection Source: NIOSH Pocket Guide (85-114)
OSHA (Bis(Dimethylthiocarbamyl)Disulfide)
50 mg/m³: Any chemical cartridge respirator with organic vapor cartridge(s) in combination with a dust, mist and fume filter. Substance reported to cause eye irritation or damage may require eye protection. Any supplied air-respirator. Substance reported to cause eye irritation or damage may require eye protection. Any self-contained breathing apparatus. Substance reported to cause eye irritation or damage may require eye protection.
125 mg/m³: Any powered air-purifying respirator with organic vapor cartridge(s) in combination with a dust, mist, and fume filter. Substance reported to cause eye irritation or damage may require eye protection. Any supplied-air respirator operated in a continuous flow mode. Substance reported to cause eye irritation or damage may require eye protection.
250 mg/m³: Any self-contained breathing apparatus with a full facepiece. Any chemical cartridge respirator with a full facepiece and organic vapor cartridge(s) in combination with a high-efficiency particulate filter. Any supplied-air respirator with a full facepiece. Any powered air-purifying respirator with a tight-fitting facepiece and organic vapor cartridge(s) in combination with a high-efficiency particulate filter. Substance reported to cause eye irritation or damage may require eye protection. Any air-purifying full facepiece respirator (gas mask) with a chin-style or front- or back-mounted organic vapor canister having a high-efficiency particulate filter.
1500 mg/m³: Any supplied-air respirator with a full facepiece and operated in a pressure-demand or other positive-pressure mode.
Emergency or Planned Entry in Unknown Concentrations or IDLH Conditions: Any self-contained breathing apparatus with a full facepiece and operated in a pressure-demand or other positive-pressure mode. Any supplied-air respirator with a full facepiece and operated in a pressure-demand or other positive-pressure mode in combination with an auxiliary self-contained breathing apparatus operated in a pressure-demand or other positive-pressure mode.

Escape: Any air-purifying full facepiece respirator (gas mask) with a chin-style or front- or back-mounted organic vapor canister having a high-efficiency particulate filter. Any appropriate escape-type self-contained breathing apparatus.

First Aid Source: HCDB.
Inhalation: Remove victim from exposure. If breathing has stopped or is difficult, give artificial respiration and call physician.
Ingestion: Call physician; induce vomiting and flow with gastric lavage; treatment thereafter is symptomatic and supportive; avoid fats, oils, and lipid solvents, which enhance absorption; rigorously prohibit ethyl alcohol in all forms for at least 10 days; inform doctor if patient has used alcohol within 48 hr.
Skin: Wash with water; if irritation persists, consult a physician.
Eyes: Wash with water; if irritation persists, consult a physician.

First Aid Source: CHRIS Manual 1991.
Inhalation: Remove victim from exposure; if breathing has stopped or is difficult, give artificial respiration and call physician.
Eyes or skin: Wash with water; if irritation persists, consult a physician.
Ingestion: Call physician; induce vomiting and follow with gastric lavage; treatment thereafter is symptomatic and supportive; avoid fats, oils, and lipid solvents, which enhance absorption; rigorously prohibit ethyl alcohol in all forms for at least 10 days; inform doctor if patient has used alcohol within 48 hr.

First Aid Source: DOT Emergency Response Guide 1993.
In case of contact with material, immediately flush eyes with running water for at least 15 min. Wash skin with soap and water. Remove and isolate contaminated clothing and shoes at the site.

--- INITIAL INCIDENT RESPONSE ---

Fire Extinguishment: Water, dry chemical, carbon dioxide. (CHRIS91)

U.S. Department of Transportation Guide to Hazardous Materials Transport Information - Publication DOT 5800.5 (1990).
DOT Shipping Name: ENVIRONMENTALLY HAZARDOUS
   SUBSTANCES, SOLID, N.O.S.
DOT ID Number: UN3077

* POTENTIAL HAZARDS *

*Health Hazards
Contact may cause burns to skin and eyes.
Fire may produce irritating or poisonous gases.
Runoff from fire control or dilution water may cause pollution.

*Fire or Explosion
Some of these materials may burn, but none of them ignites readily.

* EMERGENCY ACTION *

Keep unnecessary people away; isolate hazard area and deny entry.
Positive-pressure self-contained breathing apparatus (SCBA) and structural fire-
fighters' protective clothing will provide limited protection.
**CALL CHEMTREC AT 1-800-424-9300 FOR EMERGENCY ASSIS-
TANCE.** If water pollution occurs, notify the appropriate authorities.

*Fire
Small fires: Dry chemical, CO_2, water spray, or regular foam.
Large fires: Water spray, fog, or regular foam.
Move container from fire area if you can do it without risk.
Do not scatter spilled material with high-pressure water streams.
Dike fire-control water for later disposal.

*Spill or Leak
Stop leak if you can do it without risk.
Small dry spills: With clean shovel, place material into clean, dry container
and cover loosely; move containers from spill area.
Small spills: Take up with sand or other noncombustible absorbent material
and place into containers for later disposal.
Large spills: Dike far ahead of liquid spill for later disposal. Cover powder
spill with plastic sheet or tarp to minimize spreading.

*First Aid
In case of contact with material, immediately flush eyes with running water
for at least 15 min. Wash skin with soap and water.
Remove and isolate contaminated clothing and shoes at the site.

▼   ▼   ▼   ▼   ▼   ▼   ▼   ▼   ▼   ▼   ▼   ▼   ▼   ▼

--- IDENTIFIERS ---

Chemical Name and Synonyms: DIBENZOYL *p*-QUINONEDIOXIME

Trade Name and Synonyms: Dibenzo GMF

Chemical Family: Vulcanizing agent

Manufacturer: Uniroyal Chemical Co., Naugatuck, Connecticut

UN ID Number: NA

--- INGREDIENTS AND COMPOSITION DATA ---

Material	Cas No.	%	TLV
Dibenzoyl *p*-quinonedioxime	NA	99.9	Not established

--- PHYSICAL DATA ---

Boiling Point (°F): NA

Specific Gravity (H$_2$O = 1): 1.34

Vapor Pressure (mm Hg): NA

Percent Volatile by Volume (%): Not volatile

Vapor Density (air = 1): NA

Freezing Point: NA

Evaporation Rate: NA

pH: Not corrosive

Solubility in Water: Insoluble in water and organic solvents

Melting Point: Decomposes above 175°F

Appearance/Color/Odor: Brownish grey powder.

--- FIRE AND EXPLOSION HAZARDS ---

Flash Point: Over 175°F (Decomposes)

Flammability Limits: NA

Autoignition: NA

Extinguishing Media: Foam or water spray, dry chemical, carbon dioxide.

Special Fire Fighting Procedures: As in any fire, prevent human exposure to fire, smoke, fumes, or products of combustion. Evacuate nonessential personnel from the fire area. Firefighters should wear full-face, self-contained breathing apparatus and impervious protective clothing such as gloves, hoods, suits, and rubber boots.

Use water fog, dry chemicals, or carbon dioxide to extinguish fires involving this material. Solid streams of water should not be used because of the possibility of dispersing dust clouds of sulfur in the air, possibly causing an explosion. Cool the surrounding area with water fog to prevent ignition. Fires will rekindle before the mass is cooled down below 428°F (220°C).

Note that this product will burn in the absence of air once it is ignited. Use water to cool below the autoignition temperature. Fires involving several containers should be approached with caution because of violent burning as containers rupture.

Unusual Fire and Explosion Hazards: *Do not* use welding or cutting torch on or near any container of this material, even empty, because an explosion could occur. *Do not* use, pour, spill, or store near heat or open flame.

Hazardous Products of Combustion and Decomposition: Toxic fumes.

Hazardous Polymerization: Will not occur.

--- HEALTH HAZARD DATA ---

Effects of Overexposure: Principal routes of exposure are skin contact and inhalation. Slight irritation potential reported by the manufacturer.

Ingestion of the product in large quantities may cause mild digestive disorders, including diarrhea.

There are no data available that address medical conditions that are generally recognized as being aggravated by exposure to this product.

Emergency Response and First Aid:
Skin: Flush all affected areas with plenty of water for several minutes. Remove and clean any contaminated clothing and shoes. Seek medical attention if skin irritation occurs.
Eyes: Flush the eyes with plenty of running water for at least 15 min. Hold the eyelids apart during the flushing to ensure rinsing of the surface of the eye and lids with water. Seek medical attention if eye irritation occurs.
Inhalation: Remove to fresh air and, if indicated, give artificial respiration. If breathing is difficult, give oxygen. Call a physician. Treat immediately.
Ingestion: If swallowed, immediately give several glasses of water and induce vomiting by gagging the victim with a finger placed on the back of the victim's tongue. Give fluids until vomitus is clear. If victim is unconscious or convulsing, do not induce vomiting or give anything by mouth.

--- TOXICITY DATA ---

Animal Effects: No toxicology studies reported, but product has been in commercial use for over 25 years.

--- REACTIVITY DATA ---

Stability and Conditions to Avoid: Heat, open flames.

Materials to Avoid (Incompatibility): None reported by the manufacturer.

Hazard Decomposition Products: Unknown.

Hazardous Polymerization: Will not occur.

--- SPILL AND EMERGENCY RESPONSE ---

For Spills: Sweep spillage; wash residuals with soap and water and transfer to a closed container.

Waste Disposal Method: Same as for organic chemicals.

Empty Containers: The container for this product can present explosion or fire hazards, even when emptied. To avoid the risk of injury, do not cut, puncture, or weld on or near this container.

--- SPECIAL PROTECTION AND PRECAUTIONS ---

Respiratory Protection: Appropriate respirator selected and used in accordance with OSHA Subpart I (29 CFR 1910.134) and manufacturer's recommendations is required if airborne dust is not adequately controlled or if excessive fumes are generated from product decomposition. Self-contained breathing apparatus should be worn in high vapor and dust concentrations. Under normal use, manufacturer recommends a dust mask.

Type of Ventilation Required: General mechanical. Avoid breathing dusts and vapors. Adequate ventilation should be provided to keep mist and dust concentrations below acceptable exposure limits. Discharge from the ventilation system should comply with applicable air-pollution regulations.

Eye Protection Requirements: Chemical goggles. Eyewash fountains and safety showers should be easily accessible.

Protective Gloves and Other: Impervious gloves and clothing should be worn. Manufacturer recommends rubber gloves be used.

Storage and Handling Precautions: Containers should be stored in a cool, dry, well-ventilated area. Store away from flammable materials, sources of heat, flame, sparks, and foodstuffs. Exercise due caution to prevent damage to or leakage from the container. Avoid any conditions that might tend to create a dust explosion. Maintain good housekeeping practices to minimize dust buildup and dispersion.

Other Precautions: Use all standard practices for good personal hygiene. Completely isolate and thoroughly clean all equipment, piping, or vessels before beginning maintenance or repairs. Keep area clean. Note that this product will burn.

--- REGULATORY INFORMATION ---

Refer to DOT shipping regulations.

▼ ▼ ▼ ▼ ▼ ▼ ▼ ▼ ▼ ▼ ▼ ▼ ▼

--- IDENTIFIERS ---

Chemical Name and Synonyms: HEXAMETHYLENEDIAMINE CARBA-MATE

Trade Name and Synonyms: Diak No. 1

Chemical Family: Vulcanizing Agent

Manufacturer: E. I. du Pont de Nemours & Co., Inc., Wilmington, Delaware

UN ID Number: NA

--- INGREDIENTS AND COMPOSITION DATA ---

Material	Cas No.	%	TLV
Hexamethylenediamine carbamate	NA	100	Not established

--- PHYSICAL DATA ---

Boiling Point (°F): NA

Specific Gravity (H₂O = 1): 1.28

Vapor Pressure (mm Hg): NA

Percent Volatile by Volume (%): NA

Vapor Density (air = 1): NA

Freezing Point: NA

Evaporation Rate: NA

pH: NA

Solubility in Water: Soluble in water

Melting Point: Above 154°C (309°F)

Appearance/Color/Odor: Very fine white powder. Slight amine odor.

--- FIRE AND EXPLOSION HAZARDS ---

Flash Point: Above 149°C (300°F, open cup method)

Flammability Limits: NA

Autoignition Temperature: NA

Extinguishing Media: Foam or water spray, dry chemical, carbon dioxide, standard fire extinguishers.

Special Fire Fighting Procedures: As in any fire, prevent human exposure to fire, smoke, fumes, or products of combustion. Evacuate nonessential personnel from the fire area. Firefighters should wear full-face, self-contained breathing apparatus and impervious protective clothing such as gloves, hoods, and suits.

Unusual Fire and Explosion Hazards: *Do not* use welding or cutting torch on or near any container of this material, even empty, because an explosion could occur. *Do not* use, pour, spill, or store near heat or open flame. May form flammable dust-air mixture. Keep dust away from heat, sparks, and open flames. Use only in areas provided with grounded equipment.

Hazardous Products of Combustion and Decomposition: Toxic fumes.

Hazardous Polymerization: Will not occur.

--- HEALTH HAZARD DATA ---

Effects of Overexposure: Principal routes of exposure are skin contact and inhalation. Causes irritation to skin and eyes. Harmful if swallowed.

Emergency Response and First Aid:
Skin: Flush all affected areas with plenty of water for several minutes. Remove and clean any contaminated clothing and shoes. Seek medical attention if skin irritation occurs.
Eyes: Flush the eyes with plenty of running water for at least 15 min. Hold the eyelids apart during the flushing to ensure rinsing of the surface of the eye and lids with water. Seek medical attention if eye irritation occurs.

Inhalation: Remove to fresh air and, if indicated, give artificial respiration. If breathing is difficult, give oxygen. Call a physician. Treat immediately. *Ingestion:* If swallowed, immediately give several glasses of water and induce vomiting by gagging the victim with a finger placed on the back of the victim's tongue. Give fluids until vomitus is clear. If victim is unconscious or convulsing, do not induce vomiting or give anything by mouth.

Cancer Information: Not listed as a carcinogen by NTP (National Toxicology Program); not regulated as a carcinogen by OSHA; not evaluated by IARC (International Agency for Research on Cancer).

--- TOXICITY DATA ---

Animal Effects: No toxicology studies reported.

--- REACTIVITY DATA ---

Stability and Conditions to Avoid: Heat, open flames. Under normal use, this material is stable.

Materials to Avoid (Incompatibility): None reported by the manufacturer.

Hazard Decomposition Products: Unknown

Hazardous Polymerization: Will not occur.

--- SPILL AND EMERGENCY RESPONSE ---

For Spills: *Do not* sweep because this may create a dust hazard. Instead, soak up in oil dry or absorbent material. Flush spill area with water.

Waste Disposal Method: Dispose of in accordance with federal, state, and local regulations.

Empty Containers: The container for this product can present explosion or fire hazards, even when emptied. To avoid the risk of injury, do not cut, puncture, or weld on or near this container.

--- SPECIAL PROTECTION AND PRECAUTIONS ---

Respiratory Protection: Appropriate respirator selected and used in accordance with OSHA Subpart I (29 CFR 1910.134) and manufacturer's recommendations is required if airborne dust is not adequately controlled or if excessive fumes are generated from product decomposition. Self-contained breathing apparatus should be worn in high vapor and dust concentrations. Under normal use, manufacturer recommends a dust mask.

Type of Ventilation Required: General mechanical. Avoid breathing dusts and vapors. Adequate ventilation should be provided to keep mist and dust concentrations below acceptable exposure limits. Discharge from the ventilation system should comply with applicable air-pollution regulations.

Eye Protection Requirements: Chemical goggles. Eyewash fountains and safety showers should be easily accessible.

Protective Gloves and Other: Impervious gloves and clothing should be worn. Manufacturer recommends rubber gloves be used.

Storage and Handling Precautions: Containers should be stored in a cool, dry, well-ventilated area. Store away from flammable materials, sources of heat, flame, sparks, and foodstuffs. Exercise due caution to prevent damage to or leakage from the container. Avoid any conditions that might tend to create a dust explosion. Maintain good housekeeping practices to minimize dust buildup.

Other Precautions: Use all standard practices for good personal hygiene. Completely isolate and thoroughly clean all equipment, piping, or vessels before beginning maintenance or repairs. Keep area clean. Note that this product will burn.

--- REGULATORY INFORMATION ---

Refer to DOT shipping regulations.

▼　▼　▼　▼　▼　▼　▼　▼　▼　▼　▼　▼　▼　▼

--- IDENTIFIERS ---

Chemical Name and Synonyms: *n,n'*-DICINNAMYLIDENE-1,6-HEXANEDIAMINE

Trade Name and Synonyms: Diak No. 3

Chemical Family: Vulcanizing Agent

Manufacturer: E. I. du Pont de Nemours & Co., Inc., Wilmington, Delaware

UN ID Number: NA

--- INGREDIENTS AND COMPOSITION DATA ---

Material	Cas No.	%	TLV
Diak No. 3	NA	100	Not established

--- PHYSICAL DATA ---

Boiling Point (°F): NA

Specific Gravity (H_2O = 1): 1.09

Vapor Pressure (mm Hg): NA

Percent Volatile by Volume (%): NA

Vapor Density (air = 1): NA

Freezing Point: NA

Evaporation Rate: NA

pH: NA

Solubility in Water: Insoluble

Melting Point: 180° to 190°F

Appearance/Color/Odor: Tan powder. Cinnamon odor.

--- FIRE AND EXPLOSION HAZARDS ---

Flash Point: Over 300°F, open cup method

Flammability Limits: NA

Extinguishing Media: Foam or water spray, dry chemical, carbon dioxide, standard fire extinguishers.

Special Fire Fighting Procedures: As in any fire, prevent human exposure to fire, smoke, fumes, or products of combustion. Evacuate nonessential personnel from the fire area. Firefighters should wear full-face, self-contained breathing apparatus and impervious protective clothing such as gloves, hoods, and suits.

Unusual Fire and Explosion Hazards: *Do not* use welding or cutting torch on or near any container of this material, even empty, because an explosion could occur. *Do not* use, pour, spill, or store near heat or open flame. May form flammable dust-air mixture. Keep dust away from heat, sparks, and open flames. Use only in areas provided with grounded equipment.

Hazardous Products of Combustion and Decomposition: Toxic fumes.

Hazardous Polymerization: Will not occur.

--- HEALTH HAZARD DATA ---

Effects of Overexposure: Principal routes of exposure are skin contact and inhalation. Causes irritation to skin and eyes. Harmful if swallowed.

Emergency Response and First Aid:
Skin: Flush all affected areas with plenty of water for several minutes. Remove and clean any contaminated clothing and shoes. Seek medical attention if skin irritation occurs.
Eyes: Flush the eyes with plenty of running water for at least 15 min. Hold the eyelids apart during the flushing to ensure rinsing of the surface of the eye and lids with water. Seek medical attention if eye irritation occurs.
Inhalation: Remove to fresh air and, if indicated, give artificial respiration. If breathing is difficult, give oxygen. Call a physician. Treat immediately.
Ingestion: If swallowed, immediately give several glasses of water and induce vomiting by gagging the victim with a finger placed on the back of the victim's tongue. Give fluids until vomitus is clear. If victim is unconscious or convulsing, do not induce vomiting or give anything by mouth.

Cancer Information: Not listed as a carcinogen by NTP (National Toxicology Program); not regulated as a carcinogen by OSHA; not evaluated by IARC (International Agency for Research on Cancer).

--- TOXICITY DATA ---

Animal Effects: Lethal concentration of inhaled dust for rats is 6.2 mg/l of air for 4 hr of exposure.

--- REACTIVITY DATA ---

Stability and Conditions to Avoid: Heat, open flames. Under normal use, this material is stable.

Materials to Avoid (Incompatibility): None reported by the manufacturer.

Hazard Decomposition Products: Unknown

Hazardous Polymerization: Will not occur.

--- SPILL AND EMERGENCY RESPONSE ---

For Spills: *Do not* sweep because this may create a dust hazard. Instead, soak up in oil dry or absorbent material. Flush spill area with water.

Waste Disposal Method: Dispose of in accordance with federal, state, and local regulations.

Empty Containers: The container for this product can present explosion or fire hazards, even when emptied. To avoid the risk of injury, do not cut, puncture, or weld on or near this container.

--- SPECIAL PROTECTION AND PRECAUTIONS ---

Respiratory Protection: Respiratory Protection: Appropriate respirator selected and used in accordance with OSHA Subpart I (29 CFR 1910.134) and manufacturer's recommendations is required if airborne dust is not adequately controlled or if excessive fumes are generated from product decomposition. Self-contained breathing apparatus should be worn in high vapor and dust concentrations. Under normal use, manufacturer recommends a dust mask.

Type of Ventilation Required: General mechanical. Avoid breathing dusts and vapors. Adequate ventilation should be provided to keep mist and dust concentrations below acceptable exposure limits. Discharge from the ventilation system should comply with applicable air-pollution regulations.

Eye Protection Requirements: Chemical goggles. Eyewash fountains and safety showers should be easily accessible.

Protective Gloves and Other: Impervious gloves and clothing should be worn. Manufacturer recommends rubber gloves be used.

Storage and Handling Precautions: Containers should be stored in a cool, dry, well-ventilated area. Store away from flammable materials, sources of heat, flame, sparks, and foodstuffs. Exercise due caution to prevent damage to or leakage from the container. Avoid any conditions that might tend to create a dust explosion. Maintain good housekeeping practices to minimize dust buildup.

Other Precautions: Use all standard practices for good personal hygiene. Completely isolate and thoroughly clean all equipment, piping, or vessels before beginning maintenance or repairs. Keep area clean. Note that this product will burn.

--- REGULATORY INFORMATION ---

Refer to DOT shipping regulations.

▼   ▼   ▼   ▼   ▼   ▼   ▼   ▼   ▼   ▼   ▼   ▼   ▼   ▼

--- IDENTIFIERS ---

Chemical Name and Synonyms: POLY-DISPERSION OF LEAD OXIDES

Trade Name and Synonyms: Poly-Dispersion (TLD-90 and ERD-90)

Chemical Family: Dispersion of Lead Oxide

Manufacturer: Wyrough and Loser, Inc., Trenton, New Jersey

UN ID Number: NA

--- INGREDIENTS AND COMPOSITION DATA ---

Material	Cas No.	%	TLV
Proprietary Dispersion	NA	-	Not established

--- PHYSICAL DATA ---

Boiling Point (°F): NA

Specific Gravity (H₂O = 1): Over 3.5

Vapor Pressure (mm Hg): NA

Percent Volatile by Volume (%): NA

Vapor Density (air = 1): NA

Freezing Point: NA

Evaporation Rate: NA

pH: NA

Solubility in Water: Slight

Melting Point: NA

Appearance/Color/Odor: Yellow or Red-Rubbery Solid with little or no odor.

--- FIRE AND EXPLOSION HAZARDS ---

Flash Point: Over 400°F, open cup method

Flammability Limits: NA

Autoignition Temperature: NA

Extinguishing Media: Foam or water spray, dry chemical, carbon dioxide, standard fire extinguishers.

Special Fire Fighting Procedures: As in any fire, prevent human exposure to fire, smoke, fumes, or products of combustion. Evacuate nonessential personnel from the fire area. Firefighters should wear full-face, self-contained breathing apparatus and impervious protective clothing such as gloves, hoods, and suits.

Unusual Fire and Explosion Hazards: *Do not* use welding or cutting torch on or near any container of this material, even empty, because an explosion could occur. *Do not* use, pour, spill, or store near heat or open flame. May form flammable dust-air mixture. Keep dust away from heat, sparks, and open flames. Use only in areas provided with grounded equipment.

Hazardous Products of Combustion and Decomposition: Toxic fumes.

Hazardous Polymerization: Will not occur.

--- HEALTH HAZARD DATA ---

Effects of Overexposure: The form of the material is a dispersion in polymer matrix, and hence normal handling precludes conditions of overexposure.

Emergency Response and First Aid:
Skin: Flush all affected areas with plenty of water for several minutes. Remove and clean any contaminated clothing and shoes. Seek medical attention if skin irritation occurs.
Eyes: Flush the eyes with plenty of running water for at least 15 min. Hold the eyelids apart during the flushing to ensure rinsing of the surface of the eye and lids with water. Seek medical attention if eye irritation occurs.
Inhalation: Remove to fresh air and, if indicated, give artificial respiration. If breathing is difficult, give oxygen. Call a physician. Treat immediately.
Ingestion: If swallowed, immediately give several glasses of water and induce vomiting by gagging the victim with a finger placed on the back of the victim's tongue. Give fluids until vomitus is clear. If victim is unconscious or convulsing, do not induce vomiting or give anything by mouth.

Cancer Information: Not listed as a carcinogen by NTP (National Toxicology Program); not regulated as a carcinogen by OSHA; not evaluated by IARC (International Agency for Research on Cancer).

--- TOXICITY DATA ---

Animal Effects: None reported.

--- REACTIVITY DATA ---

Stability and Conditions to Avoid: Heat, open flames. Under normal use, this material is stable. Avoid contact with mineral acids.

Materials to Avoid (Incompatibility): None reported by the manufacturer.

Hazard Decomposition Products: Lead and lead oxide at temperatures over 800°C.

Hazardous Polymerization: Will not occur.

--- SPILL AND EMERGENCY RESPONSE ---

For Spills: No special procedures recommended.

Waste Disposal Method: Dispose of in accordance with federal, state, and local regulations.

Empty Containers: The container for this product can present explosion or fire hazards, even when emptied. To avoid the risk of injury, do not cut, puncture, or weld on or near this container.

--- SPECIAL PROTECTION AND PRECAUTIONS ---

Respiratory Protection: Wear dust and/or appropriate fume respirator.

Type of Ventilation Required: General mechanical. Avoid breathing dusts and vapors. Adequate ventilation should be provided to keep mist and dust concentrations below acceptable exposure limits. Discharge from the ventilation system should comply with applicable air-pollution regulations.

Eye Protection Requirements: Chemical goggles. Eyewash fountains and safety showers should be easily accessible.

Protective Gloves and Other: Impervious gloves and clothing should be worn. Manufacturer recommends rubber gloves be used.

Storage and Handling Precautions: Containers should be stored in a cool, dry, well-ventilated area. Store away from flammable materials, sources of heat, flame, sparks, and foodstuffs. Exercise due caution to prevent damage to or leakage from the container. Avoid any conditions that might tend to create a dust explosion. Maintain good housekeeping practices to minimize dust buildup.

Other Precautions: Use all standard practices for good personal hygiene. Completely isolate and thoroughly clean all equipment, piping, or vessels

before beginning maintenance or repairs. Keep area clean. Note that this product will burn.

--- REGULATORY INFORMATION ---

Refer to DOT shipping regulations.

▼ ▼ ▼ ▼ ▼ ▼ ▼ ▼ ▼ ▼ ▼ ▼ ▼ ▼

--- IDENTIFIERS ---

Chemical Name and Synonyms: VULCANIZED VEGETABLE OIL

Trade Name and Synonyms: Factice White 57

Chemical Family: NA

Manufacturer: American Cyanamid Co., Bound Brook, New Jersey

UN ID Number: NA

--- INGREDIENTS AND COMPOSITION DATA ---

Material	Cas No.	%	TLV
	NA		Not established

--- PHYSICAL DATA ---

Boiling Point (°F): NA

Specific Gravity (H₂O = 1): 1.28

Vapor Pressure (mm Hg): NA

Percent Volatile by Volume (%): Negligible

Vapor Density (air = 1): NA

Freezing Point: NA

Evaporation Rate: NA

pH: NA

Solubility in Water: Negligible

Melting Point: NA

Appearance/Color/Odor: White powder, slight oily odor.

--- FIRE AND EXPLOSION HAZARDS ---

Flash Point: NA

Flammability Limits: NA

Autoignition: NA

Extinguishing Media: Foam or water spray, dry chemical, carbon dioxide, standard fire extinguishers.

Special Fire Fighting Procedures: As in any fire, prevent human exposure to fire, smoke, fumes, or products of combustion. Evacuate nonessential personnel from the fire area. Firefighters should wear full-face, self-contained breathing apparatus and impervious protective clothing such as gloves, hoods, and suits. *Do not* use high-pressure water stream as air-borne dust creates an explosion hazard.

Unusual Fire and Explosion Hazards: *Do not* use welding or cutting torch on or near any container of this material, even empty, because an explosion could occur. *Do not* use, pour, spill, or store near heat or open flame. May form flammable dust-air mixture. Keep dust away from heat, sparks, and open flames. Use only in areas provided with grounded equipment. The hazard is similar to that of any organic solid including sawdust.

Hazardous Products of Combustion and Decomposition: Sulfur dioxide may be formed under fire conditions.

Hazardous Polymerization: Will not occur.

--- HEALTH HAZARD DATA ---

Effects of Overexposure: None expected.

Emergency Response and First Aid:
Skin: None recommended.
Eyes: None recommended.
Inhalation: None recommended, but author advises a minimum protection of dust mask.
Ingestion: If swallowed, immediately give several glasses of water and induce vomiting by gagging the victim with a finger placed on the back of the victim's tongue. Give fluids until vomitus is clear. If victim is unconscious or convulsing, do not induce vomiting or give anything by mouth.

Cancer Information: Not listed as a carcinogen by NTP (National Toxicology Program); not regulated as a carcinogen by OSHA; not evaluated by IARC (International Agency for Research on Cancer).

--- TOXICITY DATA ---

Animal Effects: None reported.

--- REACTIVITY DATA ---

Stability and Conditions to Avoid: Heat, open flames. Under normal use, this material is stable.

Materials to Avoid (Incompatibility): None reported by the manufacturer.

Hazard Decomposition Products: Thermal decomposition may produce carbon monoxide, carbon dioxide, and/or sulfur dioxide.

Hazardous Polymerization: Will not occur.

--- SPILL AND EMERGENCY RESPONSE ---

For Spills: Sweep up and place in a waste disposal container. Flush area with water.

Waste Disposal Method: Dispose of in accordance with federal, state, and local regulations.

Empty Containers: The container for this product can present explosion or fire hazards, even when emptied. To avoid the risk of injury, do not cut, puncture, or weld on or near this container.

--- SPECIAL PROTECTION AND PRECAUTIONS ---

Respiratory Protection: Wear dust and/or appropriate fume respirator.

Type of Ventilation Required: General mechanical. Avoid breathing dusts and vapors. Adequate ventilation should be provided to keep mist and dust concentrations below acceptable exposure limits. Discharge from the ventilation system should comply with applicable air-pollution regulations.

Eye Protection Requirements: Chemical goggles. Eyewash fountains and safety showers should be easily accessible.

Protective Gloves and Other: Impervious gloves and clothing should be worn. Manufacturer recommends rubber gloves be used.

Storage and Handling Precautions: Containers should be stored in a cool, dry, well-ventilated area. Store away from flammable materials, sources of heat, flame, sparks, and foodstuffs. Exercise due caution to prevent damage to or leakage from the container. Avoid any conditions that might tend to create a dust explosion. Maintain good housekeeping practices to minimize dust buildup.

Other Precautions: Use all standard practices for good personal hygiene. Completely isolate and thoroughly clean all equipment, piping, or vessels before beginning maintenance or repairs. Keep area clean. Note that this product will burn.

--- REGULATORY INFORMATION ---

Refer to DOT shipping regulations.

▼ ▼ ▼ ▼ ▼ ▼ ▼ ▼ ▼ ▼ ▼ ▼ ▼ ▼

--- IDENTIFIERS ---

Chemical Name and Synonyms: NONE; PROPRIETARY MIXTURE

Trade Name and Synonyms: Ferro Therm-Chek 6-V6A

Chemical Family: Barium, Cadmium, Zinc Compound

Manufacturer: Ferro Corp., Bedford, Ohio

UN ID Number: NA

--- INGREDIENTS AND COMPOSITION DATA ---

Material	Cas No.	%	TLV
Unknown	NA		Not established
Aromatic solvent	NA	5.6	Not established
Aliphatic solvent	NA	20	500 ppm
Alkyl phenol	NA	<4	Not established
Cadmium	NA	<3	Not established
Barium	NA	<5	Not established
Alkyl-aryl phosphate esters	NA	30	Not established

--- PHYSICAL DATA ---

Boiling Point (°F): >300°F

Specific Gravity (H₂O = 1): 1.03

Vapor Pressure (mm Hg): NA

Percent Volatile by Volume (%): 36

Vapor Density (air = 1): 3.9

Freezing Point: NA

Evaporation Rate: NA

pH: NA

Solubility in Water: Negligible

Melting Point: NA

Appearance/Color/Odor: Clear, amber liquid with mineral spirits odor.

--- FIRE AND EXPLOSION HAZARDS ---

Flash Point: 120°F   (TOC)

Flammability Limits: LEL: 0.9; UEL: 6.0

Autoignition Temperature: NA

Extinguishing Media: Foam or water spray, dry chemical, carbon dioxide, standard fire extinguishers.

Special Fire Fighting Procedures: As in any fire, prevent human exposure to fire, smoke, fumes, or products of combustion.   Evacuate nonessential personnel from the fire area.   Firefighters should wear full-face, self-contained breathing apparatus and impervious protective clothing such as gloves, hoods, and suits.

Unusual Fire and Explosion Hazards: *Do not* use welding or cutting torch on or near any container of this material, even empty, because an explosion could occur.   *Do not* use, pour, spill, or store near heat or open flame.

Hazardous Products of Combustion and Decomposition: Cadmium and barium will be in ash.

Hazardous Polymerization: Will not occur.

--- HEALTH HAZARD DATA ---

Effects of Overexposure: Solvent vapors have a TLV of 500 ppm.   Nonvolatile portions are toxic if ingested.   Overexposure effects include narcosis from solvent vapors.   Skin contact can cause dermatitis.

Emergency Response and First Aid:
Skin: Wipe well with cloth dampened in a little mineral oil, followed by soap and water wash.
Eyes: Flush eyes with lots of water. If irritation persists, consult a physician.
Inhalation: Give victim fresh air or oxygen. Consult physician. Material should be used in a well-ventilated area.
Ingestion: If swallowed, immediately give several glasses of water and induce vomiting by gagging the victim with a finger placed on the back of the

victim's tongue. Give fluids until vomitus is clear. If victim is unconscious or convulsing, do not induce vomiting or give anything by mouth. Immediately consult a physician to treat as barium and cadmium poisoning.

Cancer Information: Not listed as a carcinogen by NTP (National Toxicology Program); not regulated as a carcinogen by OSHA; not evaluated by IARC (International Agency for Research on Cancer).

--- TOXICITY DATA ---

Animal Effects: Numerous effects reported for heavy metal poisoning, such as barium and cadmium. Consult CHRIS, NIOSH, OSHA.

--- REACTIVITY DATA ---

Stability and Conditions to Avoid: Heat, open flames. Under normal use, this material is stable.

Materials to Avoid (Incompatibility): None reported by the manufacturer.

Hazard Decomposition Products: Thermal decomposition may produce carbon monoxide and carbon dioxide. Cadmium and barium residuals in ash.

Hazardous Polymerization: Will not occur.

--- SPILL AND EMERGENCY RESPONSE ---

For Spills: Use appropriate absorbent to stabilize and sweep into disposal container.

Waste Disposal Method: Dispose of in accordance with federal, state, and local regulations.

Empty Containers: The container for this product can present explosion or fire hazards, even when emptied. To avoid the risk of injury, do not cut, puncture, or weld on or near this container.

--- SPECIAL PROTECTION AND PRECAUTIONS ---

Respiratory Protection: Wear NIOSH-approved chemical cartridge-type respirator.

Type of Ventilation Required: General mechanical. Avoid breathing dusts and vapors. Adequate ventilation should be provided to keep mist and dust concentrations below acceptable exposure limits. Discharge from the ventilation system should comply with applicable air-pollution regulations.

Eye Protection Requirements: Chemical goggles. Eyewash fountains and safety showers should be easily accessible.

Protective Gloves and Other: Impervious gloves and clothing should be worn. Manufacturer recommends rubber gloves be used.

Storage and Handling Precautions: Containers should be stored in a cool, dry, well-ventilated area. Store away from flammable materials, sources of heat, flame, sparks, and foodstuffs. Exercise due caution to prevent damage to or leakage from the container.

Other Precautions: Use all standard practices for good personal hygiene. Completely isolate and thoroughly clean all equipment, piping, or vessels before beginning maintenance or repairs. Keep area clean. Note that this product will burn.

--- REGULATORY INFORMATION ---

None reported.

4 ACCELERATORS AND ACTIVATORS

The types of chemicals covered in this section are listed below. The information in this section is organized by chemical name, as in other parts of this book.

CHEMICAL NAME/ SYNONYMS	CHEMICAL FAMILY	TRADE NAMES
Calcium oxide, lime, quicklime	Alkaline earth	Desical P
Magnesium aluminum hydroxy carbonate	Alkaline earth	Hysafe 510
Talc, hydrous magnesium Silicate	Silicates	Mistron Compound
Azodicarbonamide-azobisformamide	Azo amide	Kempore-200
Azodicarbonamide-azobisformamide	Azo amide	Kempore 60
Diethylene glycol	Dihydroxyethyl ether	Diglycol
Cupric dimethyldithio-carbamate	Dithiocarbamate	Cumate
Zinc diethyldithio-carbamate	Dithiocarbamate	Ethyl Zimate

CHEMICAL NAME/ SYNONYMS	CHEMICAL FAMILY	TRADE NAMES
Nickel dibutyldithio-carbamate	Dithiocarbamate	NBC
Bismuth dimethyldithio-carbamate	Dithiocarbamate	Bismate
Zinc dibutyldithio-carbamate	Dithiocarbamate	Butyl Zimate
Zinc dimethyldithio-carbamate	Dithiocarbamate	Methyl Zimate
Selenium dimethyldithio-carbamate	Dithiocarbamate	Methyl Selenac
Lead dimethyldithio-carbamate	Dithiocarbamate	Methyl Ledate
Tellurium diethyldithio-carbamate	Dithiocarbamate	Ethyl Tellurac
Cadmium diethyldithio-carbamate	Dithiocarbamate	Ethyl Cadmate
Zinc salt of stearic acid	Metal soap	U.S.P. S-1271 DLG-10, DLG-20 Zinc Stearate
Di-ortho-tolylguanidine	Guanidine	DOTG Accelerator
Copper mercaptobenzo-thiazole	Thiazole	Cupsac Accelerator
Thiram	Thiuram	MOTS No. 1 Accelerator
Tetramethylthiuram	Thiuram	Monex
Hexamine	Aliphatic amine	

CHEMICAL NAME/ SYNONYMS	CHEMICAL FAMILY	TRADE NAMES
Zinc ethylphenyldithio-carbamate	Organic acid salt	
Cadmium diethyldithio-carbamate	Organic acid salt	
2-Benzothiazolethiol	Azole	Captax, MBT
2-Benzothiazolethiol, zinc salt	Organic acid salt	ZMBT
Bis(Dimethylthiocarbamyl) disulfide	Organic Sulfide	TETD
1,3-Diphenyl-2-thiourea	Rubber accelerator	A-1 Thiocarbanilide
Carbamide	Activator	BIK
Zinc dibenzyldithiocarbamate	Accelerator	Arazate
1,3-Diphenyl guanidine	Guanidine	DPG

▼   ▼   ▼   ▼   ▼   ▼   ▼   ▼   ▼   ▼   ▼   ▼   ▼   ▼

--- IDENTIFIERS ---

Name: CALCIUM OXIDE
Synonyms: Burnt Lime; Calcia; Calcium Oxide (DOT); Calx; Lime; Lime, Burned; Lime, Unslaked (DOT); Oxyde De Calcium (French); Quicklime; Quicklime (DOT); Wapniowy Tlenek (Polish); Pebble Lime
CAS: 1305-78-8; **RTECS:** EW3100000
Formula: CaO; **Mol Wt:** 56
WLN: CA O
Chemical Class: Metal oxide

See other identifiers listed below under Regulations.

--- PROPERTIES ---

Physical Description: White, gray, or yellow powder. Color depends on amount of impurities present. (NYDH)
Boiling Point: 3123.13 K; 2849.9°C; 5161.9°F
Melting Point: 2843.13 K; 2569.9°C; 4657.9°F
Flash Point: NA
Autoignition: NA
Vapor Pressure: 0 mm
UEL: NA
LEL: NA
Vapor Density: No data
Specific Gravity: 3.3 at 20°C
Density: 3.300
Water Solubility: Reacts, producing heat
Incompatibilities: Water

Reactivity with Water: Heat may cause ignition of combustibles. Material swells during reaction.
Reactivity with Common Materials: No reaction unless water present; then chief effect is that of heat liberated.
Stability during Transport: No data
Neutralizing Agents: No data
Polymerization Possibilities: No data

Toxic Fire Gases: None reported other than possible unburned vapors.
Odor Detected at (ppm): Not pertinent
Odor Description: Odorless (Source: CHRIS)
100% Odor Detection: No data

--- REGULATIONS ---

DOT hazard class: 8 CORROSIVE
DOT guide: 60
Identification number: UN1910
DOT shipping name: Calcium oxide
Packing group: III
Label(s) required: CORROSIVE
Packaging exceptions: 173.154
Nonbulk packaging: 173.213
Bulk packaging: 173.240

Quantity limitations:
Passenger air/rail: 25 kg
Cargo aircraft only: 100 kg
Vessel stowage: A

STCC Number: 4944515

Clean Water Act Sect. 307: No
Clean Water Act Sect. 311: No
Clean Air Act: Not listed
EPA Waste Number: None
CERCLA Ref: Not listed
RQ Designation: Not listed
SARA TPQ Value: Not listed
SARA Sect. 312 Categories:
Acute toxicity: adverse effect to target organs.
Reactive hazard: water reactive.

U.S. Postal Service Mailability:
Hazard class: ORM-B
Mailability: Domestic service and air transportation; shipper's declaration
Max per parcel: 25 lb; 5 lb

NFPA Codes:
Health Hazard (blue): (1) Slightly hazardous to health. As a precaution wear self-contained breathing apparatus.
Flammability (red): (0) This material does not readily burn.
Reactivity (yellow): (1) Normally stable, but may become unstable at elevated temperature and pressures.
Special: Unspecified.

--- TOXICITY DATA ---

Short-term Toxicity: *Inhalation:* May cause sneezing, coughing, inflammation of respiratory system, bronchitis, pneumonia, and mouth and nose sores. May cause both heat burns and chemical burns to the nose and throat. *Skin:* May cause both heat and chemical burns, sores, and scarring. Severity of symptoms increased if skin is wet. *Eyes:* May cause both heat and chemical burns, sores, scarring, holes, irritation, and permanent injury. *Ingestion:* May cause both heat and chemical burns to the mouth, throat, and stomach. Estimated lethal dose is 36 g (1.2 oz). (NYDH)

Long-term Toxicity: May cause thickening, scaling, and cracking of the skin, brittleness and cracking of fingernails; eye irritation, bronchitis and pneumonia. (NYDH)

Target Organs: Respiratory system, skin, eyes.

Symptoms: Causes burns on mucous membrane and skin. Inhalation of dust causes sneezing. (Source: CHRIS)

Conc IDLH: Unknown

ACGIH TLV: TLV = 2 mg/m^3
ACGIH STEL: Not listed

OSHA PEL: Transitional Limits: PEL = 5 mg/m^3; Final Rule Limits: TWA = 5 mg/m^3. The final rule limit TWA of 5 mg/m^3 is not in effect as a RES.

MAK Information: 5 calculated as total dust mg/m^3; Local irritant: Peak = 2xMAK for 5 min., 8 times per shift.

Carcinogen: N

Carcinogen Lists: *IARC*: Not listed; *MAK*: Not listed; *NIOSH*: Not listed; *NTP*: Not listed; *ACGIH*: Not listed; *OSHA*: Not listed.

LD50 Value: No LD50 in RTECS 1992

Reproductive Toxicity (1992 RTECS): This chemical has no known mammalian reproductive toxicity.

--- PROTECTION AND FIRST AID ---

Protection Suggested from the CHRIS Manual: Protective gloves, goggles, and any type of respirator prescribed for fine dust.

NIOSH Pocket Guide to Chemical Hazards:
  **Wear Appropriate Equipment to Prevent: Reasonable probability of skin contact.

  **Wear Eye Protection to Prevent: Any possibility of eye contact.

**Exposed Personnel Should Wash: Promptly when skin becomes contaminated and at the end of each work shift.

**Work Clothing Should Be Changed Daily: If there is any reasonable possibility that the clothing may be contaminated.

**Remove Clothing: Promptly remove nonimpervious clothing that becomes contaminated.

**The Following Equipment Should Be Made Available: Eyewash, quick drench.

**Reference: NIOSH

Recommended Respiration Protection Source: NIOSH Pocket Guide (85-114) ACGIH (Calcium oxide)

10 mg/m³: Any dust and mist respirator.

20 mg/m³: Any dust and mist respirator except single-use and quarter-mask respirators. Any supplied-air respirator. Any self-contained breathing apparatus.

50 mg/m³: Any powered air-purifying respirator with a high-efficiency particulate filter. Any supplied-air respirator operated in a continuous-flow mode.

100 mg/m³: Any air-purifying full facepiece respirator with a high-efficiency particulate filter. Any self-contained breathing apparatus with a full facepiece. Any supplied-air respirator with a full facepiece. Any powered air-purifying respirator with a tight-fitting facepiece and a high-efficiency particulate filter.

250 mg/m³: Any supplied-air respirator with a half-mask and operated in a pressure demand or other positive-pressure mode.

Emergency or planned entry in unknown concentrations or IDLH conditions: Any self-contained breathing apparatus with a full facepiece and operated in a pressure-demand or other positive-pressure mode. Any supplied-air respirator with a full facepiece and operated in a pressure-demand or other positive-pressure mode in combination with an auxiliary self-contained breathing apparatus operated in a pressure-demand or other positive-pressure mode.

Escape: Any air-purifying full facepiece respirator with a high-efficiency particulate filter. Any appropriate escape-type self-contained breathing apparatus.

First Aid Source: CHRIS Manual 1991.
Ingestion: If victim is conscious, have him drink water or milk. *Do not* induce vomiting.
Skin and Eyes: Flush with water and seek medical help.

First Aid Source: DOT Emergency Response Guide 1993.
Move victim to fresh air; call emergency medical care. In case of contact with material, immediately flush skin or eyes with running water for at least 15 min. Remove and isolate contaminated clothing and shoes at the site. Keep victim quiet and maintain normal body temperature.

--- INITIAL INCIDENT RESPONSE ---

Fire Extinguishment: Extinguish adjacent fires with dry chemical or carbon dioxide. *Note:* Do not use water on adjacent fires. (CHRIS 91)

U.S. Department of Transportation Guide to Hazardous Materials Transport Information - Publication DOT 5800.5 (1990).
DOT Shipping Name: CALCIUM OXIDE
DOT ID Number: UN1910

ERG90   GUIDE 60
* POTENTIAL HAZARDS *

*Health Hazards
Contact causes burns to skin and eyes.
If inhaled, may be harmful.
Fire may produce irritating or poisonous gases.
Runoff from fire control or dilution water may cause pollution.

*Fire or Explosion
Some of these materials may burn, but none of them ignites readily.
Flammable/poisonous gases may accumulate in tanks and hopper cars.
Some of these materials may ignite combustibles (wood, paper, oil, etc.).

* EMERGENCY ACTION *

Keep unnecessary people away; isolate hazard area and deny entry.
Stay upwind; keep out of low areas.
Positive-pressure self-contained breathing apparatus (SCBA) and structural firefighters' protective clothing will provide limited protection.

**CALL CHEMTREC AT 1-800-424-9300 FOR EMERGENCY ASSIS-
TANCE.** If water pollution occurs, notify the appropriate authorities.

*Fire
Some of these materials may react violently with water.
Small fires: Dry chemical, CO_2, water spray, or regular foam.
Large fires: Water spray, fog, or regular foam.
Move container from fire area if you can do it without risk.
Apply cooling water to sides of containers that are exposed to flames until
well after fire is out. Stay away from ends of tanks.

*Spill or Leak
Do not touch spilled material; stop leak if you can do it without risk.
Small spills: Take up with sand or other noncombustible absorbent material
and place into containers for later disposal.
Small dry spills: With clean shovel, place material into clean, dry container
and cover loosely; move containers from spill area.
Large spills: Dike far ahead of liquid spill for later disposal.

*First Aid
Move victim to fresh air; call emergency medical care.
In case of contact with material, immediately flush skin or eyes with running
water for at least 15 min.
Remove and isolate contaminated clothing and shoes at the site.
Keep victim quiet and maintain normal body temperature.

▼   ▼   ▼   ▼   ▼   ▼   ▼   ▼   ▼   ▼   ▼   ▼   ▼   ▼

--- IDENTIFIERS ---

Chemical Name and Synonyms: MAGNESIUM ALUMINUM HYDROXY
CARBONATE

Trade Name and Synonyms: Hysafe 510

Chemical Family: Alkaline Earth

Manufacturer: Havre de Grace, Maryland

UN ID Number: NA

--- INGREDIENTS AND COMPOSITION DATA ---

Material	Cas No.	%	TLV
Magnesium Aluminum Hydroxy Carbonate	11097-59-9		Not established

--- PHYSICAL DATA ---

Boiling Point (°F): NA

Specific Gravity (H$_2$O = 1): 2.09

Vapor Pressure (mm Hg): NA

Percent Volatile by Volume (%): NA

Vapor Density (air = 1): NA

Freezing Point: NA

Evaporation Rate: NA

pH: NA

Solubility in Water: Insoluble

Melting Point: NA

Appearance/Color/Odor: White powder, no odor.

--- FIRE AND EXPLOSION HAZARDS ---

Flash Point: None (Thermally degrades at temperatures >300°C)

Flammability Limits: NA

Autoignition Temperature: NA

Extinguishing Media: Foam or water spray, dry chemical, carbon dioxide, standard fire extinguishers.

Special Fire Fighting Procedures: As in any fire, prevent human exposure to fire, smoke, fumes, or products of combustion. Evacuate nonessential personnel from the fire area. Firefighters should wear full-face, self-contained breathing apparatus and impervious protective clothing such as gloves, hoods, and suits.

Unusual Fire and Explosion Hazards: None reported by manufacturer.

Hazardous Products of Combustion and Decomposition: Reacts with acids, forming carbon dioxide. Manufacturer provides no information on products of combustion.

Hazardous Polymerization: Will not occur.

--- HEALTH HAZARD DATA ---

Effects of Overexposure: This is an experimental product and no health data have been reported by the manufacturer. It has been noted that prolonged contact with this material may cause irritation to skin and mucous membranes. There are no other known medical conditions aggravated by prolonged exposure at this time.

Emergency Response and First Aid:
Skin: Wash with lots of clean water for at least 15 min.
Eyes: Flush eyes with lots of water. If irritation persists, consult a physician.
Inhalation: Give victim fresh air or oxygen. Consult physician. Material should be used in a well-ventilated area.
Ingestion: If swallowed, immediately give several glasses of water and induce vomiting by gagging the victim with a finger placed on the back of the victim's tongue. Give fluids until vomitus is clear. If victim is unconscious or convulsing, do not induce vomiting or give anything by mouth.

Cancer Information: Not listed as a carcinogen by NTP (National Toxicology Program); not regulated as a carcinogen by OSHA; not evaluated by IARC (International Agency for Research on Cancer).

--- TOXICITY DATA ---

Animal Effects: No toxicity data could be found on this product.

--- REACTIVITY DATA ---

Stability and Conditions to Avoid: Heat, open flames. Under normal use, this material is stable. Product will react with acids, producing carbon dioxide gas.

Materials to Avoid (Incompatibility): Acids.

Hazard Decomposition Products: Thermal decomposition may produce carbon monoxide and carbon dioxide.

Hazardous Polymerization: Will not occur.

--- SPILL AND EMERGENCY RESPONSE ---

For Spills: Use appropriate absorbent to stabilize and sweep into disposal container.

Waste Disposal Method: Sweep up and remove as for any inert solid material, such as sand or soil.

Empty Containers: The container for this product can present explosion or fire hazards, even when emptied. To avoid the risk of injury, do not cut, puncture, or weld on or near this container.

--- SPECIAL PROTECTION AND PRECAUTIONS ---

Respiratory Protection: Dust respirator as appropriate.

Type of Ventilation Required: General mechanical. Avoid breathing dusts and vapors. Adequate ventilation should be provided to keep mist and dust concentrations below acceptable exposure limits. Discharge from the ventilation system should comply with applicable air-pollution regulations.

Eye Protection Requirements: Chemical goggles. Eyewash fountains and safety showers should be easily accessible.

Protective Gloves and Other: Impervious gloves and clothing should be worn, although manufacturer makes no specific recommendation.

Storage and Handling Precautions: Containers should be stored in a cool, dry, well-ventilated area. Store away from flammable materials, sources of heat, flame, sparks, and foodstuffs. Exercise due caution to prevent damage to or leakage from the container. Do not mix or expose to acids.

Other Precautions: Use all standard practices for good personal hygiene. Completely isolate and thoroughly clean all equipment, piping, or vessels before beginning maintenance or repairs. Keep area clean. As with any powder or dustlike material, improper handling can lead to spontaneous combustion (dust explosions).

--- REGULATORY INFORMATION ---

None reported.

▼   ▼   ▼   ▼   ▼   ▼   ▼   ▼   ▼   ▼   ▼   ▼   ▼   ▼

--- IDENTIFIERS ---

Name: TALC (NONASBESTOS FORM)
Synonyms: Hydrous Magnesium Silicate; Steatite Talc; Non-Fibrous Talc; Non-Asbestiform Talc
CAS: 14807-96-6; **RTECS:** VV8790000
Formula: $H_2O_3Si.3/4$ Mg; **Mol Wt:** 295
WLN: .MG4.SI-O3*3

See other identifiers listed below under Regulations.

--- PROPERTIES ---

Physical Description: Odorless solid
Boiling Point: NA
Melting Point: 1173.15 K; 900°C; 1652°F
Flash Point: NA
Autoignition: NA
Vapor Pressure: About 0 mm
UEL: NA
LEL: NA
Vapor Density: No data
Specific Gravity: No data
Water Solubility: INSOL
Incompatibilities: None hazardous

Reactivity with Water: No data on water reactivity
Reactivity with Common Materials: No data
Stability during Transport: No data
Neutralizing Agents: No data
Polymerization Possibilities: No data

Toxic Fire Gases: None reported other than possible unburned vapors.
Odor Detected at (ppm): Unknown
Odor Description: No data
100% Odor Detection: No data

--- REGULATIONS ---

DOT hazard class: Not given
Packaging exceptions: 173.
Nonbulk packaging: 173.
Bulk packaging: 173.

STCC Number: Not listed

Clean Water Act Sect. 307: No
Clean Water Act Sect. 311: No
Clean Air Act: Not listed
EPA Waste Number: None
CERCLA Ref: Not listed
RQ Designation: Not listed
SARA TPQ Value: Not listed
SARA Sect. 312 Categories:
  Acute toxicity: irritant.

U.S. Postal Service Mailability: Not given

NFPA Codes:
  Health Hazard (blue): Unspecified
  Flammability (red): Unspecified
  Reactivity (yellow): Unspecified
  Special: Unspecified

--- TOXICITY DATA ---

Short-term Toxicity: Unknown

Long-term Toxicity: Unknown

Symptoms: Fibriotic Pneumoconiosis. (Source: NIOSHP)

Conc IDLH: Unknown

ACGIH TLV: TLV = 2 mg/m³ respirable dust (no fibers)
ACGIH STEL: Not listed

OSHA PEL: Transitional Limits: PEL = 20 MPPCF ppm; Final Rule Limits: TWA = (respirable dust), 2 mg/m³.

MAK Information: 2 calculated as fine dust, mg/m³

Carcinogen: N; **Status:** See below

Carcinogen Lists: *IARC*: Not classified as to human carcinogenicity or probably not carcinogenic to humans; *MAK*: Not listed; *NIOSH*: Not listed; *NTP*: Not listed; *ACGIH*: Not listed; *OSHA*: Not listed.

LD50 Value: No LD50 in RTECS 1992

Reproductive Toxicity (1992 RTECS): This chemical has no known mammalian reproductive toxicity.

--- PROTECTION AND FIRST AID ---

Recommended Respiration Protection Source: NIOSH Pocket Guide (85-114) ACGIH (Talc (Nonasbestos Form))
10 mg/m³: Any dust and mist respirator.
20 mg/m³: Any dust and mist respirator except single-use and quarter-mask respirators. Any supplied-air respirator. Any self-contained breathing apparatus.
50 mg/m³: Any powered air-purifying respirator with a dust and mist filter. Any supplied-air respirator operated in a continuous-flow mode.
100 mg/m³: Any air-purifying full facepiece respirator with a high-efficiency particulate filter. Any powered air-purifying respirator with a tight-fitting facepiece and a high-efficiency particulate filter. Any self-contained breathing apparatus with a full facepiece. Any supplied-air respirator with a full facepiece. Any supplied-air respirator with a tight-fitting facepiece operated in a continuous-flow mode.

2000 mg/m³: Any supplied-air respirator with a half-mask and operated in a pressure-demand or other positive-pressure mode.

4000 mg/m³: Any supplied-air respirator with a full facepiece and operated in a pressure-demand or other positive-pressure mode.

Emergency or planned entry in unknown concentrations or IDLH conditions: Any self-contained breathing apparatus with a full facepiece and operated in a pressure-demand or other positive-pressure mode. Any supplied-air respirator with a full facepiece and operated in a pressure-demand or other positive-pressure mode in combination with an auxiliary self-contained breathing apparatus operated in a pressure-demand or other positive-pressure mode.

Escape: Any air-purifying full facepiece respirator with a high-efficiency particulate filter. Any appropriate escape-type self-contained breathing apparatus.

First Aid Source: NIOSH
Inhalation: None given
Ingestion: None given
Skin: None given
Eyes: Irrigate immediately

--- INITIAL INCIDENT RESPONSE ---

No DOT Guide information for this product.

▼ ▼ ▼ ▼ ▼ ▼ ▼ ▼ ▼ ▼ ▼ ▼ ▼ ▼

--- IDENTIFIERS ---

Chemical Name and Synonyms: AZODICARBONAMIDE-AZOBISFOR-MAMIDE, $C_2H_4O_2N_4$

Trade Name and Synonyms: Kempore SDA-200

Chemical Family: Azo amide

Manufacturer: Stepan Chemical Co., Wilmington, Massachusetts

UN ID Number: NA

--- INGREDIENTS AND COMPOSITION DATA ---

Material	Cas No.	%	TLV
Azo amide	123-77-3		Not established

--- PHYSICAL DATA ---

Boiling Point (°F): 447.8°F

Specific Gravity (H$_2$O = 1): NA

Vapor Pressure (mm Hg): 1.2 at 200°C

Percent Volatile by Volume (%): NA

Vapor Density (air = 1): 16

Freezing Point: NA

Evaporation Rate: NA

pH: NA

Solubility in Water: Negligible

Melting Point: NA

Appearance/Color/Odor: No description given by manufacturer's MSDS.

--- FIRE AND EXPLOSION HAZARDS ---

Flash Point: 170°F (TCC)

Flammability Limits: NA

Autoignition: NA

Extinguishing Media: Foam or water spray, dry chemical, carbon dioxide, standard fire extinguishers.

Special Fire Fighting Procedures: As in any fire, prevent human exposure to fire, smoke, fumes, or products of combustion. Evacuate nonessential personnel from the fire area. Firefighters should wear full-face, self-contained breathing apparatus and impervious protective clothing such as gloves, hoods, and suits.

Unusual Fire and Explosion Hazards: None reported by manufacturer.

Hazardous Products of Combustion and Decomposition: Cyanuric acid.

Hazardous Polymerization: Will not occur.

--- HEALTH HAZARD DATA ---

Effects of Overexposure: May cause allergic reactions in extra-sensitive persons. Material is irritating to the eyes if contact is made.

Emergency Response and First Aid:
Skin: Wash with lots of clean water for at least 15 min.
Eyes: Flush eyes with lots of water. If irritation persists, consult a physician.
Inhalation: Give victim fresh air or oxygen. Consult physician. Material should be used in a well-ventilated area.
Ingestion: If swallowed, immediately give several glasses of water and induce vomiting by gagging the victim with a finger placed on the back of the victim's tongue. Give fluids until vomitus is clear. If victim is unconscious or convulsing, do not induce vomiting or give anything by mouth.

Cancer Information: Not listed as a carcinogen by NTP (National Toxicology Program); not regulated as a carcinogen by OSHA; not evaluated by IARC (International Agency for Research on Cancer).

--- TOXICITY DATA ---

Animal Effects: No toxicity data could be found on this product. Manufacturer states that the product is nontoxic.

--- REACTIVITY DATA ---

Stability and Conditions to Avoid: Heat, open flames. Under normal use, this material is stable. Product will react with oxidizing agents.

Materials to Avoid (Incompatibility): Oxidizing agents.

Hazard Decomposition Products: Thermal decomposition may produce carbon monoxide and carbon dioxide.

Hazardous Polymerization: Will not occur.

--- SPILL AND EMERGENCY RESPONSE ---

For Spills: Sweep up and discard in an appropriate container.

Waste Disposal Method: Sweep up and remove as for any inert solid material, such as sand or soil.

Empty Containers: The container for this product can present explosion or fire hazards, even when emptied.   To avoid the risk of injury, do not cut, puncture, or weld on or near this container.

--- SPECIAL PROTECTION AND PRECAUTIONS ---

Respiratory Protection: Dust respirator as appropriate.

Type of Ventilation Required: General mechanical. Avoid breathing dusts and vapors. Adequate ventilation should be provided to keep mist and dust concentrations below acceptable exposure limits.   Discharge from the ventilation system should comply with applicable air-pollution regulations.

Eye Protection Requirements: Chemical goggles. Eyewash fountains and safety showers should be easily accessible.

Protective Gloves and Other: Manufacturer makes no specific recommendation.

Storage and Handling Precautions: Containers should be stored in a cool, dry, well-ventilated area. Store away from flammable materials, sources of heat, flame, sparks, and foodstuffs. Exercise due caution to prevent damage to or leakage from the container. Do not mix or expose to acids.

Other Precautions: Use all standard practices for good personal hygiene. Completely isolate and thoroughly clean all equipment, piping, or vessels before beginning maintenance or repairs. Keep area clean. As with any powder or dustlike material, improper handling can lead to spontaneous combustion (dust explosions).

--- REGULATORY INFORMATION ---

None reported.

▼ ▼ ▼ ▼ ▼ ▼ ▼ ▼ ▼ ▼ ▼ ▼ ▼ ▼

--- IDENTIFIERS ---

Chemical Name and Synonyms: AZODICARBONAMIDE-AZOBISFOR-MAMIDE, $C_2H_4O_2N_4$

Trade Name and Synonyms: Kempore 60

Chemical Family: Azo amide

Manufacturer: Stepan Chemical Co., Wilmington, Massachusetts

UN ID Number: NA

--- INGREDIENTS AND COMPOSITION DATA ---

Material	Cas No.	%	TLV
Azo amide	123-77-3		Not established

--- PHYSICAL DATA ---

Decomposition Point (°C): 195 to 200

Specific Gravity (H_2O = 1): 1.65

Vapor Pressure (mm Hg): None

Percent Volatile by Volume (%): NA

Vapor Density (air = 1): NA

Freezing Point: NA

Evaporation Rate: NA

pH: NA

Solubility in Water: Insoluble

Melting Point: NA

Appearance/Color/Odor: Yellow, odorless powder.

--- FIRE AND EXPLOSION HAZARDS ---

Flash Point: NA (Material does not support combustion and is self-extinguishing when the source of flame is removed).

Flammability Limits: NA

Autoignition Temperature: NA

Extinguishing Media: Foam or water spray, dry chemical, carbon dioxide, standard fire extinguishers.

Special Fire Fighting Procedures: As in any fire, prevent human exposure to fire, smoke, fumes, or products of combustion. Evacuate nonessential personnel from the fire area. Firefighters should wear full-face, self-contained breathing apparatus and impervious protective clothing such as gloves, hoods, and suits.

Unusual Fire and Explosion Hazards: None reported by manufacturer.

Hazardous Products of Combustion and Decomposition: Cyanuric acid.

Hazardous Polymerization: Will not occur.

--- HEALTH HAZARD DATA ---

Effects of Overexposure: May cause allergic reactions in extra-sensitive persons. Material is irritating to the eyes if contact is made.

Emergency Response and First Aid:
Skin: Wash with lots of clean water for at least 15 min.
Eyes: Flush eyes with lots of water. If irritation persists, consult a physician.
Inhalation: Give victim fresh air or oxygen. Consult physician. Material should be used in a well-ventilated area.
Ingestion: If swallowed, immediately give several glasses of water and induce vomiting by gagging the victim with a finger placed on the back of the

victim's tongue. Give fluids until vomitus is clear. If victim is unconscious or convulsing, do not induce vomiting or give anything by mouth.

Cancer Information: Not listed as a carcinogen by NTP (National Toxicology Program); not regulated as a carcinogen by OSHA; not evaluated by IARC (International Agency for Research on Cancer).

--- TOXICITY DATA ---

Animal Effects: No toxicity data could be found on this product. Manufacturer states that the product is nontoxic.

--- REACTIVITY DATA ---

Stability and Conditions to Avoid: Heat, open flames. Under normal use, this material is stable. Product will react with oxidizing agents.

Materials to Avoid (Incompatibility): Oxidizing agents.

Hazard Decomposition Products: Thermal decomposition may produce carbon monoxide and carbon dioxide.

Hazardous Polymerization: Will not occur.

--- SPILL AND EMERGENCY RESPONSE ---

For Spills: Sweep up and discard in an appropriate container.

Waste Disposal Method: Sweep up and remove as for any inert solid material, such as sand or soil.

Empty Containers: The container for this product can present explosion or fire hazards, even when emptied. To avoid the risk of injury, do not cut, puncture, or weld on or near this container.

--- SPECIAL PROTECTION AND PRECAUTIONS ---

Respiratory Protection: Dust respirator as appropriate.

Type of Ventilation Required: General mechanical. Avoid breathing dusts and vapors. Adequate ventilation should be provided to keep mist and dust concentrations below acceptable exposure limits. Discharge from the ventilation system should comply with applicable air-pollution regulations.

Eye Protection Requirements: Chemical goggles. Eyewash fountains and safety showers should be easily accessible.

Protective Gloves and Other: Manufacturer makes no specific recommendation.

Storage and Handling Precautions: Containers should be stored in a cool, dry, well-ventilated area. Store away from flammable materials, sources of heat, flame, sparks, and foodstuffs. Exercise due caution to prevent damage to or leakage from the container. Do not mix or expose to acids.

Other Precautions: Use all standard practices for good personal hygiene. Completely isolate and thoroughly clean all equipment, piping, or vessels before beginning maintenance or repairs. Keep area clean. As with any powder or dustlike material, improper handling can lead to spontaneous combustion (dust explosions).

--- REGULATORY INFORMATION ---

None reported.

▼ ▼ ▼ ▼ ▼ ▼ ▼ ▼ ▼ ▼ ▼ ▼ ▼ ▼

--- IDENTIFIERS ---

Name: GLYCOL ETHER
Synonyms: Bis(2-Hydroxyethyl) Ether; Brecolane NDG; Deactivator E; Deactivator H; DEG; Dicol; Diglycol; beta,beta'-Dihydroxydiethyl Ether; Dissolvant APV; Ethanol 2,2'-Oxydi-; Ethylene Diglycol; Glycol Ether; Glycol Ethyl Ether; 3-Oxapentane-1,5-Diol; 3-Oxa-1,5-Pentanediol; 2,2'-Oxybisethanol; 2,2'-Oxydiethanol; TL4N; Diethylene Glycol; 2,2-Dihydroxydiethyl Ether
CAS: 111-46-6; **RTECS:** ID5950000
Formula: $C_4H_{10}O_3$; **Mol Wt:** 106.14
WLN: Q2O2Q
Chemical Class: Glycol ether

See other identifiers listed below under Regulations.

--- PROPERTIES ---

Boiling Point: 518.95 K; 245.8°C; 474.4°F
Melting Point: 265.15 K; -8°C; 17.6°F
Flash Point: 416.15 K; 143°C; 289.4°F
Autoignition: 418.15 K; 145°C; 293°F
Critical Temp: 681 K; 407.85°C; 766.13°F
Critical Press: 4.7 kN/m²; 46.3 atm; 680 psia
Heat of Vap: 270 Btu/lb; 149.95 cal/g; 6.274x E5 J/kg
Heat of Comb: -9617 Btu/lb; -5346 cal/g; -223x E5 J/kg
Vapor Pressure: 1 mm at 91.8°C
UEL: 10.8%
LEL: 1.6%
Vapor Density: 3.66 (air = 1)
Evaporation Rate: <0.01 (*n*-butyl acetate = 1)
Specific Gravity: No data
Density: 1.118 at 20/20
Water Solubility: Freely miscible

Reactivity with Water: No data on water reactivity
Reactivity with Common Materials: No data
Stability during Transport: No data
Neutralizing Agents: No data
Polymerization Possibilities: No data

Toxic Fire Gases: Acrid smoke and in fumes
Odor Detected at (ppm): Not pertinent
Odor Description: Practically odorless (Source: CHRIS)
100% Odor Detection: No data

--- REGULATIONS ---

DOT hazard class: Not given
Packaging exceptions: 173.
Nonbulk packaging: 173.
Bulk packaging: 173.

STCC Number: Not listed

Clean Water Act Sect. 307: No
Clean Water Act Sect. 311: No
Clean Air Act: Not listed

EPA Waste Number: None
CERCLA Ref: N
RQ Designation: Not listed
SARA TPQ Value: Not listed
SARA Sect. 312 Categories:
Acute toxicity: irritant.
Acute toxicity: adverse effect to target organs.
Chronic toxicity: adverse effect to target organs after long periods of exposure.
Chronic toxicity: reproductive toxin.

U.S. Postal Service Mailability: Not given

NFPA Codes:
Health Hazard (blue): (1) Slightly hazardous to health. As a precaution, wear self-contained breathing apparatus.
Flammability (red): (1) This material must be preheated before ignition can occur.
Reactivity (yellow): (0) Stable even under fire conditions.
Special: Unspecified.

--- TOXICITY DATA ---

Short-term Toxicity: Unknown

Long-term Toxicity: Unknown

Target Organs: Skin, eyes, liver, kidneys

Symptoms: Ingestion of large amounts may cause degeneration of kidney and liver and cause death. Liquid may cause slight skin irritation. (Source: CHRIS)

Conc IDLH: Unknown

NIOSH REL: Not given

ACGIH TLV: Not listed
ACGIH STEL: Not listed
WEEL (ACGIH): 50 ppm, total; 10 mg/m^3, aerosol only (8-hr TWA)

OSHA PEL: Not in Table Z-1-A

MAK Information: Not listed

Carcinogen: N; **Status:** See below

Carcinogen Lists: *IARC*: Not listed; *MAK*: Not listed; *NIOSH*: Not listed; *NTP*: Not listed; *ACGIH*: Not listed; *OSHA*: Not listed.

Human Toxicity Data: (Source: NIOSH RTECS)
   * orl-hmn LDLo:1 g/kg JIHTAB 21, 173, 39

LD50 Value: orl-rat LD50:12565 mg/kg

Other Species Toxicity Data: (Source: NIOSH RTECS 1992)
   orl-rat LD50:12565 mg/kg
   ipr-rat LD50:7700 mg/kg
   scu-rat LD50:18800 mg/kg
   ivn-rat LD50:6565 mg/kg
   ims-rat LDLo:7826 mg/kg
   unr-rat LD50:15650 mg/kg
   orl-mus LD50:23700 mg/kg
   ihl-mus LCLo:130 mg/m^3/2H
   ipr-mus LD50:9719 mg/kg
   scu-mus LDLo:5 g/kg
   unr-mus LD50:13300 mg/kg
   orl-dog LD50:9 g/kg
   orl-cat LD50:3300 mg/kg
   orl-rbt LD50:4400 mg/kg
   skn-rbt LD50:11890 mg/kg
   ivn-rbt LDLo:2236 mg/kg
   ims-rbt LDLo:4472 mg/kg
   unr-rbt LD50:2688 mg/kg
   orl-gpg LD50:7800 mg/kg
   unr-gpg LD50:14 g/kg

Irritation Data: (Source: NIOSH RTECS 1992)
   skn-hmn 112 mg/3D-I MLD
   eye-rbt 50 mg MLD

Reproductive Toxicity (1992 RTECS): This chemical is a mammalian reproductive toxin.

Reproductive Toxicity Data (1992 RTECS):
orl-rat TDLo:50 g/kg (1-20D preg) OYYAA2, 27, 801, 84
Specific Developmental Abnormalities
Musculoskeletal system

orl-rat TDLo:76420 mg/kg (6-15D preg) EPASR*8EHQ-0291-1175
Effects on Embryo or Fetus
Fetotoxicity (except death, e.g., stunted fetus)

orl-rat TDLo:38212 mg/kg (6-15D preg) EPASR*8EHQ-0291-1175
Specific Developmental Abnormalities
Musculoskeletal system

orl-mus TDLo:343 g/kg (multigeneration) FAATDF 14, 622, 90
Maternal Effects
Parturition
Effects on Embryo or Fetus
Fetal death
Effects on Newborn
Sex ratio

--- PROTECTION AND FIRST AID ---

First Aid Source: CHRIS Manual 1991.
Inhalation: No problem likely. If any ill effects do develop, get medical attention.
Ingestion: Induce vomiting if ingested. No known antidote; treat symptomatically.
Eye and skin: Flush with water. If any ill effects occur, get medical attention.

--- INITIAL INCIDENT RESPONSE ---

Fire Extinguishment: Alcohol foam, carbon dioxide, dry chemical. *Note:* Water or foam may cause frothing. (CHRIS 91)

No DOT Guide information for this product.

▼ ▼ ▼ ▼ ▼ ▼ ▼ ▼ ▼ ▼ ▼ ▼ ▼ ▼

--- IDENTIFIERS ---

Chemical Name and Synonyms: CUPRIC DIMETHYLDITHIOCAR-
BAMATE, Cu[SC(S)N(CH$_3$)$_2$]$_2$

Trade Name and Synonyms: Cumate

Chemical Family: Dithiocarbamate

Manufacturer: R. T. Vanderbilt Co., Inc., Norwalk, Connecticut

UN ID Number: NA

--- INGREDIENTS AND COMPOSITION DATA ---

Material	Cas No.	%	TLV
Cupric dimethyl-dithiocarbamate	NA	98 min.	Acute oral LD50 (mice), 11,383 mg/kg

--- PHYSICAL DATA ---

Boiling Point (°F): NA

Specific Gravity (H$_2$O = 1): 1.75

Vapor Pressure (mm Hg): NA

Percent Volatile by Volume (%): NA

Vapor Density (air = 1): NA

Freezing Point: NA

Evaporation Rate: NA

pH: NA

Solubility in Water: Negligible

Melting Point: NA

Appearance/Color/Odor: Dark brown powder.   No characteristic odor described.

--- FIRE AND EXPLOSION HAZARDS ---

Flash Point: NA

Flammability Limits: NA

Autoignition: NA

Extinguishing Media: Foam, dry chemical, carbon dioxide.

Special Fire Fighting Procedures: As in any fire, prevent human exposure to fire, smoke, fumes, or products of combustion.   Evacuate nonessential personnel from the fire area.   Firefighters should wear full-face, self-contained breathing apparatus and impervious protective clothing such as gloves, hoods, and suits.

Unusual Fire and Explosion Hazards: None reported by manufacturer.

Hazardous Products of Combustion and Decomposition: CuO, amines, CS_2, CO_2.

Hazardous Polymerization: Will not occur.

--- HEALTH HAZARD DATA ---

Effects of Overexposure: May cause allergic reactions in extra-sensitive persons.   Material is irritating to the eyes if contact is made.

Emergency Response and First Aid:
Skin: Wash with lots of clean water for at least 15 min.
Eyes: Flush eyes with lots of water.   If irritation persists, consult a physician.
Inhalation: Give victim fresh air or oxygen.   Consult physician.   Material should be used in a well-ventilated area.
Ingestion: If swallowed, immediately give several glasses of water and induce vomiting by gagging the victim with a finger placed on the back of the victim's tongue.   Give fluids until vomitus is clear.   If victim is unconscious or convulsing, do not induce vomiting or give anything by mouth.

Cancer Information: Not listed as a carcinogen by NTP (National Toxicology Program); not regulated as a carcinogen by OSHA; not evaluated by IARC (International Agency for Research on Cancer).

--- TOXICITY DATA ---

Animal Effects: Acute oral LD50 in mice reported to be 11,383 mg/kg.

--- REACTIVITY DATA ---

Stability and Conditions to Avoid: Heat, open flames. Under normal use, this material is stable. Product will react with acids.

Materials to Avoid (Incompatibility): Acids, acidic conditions.

Hazard Decomposition Products: Thermal decomposition may produce carbon monoxide, carbon dioxide, CuO, amines, and CS_2.

Hazardous Polymerization: Will not occur.

--- SPILL AND EMERGENCY RESPONSE ---

For Spills: Sweep up and discard in an appropriate container. Wash area with soap and water.

Waste Disposal Method: As per organic compounds.

Empty Containers: The container for this product can present explosion or fire hazards, even when emptied. To avoid the risk of injury, do not cut, puncture, or weld on or near this container.

--- SPECIAL PROTECTION AND PRECAUTIONS ---

Respiratory Protection: Dust respirator as appropriate.

Type of Ventilation Required: General mechanical. Avoid breathing dusts and vapors. Adequate ventilation should be provided to keep mist and dust concentrations below acceptable exposure limits. Discharge from the ventilation system should comply with applicable air-pollution regulations.

Eye Protection Requirements: Chemical goggles. Eyewash fountains and safety showers should be easily accessible.

Protective Gloves and Other: Rubber gloves and apron.

Storage and Handling Precautions: Containers should be stored in a cool, dry, well-ventilated area. Store away from flammable materials, sources of heat, flame, sparks, and foodstuffs. Exercise due caution to prevent damage to or leakage from the container. Do not mix or expose to acids.

Other Precautions: Use all standard practices for good personal hygiene. Completely isolate and thoroughly clean all equipment, piping, or vessels before beginning maintenance or repairs. Keep area clean. As with any powder or dustlike material, improper handling can lead to spontaneous combustion (dust explosions).

--- REGULATORY INFORMATION ---

None reported.

▼ ▼ ▼ ▼ ▼ ▼ ▼ ▼ ▼ ▼ ▼ ▼ ▼ ▼

--- IDENTIFIERS ---

Chemical Name and Synonyms: ZINC DIETHYLDITHIOCARBAMATE, ZDEC, ZINC, BIS(DIETHYLCARBAMODITHIOATO-S,S')-

Trade Name and Synonyms: Ethyl Zimate

Chemical Family: NA

Manufacturer: R. T. Vanderbilt Co., Inc., Norwalk, Connecticut

UN ID Number: NA

--- INGREDIENTS AND COMPOSITION DATA ---

Material	Cas No.	%	TLV
ZDEC	14324-55-1		OSHA TWA: 5 mg/m^3 (respirable dust) 15 mg/m^3 (total dust) ACGIH TWA: 10 mg/m^3 (total dust)

--- PHYSICAL DATA ---

Boiling Point (°F): NA

Specific Gravity (H₂O = 1): 1.48

Vapor Pressure (mm Hg): NA

Percent Volatile by Volume (%): NA

Vapor Density (air = 1): NA

Freezing Point: NA

Evaporation Rate: NA

pH: NA

Solubility in Water: Negligible

Melting Point: NA

Appearance/Color/Odor: White powder with no distinctive odor.

--- FIRE AND EXPLOSION HAZARDS ---

Flash Point: NA

Flammability Limits: NA

Autoignition Temperature: NA

Extinguishing Media: Foam, dry chemical, carbon dioxide.

Special Fire Fighting Procedures: As in any fire, prevent human exposure to fire, smoke, fumes, or products of combustion. Evacuate nonessential personnel from the fire area. Firefighters should wear full-face, self-contained breathing apparatus and impervious protective clothing such as gloves, hoods, and suits.

Unusual Fire and Explosion Hazards: None reported by manufacturer.

Hazardous Products of Combustion and Decomposition: When exposed to flames, emits acrid fumes. Oxides of carbon, sulfur, nitrogen, and zinc upon combustion.

Hazardous Polymerization: Will not occur.

--- HEALTH HAZARD DATA ---

Effects of Overexposure: May irritate or sensitize skin. Do not ingest alcoholic beverages immediately before or after handling (Antabuse effect). May cause eye, skin, and upper respiratory irritation with prolonged exposure to the dust. Continuous skin contact could lead to dermatitis and possible skin sensitization.

Emergency Response and First Aid:
Skin: Wash with lots of clean water for at least 15 min.
Eyes: Flush eyes with lots of water. If irritation persists, consult a physician.
Inhalation: Give victim fresh air or oxygen. Consult physician. Material should be used in a well-ventilated area.
Ingestion: If swallowed, immediately give several glasses of water and induce vomiting by gagging the victim with a finger placed on the back of the victim's tongue. Give fluids until vomitus is clear. If victim is unconscious or convulsing, do not induce vomiting or give anything by mouth.

Cancer Information: Not listed as a carcinogen by NTP (National Toxicology Program); not regulated as a carcinogen by OSHA; not evaluated by IARC (International Agency for Research on Cancer).

--- TOXICITY DATA ---

Animal Effects:
Acute oral LD50 3530 mg/kg in rats.
Acute dermal LD50 >3160 mg/kg in rabbits.
Chronic effects are not known.

--- REACTIVITY DATA ---

Stability and Conditions to Avoid: To prevent formation of suspect carcinogenic nitrosamines, do not use with nitrosating agents. Under dusty conditions, static electricity may cause an explosion. Avoid contact with skin and eyes.

Materials to Avoid (Incompatibility): Acids, acidic conditions, nitrosating agents.

Hazard Decomposition Products: See above sections.

Hazardous Polymerization: Will not occur.

--- SPILL AND EMERGENCY RESPONSE ---

For Spills: Sweep up and discard in an appropriate container. Wash area with soap and water.

Waste Disposal Method: As per organic compounds.

Empty Containers: The container for this product can present explosion or fire hazards, even when emptied. To avoid the risk of injury, do not cut, puncture, or weld on or near this container.

▼   ▼   ▼   ▼   ▼   ▼   ▼   ▼   ▼   ▼   ▼   ▼   ▼   ▼

--- IDENTIFIERS ---

Chemical Name and Synonyms: NICKEL DIBUTYLDITHIOCARBAMATE

Trade Name and Synonyms: NBC

Chemical Family: Rubber Accelerator

Manufacturer: E. I. du Pont de Nemours & Co., Inc., Wilmington, Delaware

UN ID Number: NA

--- INGREDIENTS AND COMPOSITION DATA ---

Material	Cas No.	%	TLV
NBC	NA		Not established

--- PHYSICAL DATA ---

Boiling Point (°F): NA

Specific Gravity (H_2O = 1): 1.26

Vapor Pressure (mm Hg): NA

Percent Volatile by Volume (%): NA

Vapor Density (air = 1): NA

Freezing Point: NA

Evaporation Rate: NA

pH: NA

Solubility in Water: Negligible

Melting Point (°F): 187

Appearance/Color/Odor: Dark green flakes with mild odor.

--- FIRE AND EXPLOSION HAZARDS ---

Flash Point: 505°F (263°C) open cup

Flammability Limits: NA

Autoignition Temperature: NA

Extinguishing Media: Foam, dry chemical, carbon dioxide, water, regular fire extinguishers.

Special Fire Fighting Procedures: As in any fire, prevent human exposure to fire, smoke, fumes, or products of combustion. Evacuate nonessential personnel from the fire area. Firefighters should wear full-face, self-contained breathing apparatus and impervious protective clothing such as gloves, hoods, and suits.

Unusual Fire and Explosion Hazards: None reported by manufacturer.

Hazardous Products of Combustion and Decomposition: When exposed to flames, emits acrid fumes. Oxides of carbon, sulfur, nitrogen, and zinc upon combustion.

Hazardous Polymerization: Will not occur.

--- HEALTH HAZARD DATA ---

Effects of Overexposure: May irritate or sensitize skin, although generally not considered a skin sensitizer.

Emergency Response and First Aid:
Skin: Wash with lots of clean water for at least 15 min.
Eyes: Flush eyes with lots of water. If irritation persists, consult a physician.
Inhalation: Give victim fresh air or oxygen. Consult physician. Material should be used in a well-ventilated area.
Ingestion: If swallowed, immediately give several glasses of water and induce vomiting by gagging the victim with a finger placed on the back of the victim's tongue. Give fluids until vomitus is clear. If victim is unconscious or convulsing, do not induce vomiting or give anything by mouth.

Cancer Information: Not listed as a carcinogen by NTP (National Toxicology Program); not regulated as a carcinogen by OSHA; not evaluated by IARC (International Agency for Research on Cancer).

--- TOXICITY DATA ---

Animal Effects: Approximate lethal dose (ADL) is greater than 17,000 mg/kg body weight for male rat.

--- REACTIVITY DATA ---

Stability and Conditions to Avoid: None reported by manufacturer other than to minimize skin exposure.

Materials to Avoid (Incompatibility): None reported by manufacturer.

Hazard Decomposition Products: None reported.

Hazardous Polymerization: Will not occur.

--- SPILL AND EMERGENCY RESPONSE ---

For Spills: Sweep up and discard in an appropriate container. Wash area with soap and water.

Waste Disposal Method: As per organic compounds.

Empty Containers: The container for this product can present explosion or fire hazards, even when emptied. To avoid the risk of injury, do not cut, puncture, or weld on or near this container. Dusty environments can cause conditions of spontaneous combustion.

--- SPECIAL PROTECTION AND PRECAUTIONS ---

Respiratory Protection: Dust respirator as appropriate.

Type of Ventilation Required: General mechanical. Avoid breathing dusts and vapors. Adequate ventilation should be provided to keep mist and dust concentrations below acceptable exposure limits. Discharge from the ventilation system should comply with applicable air-pollution regulations.

Eye Protection Requirements: Chemical goggles. Eyewash fountains and safety showers should be easily accessible.

Protective Gloves and Other: Rubber or PVC gloves.

Storage and Handling Precautions: Containers should be stored in a cool, dry, well-ventilated area. Store away from flammable materials, sources of heat, flame, sparks, and foodstuffs. Exercise due caution to prevent damage to or leakage from container.

Other Precautions: Use all standard practices for good personal hygiene. Completely isolate and thoroughly clean all equipment, piping, or vessels before beginning maintenance or repairs. Keep area clean. As with any powder or dustlike material, improper handling can lead to spontaneous combustion (dust explosions).

--- REGULATORY INFORMATION ---

None reported.

▼   ▼   ▼   ▼   ▼   ▼   ▼   ▼   ▼   ▼   ▼   ▼   ▼   ▼

--- IDENTIFIERS ---

Chemical Name and Synonyms: BISMUTH DIMETHYLDITHIOCAR-
BAMATE

Trade Name and Synonyms: Bismate Powder

Chemical Family: Rubber Accelerator

Manufacturer: R. T. Vanderbilt Co., Inc., Norwalk, Connecticut

UN ID Number: NA

--- INGREDIENTS AND COMPOSITION DATA ---

Material	Cas No.	%	TLV
Bismuth dimethyl- dithiocarbamate	21260-46-8	35 to 38	OSHA TWA: 5 mg/m³ (respirable dust) 15 mg/m³ (total dust)
Mixed petroleum Process oil	64741-96-4 64741-97-5		Not established
Proprietary ingredient	NJTSR No. 800983-5004P		Not established

--- PHYSICAL DATA ---

Boiling Point (°F): NA

Specific Gravity (H₂O = 1): 2.04

Vapor Pressure (mm Hg): NA

Percent Volatile by Volume (%): NA

Vapor Density (air = 1): NA

Freezing Point: NA

Evaporation Rate: NA

pH: NA

Solubility in Water: Negligible

Melting Point (°C): Above 230 with decomposition

Fineness (100 mesh): 99.9% minimum

Appearance/Color/Odor: Lemon yellow powder.

--- FIRE AND EXPLOSION HAZARDS ---

Flash Point: NA

Flammability Limits: NA

Autoignition Temperature: NA

Extinguishing Media: Foam, dry chemical, carbon dioxide.

Special Fire Fighting Procedures: As in any fire, prevent human exposure to fire, smoke, fumes, or products of combustion. Evacuate nonessential personnel from the fire area. Firefighters should wear full-face, self-contained breathing apparatus and impervious protective clothing such as gloves, hoods, and suits.

Unusual Fire and Explosion Hazards: Decomposition products may be hazardous (see data below).

Hazardous Products of Combustion and Decomposition: Bismuth oxide, carbon and sulfur dioxide, and carbon disulfide at decomposition temperatures.

Hazardous Polymerization: Will not occur.

--- HEALTH HAZARD DATA ---

Effects of Overexposure: None listed in available literature.

Emergency Response and First Aid:
Skin: Wash with lots of clean water for at least 15 min.
Eyes: Flush eyes with lots of water. If irritation persists, consult a physician.

Inhalation: Give victim fresh air or oxygen. Consult physician. Material should be used in a well-ventilated area.

Ingestion: If swallowed, immediately give several glasses of water and induce vomiting by gagging the victim with a finger placed on the back of the victim's tongue. Give fluids until vomitus is clear. If victim is unconscious or convulsing, do not induce vomiting or give anything by mouth.

Cancer Information: Not listed as a carcinogen by NTP (National Toxicology Program); not regulated as a carcinogen by OSHA; not evaluated by IARC (International Agency for Research on Cancer).

--- TOXICITY DATA ---

Animal Effects:
RTEC: Scu-mus TDLo 1000 mg/kg. National Technical Information Service Reference. This effect is listed as inconclusive due to insufficient study data or method. NCI carcinogenesis study reflects no significant increase in tumors.

Manufacturer Studies: Toxicity >3000 mg/kg acute oral rat (not OSHA Toxic). Human patch test, negative.

--- REACTIVITY DATA ---

Stability and Conditions to Avoid: None reported by manufacturer other than to minimize skin exposure.

Materials to Avoid (Incompatibility): Acids and oxidizing agents.

Hazard Decomposition Products: Bismuth oxide, carbon and sulfur dioxide, and carbon disulfide.

Hazardous Polymerization: Will not occur.

--- SPILL AND EMERGENCY RESPONSE ---

For Spills: Vacuum or moisten material prior to sweeping. Place in closed container for disposal.

Waste Disposal Method: Not classified as a RCRA hazardous waste. Dispose of according to applicable environmental regulations.

Empty Containers: The container for this product can present explosion or fire hazards, even when emptied. To avoid the risk of injury, do not cut, puncture, or weld on or near this container. Dusty environments can cause spontaneous combustion conditions.

--- SPECIAL PROTECTION AND PRECAUTIONS ---

Respiratory Protection: Dust respirator as appropriate.

Type of Ventilation Required: General mechanical. Avoid breathing dusts and vapors. Adequate ventilation should be provided to keep mist and dust concentrations below acceptable exposure limits. Discharge from the ventilation system should comply with applicable air-pollution regulations.

Eye Protection Requirements: Chemical goggles. Eyewash fountains and safety showers should be easily accessible.

Protective Gloves and Other: Not typically recommended.

Storage and Handling Precautions: Containers should be stored in a cool, dry, well-ventilated area. Store away from flammable materials, sources of heat, flame, sparks, and foodstuffs. Exercise due caution to prevent damage to or leakage from the container.

Other Precautions: Use all standard practices for good personal hygiene. Completely isolate and thoroughly clean all equipment, piping, or vessels before beginning maintenance or repairs. Keep area clean. As with any powder or dustlike material, improper handling can lead to spontaneous combustion (dust explosions).

--- REGULATORY INFORMATION ---

None reported.

▼   ▼   ▼   ▼   ▼   ▼   ▼   ▼   ▼   ▼   ▼   ▼   ▼   ▼

--- IDENTIFIERS ---

Chemical Name and Synonyms: ZINC DIBUTYLDITHIOCARBAMATE

Trade Name and Synonyms: Butyl Zimate

Chemical Family: Rubber Accelerator

Manufacturer: R. T. Vanderbilt Co., Inc., Norwalk, Connecticut

UN ID Number: NA

--- INGREDIENTS AND COMPOSITION DATA ---

Material	Cas No.	%	TLV
Zinc compound containing zinc dibutyldithiocarbamate	136-23-2	13 to 15% zinc	Not established

--- PHYSICAL DATA ---

Boiling Point (°F): NA

Specific Gravity (H₂O = 1): 1.21

Vapor Pressure (mm Hg): NA

Percent Volatile by Volume (%): NA

Vapor Density (air = 1): NA

Freezing Point: NA

Evaporation Rate: NA

pH: NA

Solubility in Water: Negligible

Melting Point (°C): NA

Appearance/Color/Odor: White to cream color.

--- FIRE AND EXPLOSION HAZARDS ---

Flash Point: NA

Flammability Limits: NA

Autoignition Temperature: NA

Extinguishing Media: Foam, dry chemical, carbon dioxide.

Special Fire Fighting Procedures: As in any fire, prevent human exposure to fire, smoke, fumes, or products of combustion. Evacuate nonessential personnel from the fire area. Firefighters should wear full-face, self-contained breathing apparatus and impervious protective clothing such as gloves, hoods, and suits. Manufacturer recommends positive-pressure self-contained breathing apparatus.

Unusual Fire and Explosion Hazards: Decomposition products may be hazardous.

Hazardous Products of Combustion and Decomposition: Oxides of sulfur, nitrogen, carbon, and zinc at combustion temperatures.

Hazardous Polymerization: Will not occur.

--- HEALTH HAZARD DATA ---

Effects of Overexposure: None listed in available literature.

Emergency Response and First Aid:
Skin: Wash with lots of clean water for at least 15 min.
Eyes: Flush eyes with lots of water. If irritation persists, consult a physician.
Inhalation: Give victim fresh air or oxygen. Consult physician. Material should be used in a well-ventilated area.
Ingestion: None reported by manufacturer.

Cancer Information: Not listed as a carcinogen by NTP (National Toxicology Program); not regulated as a carcinogen by OSHA; not evaluated by IARC (International Agency for Research on Cancer).

--- TOXICITY DATA ---

Animal Effects: For Butyl Zimate: acute oral LD50 (rats) >16,000 mg/kg. In a 92-day study, 50 mg/kg per day produced no effect in rats. In a 17-week study, the no-untoward-effect level was 500 ppm in the diet of rats, providing an intake of between 41 and 47 mg/kg/day. Ames test was negative. There are no known medical conditions that are aggravated by exposure.

--- REACTIVITY DATA ---

Stability and Conditions to Avoid: None reported by manufacturer other than to minimize skin exposure.

Materials to Avoid (Incompatibility): Acids and oxidizing agents.

Hazard Decomposition Products: Oxides of sulfur, nitrogen, carbon, and zinc at combustion temperatures.

Hazardous Polymerization: Will not occur.

--- SPILL AND EMERGENCY RESPONSE ---

For Spills: Sweep up. Wash residuals with soap and water. Transfer to a closed container.

Waste Disposal Method: As per any organic chemical.

Empty Containers: The container for this product can present explosion or fire hazards, even when emptied. To avoid the risk of injury, do not cut, puncture, or weld on or near this container.

--- SPECIAL PROTECTION AND PRECAUTIONS ---

Respiratory Protection: Dust respirator as appropriate.

Type of Ventilation Required: General mechanical. Avoid breathing dusts and vapors. Adequate ventilation should be provided to keep mist and dust concentrations below acceptable exposure limits. Discharge from the ventilation system should comply with applicable air-pollution regulations.

Eye Protection Requirements: Chemical goggles. Eyewash fountains and safety showers should be easily accessible.

Protective Gloves and Other: Not typically recommended.

Storage and Handling Precautions: Containers should be stored in a cool, dry, well-ventilated area. Store away from flammable materials, sources of heat, flame, sparks, and foodstuffs. Exercise due caution to prevent damage to or leakage from the container.

Other Precautions: Use all standard practices for good personal hygiene. Completely isolate and thoroughly clean all equipment, piping, or vessels before beginning maintenance or repairs. Keep area clean. As with any powder or dustlike material, improper handling can lead to spontaneous combustion (dust explosions).

--- REGULATORY INFORMATION ---

None reported.

▼ ▼ ▼ ▼ ▼ ▼ ▼ ▼ ▼ ▼ ▼ ▼ ▼ ▼

--- IDENTIFIERS ---

Chemical Name and Synonyms: ZINC DIMETHYLDITHIOCARBAMATE

Trade Name and Synonyms: Methyl Zimate

Chemical Family: Rubber Accelerator

Manufacturer: R. T. Vanderbilt Co., Inc., Norwalk, Connecticut

UN ID Number: NA

--- INGREDIENTS AND COMPOSITION DATA ---

Material	Cas No.	%	TLV
Zinc compound containing zinc dimethyldithiocarbamate	137-30-4	19.5 to 23% zinc	Not established

OSHA - TWA/PEL	5 mg/m^3
ACGIH - TWA/TLV	5 mg/m^3
- STEL	10 mg/m^3

--- PHYSICAL DATA ---

Boiling Point (°F): NA

Specific Gravity (H₂O = 1): 1.71

Vapor Pressure (mm Hg): NA

Percent Volatile by Volume (%): NA

Vapor Density (air = 1): NA

Freezing Point: NA

Evaporation Rate: NA

pH: NA

Solubility in Water: Negligible

Melting Point (°C): NA

Appearance/Color/Odor: White powder, slight characteristic odor.

--- FIRE AND EXPLOSION HAZARDS ---

Flash Point: NA

Flammability Limits: NA

Autoignition Temperature: NA

Extinguishing Media: Foam, dry chemical, carbon dioxide, water.

Special Fire Fighting Procedures: As in any fire, prevent human exposure to fire, smoke, fumes, or products of combustion. Evacuate nonessential personnel from the fire area. Firefighters should wear full-face, self-contained breathing apparatus and impervious protective clothing such as gloves, hoods, and suits. Manufacturer recommends positive-pressure self-contained breathing apparatus.

Unusual Fire and Explosion Hazards: Decomposition products may be hazardous.

Hazardous Products of Combustion and Decomposition: Oxides of sulfur, nitrogen, carbon, and zinc at combustion temperatures.

Hazardous Polymerization: Will not occur.

--- HEALTH HAZARD DATA ---

Effects of Overexposure: May cause eye, skin, and upper respiratory tract irritation on prolonged exposure to dust. Continuous skin contact could lead to dermatitis and possible skin sensitization.

Do not ingest alcoholic beverages immediately before or after handling this material (Antabase effect).

MAY BE FATAL IF INHALED. May irritate eyes and skin. May sensitize skin.

Emergency Response and First Aid:
Skin: Wash with lots of clean water for at least 15 min.
Eyes: Flush eyes with lots of water. If irritation persists, consult a physician.
Inhalation: Give victim fresh air or oxygen. Consult physician. Material should be used in a well-ventilated area.
Ingestion: If swallowed, drink promptly a large quantity of milk, egg whites, or gelatin solution, or if these are not available, large quantities of water. Consult a physician immediately.

Cancer Information: IARC Group 3. See Toxicity Data below.

--- TOXICITY DATA ---

Animal Effects:
For Ziram:
Acute oral LD50 (rats): 1400 mg/kg and in second study 320 mg/kg.
Acute dermal LD50 (rabbits): >2000 mg/kg (low toxicity).
Eye irritation (rabbits): Irritant. Severe ocular lesions were produced.

Skin irritation (rabbits): slight.

Inhalation LC50 (rats): 81 mg/m^3/4 hr (highly toxic). Four groups of 10 rats were exposed to 21, 63, 90 and 124 mg/m^3 of ziram for 4 hr. The animals were observed for 14 days. All rats that died spontaneously exhibited lung discoloration and effects of the digestive tract, in particular distended stomachs and intestinal tracts. Similar symptoms were observed in the surviving rats, but some appeared normal.

A teratogenic study using rats indicated that under the test conditions the product is not teratogenic at dose levels as high as 140 mg/kg/day. A rabbit study found no teratogenic effects at dose levels up to 15 mg/kg/day (max. tested).

A reproductive study indicated no adverse effects in three generations at 29.6 mg/kg/day for male rats and 33.8 mg/kg/day for female rats.

Mutagenicity: L5178Y mouse, lymphoma: negative.

A mutagenic study (Salmonella microsome) showed positive results at 5 mg/plate. In a second bacterial mutation study with five dose levels of 0.5, 1.5, 5, 15, and 50 mg/plate, ziram was mutagenic only at dose levels above 50 mg/plate.

Ziram failed to show evidence of DNA-damaging activity in cultured rat hepatocytes in vitro.

IARC Group 3 (not classifiable as to their carcinogenicity to humans). Inadequate evidence in animals.

Ziram failed to show clastogenic activity in Chinese Hamster ovary cells, in vitro, with or without metabolic activation.

NTP conducted a carcinogenesis bioassay of ziram (NTP Technical Reports Series No. 238). C-cell carcinomas of the thyroid gland were reported in 2 out of 49 male F344/N rats at a daily dose of 11 mg/kg and in 7 out of 49 at a daily dose of 22 mg/kg. Female B6C3F1 mice showed some lung adenomas, but the study was complicated by an intercurrent Sendai virus infection. No carcinomas were shown under the same test conditions for female rats and male mice.

NTP does not list ziram in its Fifth Annual Report on Carcinogens (1989). Ziram is not included in the chemicals proposed for addition to the Sixth Annual Report on Carcinogens. There is no known human carcinogen association after more than 30 years of application.

A long-term feeding study (2 yr) showed no carcinogenic response in rats fed a daily diet of 0.025% ziram (250 ppm).

A short-term feeding study (90 days) showed no liver, lung, spleen, or kidney damage in rats fed a daily diet of 20 mg of ziram.

In a study comparing seven dithiocarbamate fungicides for skin sensitization, ziram was ranked as a moderate allergic skin sensitizer, being less active in this regard than most of the other dithiocarbamates tested.

Medical Conditions Generally Aggravated by Exposure: Unknown.

--- REACTIVITY DATA ---

Stability and Conditions to Avoid: None reported by manufacturer other than to minimize skin exposure.

Materials to Avoid (Incompatibility): Acids and oxidizing agents.

Hazard Decomposition Products: Oxides of sulfur, nitrogen, carbon, and zinc at combustion temperatures.

Hazardous Polymerization: Will not occur.

--- SPILL AND EMERGENCY RESPONSE ---

For Spills: Sweep up. Wash residuals with soap and water. Transfer to a closed container.

Waste Disposal Method: As per any organic chemical.

Empty Containers: The container for this product can present explosion or fire hazards, even when emptied. To avoid the risk of injury, do not cut, puncture, or weld on or near this container.

--- SPECIAL PROTECTION AND PRECAUTIONS ---

Respiratory Protection: NIOSH-approved chemical cartridge for organic vapors and acid gases, in combination with a dust prefilter.

Type of Ventilation Required: General mechanical. Avoid breathing dusts and vapors. Adequate ventilation should be provided to keep mist and dust concentrations below acceptable exposure limits. Discharge from the ventilation system should comply with applicable air-pollution regulations. Local exhaust ventilation is recommended with a capture velocity of 150 to 200 fpm.

Eye Protection Requirements: Chemical goggles. Eyewash fountains and safety showers should be easily accessible.

Protective Gloves and Other: Rubber gloves, goggles.

Storage and Handling Precautions: Containers should be stored in a cool, dry, well-ventilated area. Store away from flammable materials, sources of heat, flame, sparks, and foodstuffs. Exercise due caution to prevent damage to or leakage from the container.

Prevent inhalation of dust. Prevent contact with eyes and skin. Under dusty conditions, static electricity may cause an explosion. Keep containers closed. Store in a cool, dry place.

Use with adequate ventilation. Wash thoroughly after handling before eating and smoking. Work clothing should be frequently laundered and not worn away from work premises.

Other Precautions: Use all standard practices for good personal hygiene. Completely isolate and thoroughly clean all equipment, piping, or vessels before beginning maintenance or repairs. Keep area clean. As with any powder or dust-like material, improper handling can lead to spontaneous combustion (dust explosions).

--- REGULATORY INFORMATION ---

None reported.

▼ ▼ ▼ ▼ ▼ ▼ ▼ ▼ ▼ ▼ ▼ ▼ ▼ ▼

--- IDENTIFIERS ---

Name: SELENIUM, TETRAKIS(DIMETHYLDITHIOCARBAMATO)-
Synonyms: Methyl Selenac; Selenium Dimethyldithiocarbamate; Tetrakis-(Dimethylcarbamodithioato-S,S')Selenium
CAS: 144-34-3; **RTECS:** VT0780000
Formula: $C_{12}H_{24}N_4S_8$. Se; **Mol Wt:** 559.84
WLN: 1N1

See other identifiers listed below under Regulations.

--- PROPERTIES ---

Boiling Point: NA
Melting Point: NA
Flash Point: NA
Autoignition: NA
UEL: NA
LEL: NA
Vapor Density: No data
Specific Gravity: No data

Reactivity with Water: No data on water reactivity
Reactivity with Common Materials: No data
Stability during Transport: No data
Neutralizing Agents: No data
Polymerization Possibilities: No data

Toxic Fire Gases: Oxides of nitrogen
Odor Detected at (ppm): Unknown
Odor Description: No data
100% Odor Detection: No data

--- REGULATIONS ---

DOT hazard class: Not given
Packaging exceptions: 173.
Nonbulk packaging: 173.
Bulk packaging: 173.

STCC Number: Not listed

Clean Water Act Sect. 307: Yes
Clean Water Act Sect. 311: No
Clean Air Act: CAA '90 by category
EPA Waste Number: D010
CERCLA Ref: Not listed
RQ Designation: Not listed
SARA TPQ Value: Not listed
SARA Sect. 312 Categories: No category

Listed in SARA Sect. 313: Yes
De Minimus Concentration: 1.0%

U.S. Postal Service Mailability: Not given

--- TOXICITY DATA ---

Short-term Toxicity: Unknown

Long-term Toxicity: Unknown

Conc IDLH: Unknown

NIOSH REL: Not given

ACGIH TLV: TLV = 0.2 mg/m^3 as selenium
ACGIH STEL: As selenium

OSHA PEL: Transitional Limits: PEL = 0.2 mg/m^3; Final Rule Limits: TWA = 0.2 mg/m^3.

MAK Information: Not listed

Carcinogen: N; **Status:** See below

Carcinogen Lists: *IARC*: Not classified as to human carcinogenicity or probably not carcinogenic to humans; *MAK*: Not listed; *NIOSH*: Not listed; *NTP*: Not listed; *ACGIH*: Not listed; *OSHA*: Not listed.

LD50 Value: No LD50 in RTECS 1992

Reproductive Toxicity (1992 RTECS): This chemical has no known mammalian reproductive toxicity.

--- INITIAL INCIDENT RESPONSE ---

No DOT Guide information for this product.

▼   ▼   ▼   ▼   ▼   ▼   ▼   ▼   ▼   ▼   ▼   ▼   ▼   ▼

--- IDENTIFIERS ---

Chemical Name and Synonyms: LEAD DIMETHYLDITHIOCARBAMATE

Trade Name and Synonyms: Methyl Ledate

Chemical Family: Metal Dithiocarbamate

Manufacturer: R. T. Vanderbilt Co., Inc., Norwalk, Connecticut

UN ID Number: NA

--- INGREDIENTS AND COMPOSITION DATA ---

Material	Cas No.	%	TLV
Lead dimethyldithio-carbamate	NA	95	Not established

--- PHYSICAL DATA ---

Boiling Point (°F): NA

Specific Gravity (H$_2$O = 1): 2.41

Vapor Pressure (mm Hg): NA

Percent Volatile by Volume (%): NA

Vapor Density (air = 1): NA

Freezing Point: NA

Evaporation Rate: NA

pH: NA

Solubility in Water: Negligible

Melting Point (°C): NA

Appearance/Color/Odor: White powder, no odor reported.

--- FIRE AND EXPLOSION HAZARDS ---

Flash Point: NA

Flammability Limits: NA

Autoignition Temperature: NA

Extinguishing Media: Foam, dry chemical, carbon dioxide.

Special Fire Fighting Procedures: As in any fire, prevent human exposure to fire, smoke, fumes, or products of combustion. Evacuate nonessential personnel from the fire area. Firefighters should wear full-face, self-contained breathing apparatus and impervious protective clothing such as gloves, hoods, and suits. Manufacturer recommends positive-pressure self-contained breathing apparatus.

Unusual Fire and Explosion Hazards: Decomposition products may be hazardous.

Hazardous Products of Combustion and Decomposition: Oxides of lead, sulfur, and nitrogen at decomposition temperatures.

Hazardous Polymerization: Will not occur.

--- HEALTH HAZARD DATA ---

Effects of Overexposure: Inhalation of dust is harmful; may affect the central nervous system and cause neuromuscular, genitourinary, and hematopoietic complications.

Emergency Response and First Aid:
Skin: Wash with lots of clean water for at least 15 min.
Eyes: Flush eyes with lots of water. If irritation persists, consult a physician.
Inhalation: Give victim fresh air or oxygen. Consult physician regarding the necessity of blood and urine tests. Material should be used in a well-ventilated area.
Ingestion: None recommended. Consult physician immediately.

Cancer Information: IARC Group 3.

--- TOXICITY DATA ---

Animal Effects: See RTECS studies for lead poisoning.

--- REACTIVITY DATA ---

Stability and Conditions to Avoid: None reported by manufacturer other than to minimize skin exposure and inhalation.

Materials to Avoid (Incompatibility): Acids and oxidizing agents.

Hazard Decomposition Products: Highly toxic oxides of lead, nitrogen, and sulfur.

Hazardous Polymerization: Will not occur.

--- SPILL AND EMERGENCY RESPONSE ---

For Spills: Sweep up. Wash residuals with soap and water. Transfer to a closed container.

Waste Disposal Method: According to RCRA 40 CFR Section 261.24.

Empty Containers: The container for this product can present explosion or fire hazards, even when emptied. To avoid the risk of injury, do not cut, puncture, or weld on or near this container.

--- SPECIAL PROTECTION AND PRECAUTIONS ---

Respiratory Protection: Dust mask.

Type of Ventilation Required: General mechanical. Avoid breathing dusts and vapors. Adequate ventilation should be provided to keep mist and dust concentrations below acceptable exposure limits. Discharge from the ventilation system should comply with applicable air-pollution regulations. Local exhaust ventilation is recommended with a capture velocity of 150 to 200 fpm.

Eye Protection Requirements: Chemical goggles. Eyewash fountains and safety showers should be easily accessible.

Protective Gloves and Other: Rubber gloves, goggles.

Storage and Handling Precautions: Containers should be stored in a cool, dry, well-ventilated area. Store away from flammable materials, sources of heat, flame, sparks, and foodstuffs. Exercise due caution to prevent damage to or leakage from the container.

Prevent inhalation of dust. Prevent contact with eyes and skin. Under dusty conditions, static electricity may cause an explosion. Keep containers closed. Store in a cool, dry place.

Use with adequate ventilation. Wash thoroughly after handling before eating and smoking. Work clothing should be frequently laundered and not worn away from work premises.

Other Precautions: Use all standard practices for good personal hygiene. Completely isolate and thoroughly clean all equipment, piping, or vessels before beginning maintenance or repairs. Keep area clean. As with any powder or dust-like material, improper handling can lead to spontaneous combustion (dust explosions).

--- REGULATORY INFORMATION ---

RCRA 40 CFR Section 261.24.

▼ ▼ ▼ ▼ ▼ ▼ ▼ ▼ ▼ ▼ ▼ ▼ ▼ ▼

--- IDENTIFIERS ---

Chemical Name and Synonyms: TELLURIUM DIETHYLDITHIO-CARBAMATE AND MIXED PETROLEUM PROCESS OIL

Trade Name and Synonyms: Ethyl Tellurac

Chemical Family: Rubber Accelerator

Manufacturer: R. T. Vanderbilt Co., Inc., Norwalk, Connecticut

UN ID Number: NA

--- INGREDIENTS AND COMPOSITION DATA ---

Material	Cas No.	%	TLV
Tellurium diethyldithio- carbamate	20941-65-5		Not established
Mixed petroleum Process oil	64741-96-4 and 64741-97-5		

For Tellurium:
ACGIH and OSHA TWA 0.10 mg/m^3/8 hr
For Particulates (NOC):
OSHA TWA (total dust)   5 mg/m^3
ACGIH TWA (total dust)   10 mg/m^3

--- PHYSICAL DATA ---

Boiling Point (°F): NA

Specific Gravity (H$_2$O = 1): 1.44

Vapor Pressure (mm Hg): NA

Percent Volatile by Volume (%): NA

Vapor Density (air = 1): NA

Freezing Point: NA

Evaporation Rate: NA

pH: NA

Solubility in Water: Negligible

Melting Point: NA

Appearance/Color/Odor: Orange yellow powder; no characteristic odor reported.

--- FIRE AND EXPLOSION HAZARDS ---

Flash Point: NA

Flammability Limits: NA

Autoignition Temperature: NA

Extinguishing Media: Foam, dry chemical, carbon dioxide.

Special Fire Fighting Procedures: As in any fire, prevent human exposure to fire, smoke, fumes, or products of combustion. Evacuate nonessential personnel from the fire area. Firefighters should wear full-face, self-contained breathing apparatus and impervious protective clothing such as gloves, hoods, and suits. Manufacturer recommends positive-pressure self-contained breathing apparatus.

Unusual Fire and Explosion Hazards: Dust may cause spontaneous combustion.

Hazardous Products of Combustion and Decomposition: Oxides of tellurium, nitrogen, sulfur, and carbon at combustion temperatures.

Hazardous Polymerization: Will not occur.

--- HEALTH HAZARD DATA ---

Effects of Overexposure: May cause irritation to eyes. Garlic odor of breath and sweat. Possible (but very remote) symptoms are somnolence, anorexia, and nausea.

Emergency Response and First Aid:
Skin: Wash with lots of clean water for at least 15 min.
Eyes: Flush eyes with lots of water. If irritation persists, consult a physician.
Inhalation: Expose to fresh air. Keep warm and quiet. Give artificial respiration if necessary.
Ingestion: None recommended. Consult physician immediately.

Cancer Information: IARC Group 3.

--- TOXICITY DATA ---

Animal Effects:
Acute dermal LD50 (rabbits): >16,000 mg/kg
Acute oral LD50 (mice): >17,000 mg/kg
Acute oral LD50 (rats): >5000 mg/kg
Acute inh. LC50 (rats): 0.51 mg/L (maximum attenuation)

When excess tellurium exists, a garlic odor to the breath, dryness of the mouth, anorexia, and occasional nausea may be experienced (even at Te levels below the established standard).

Slightly irritating to the eyes of rabbits under unrinsed test conditions. Guinea pig sensitization and rabbit skin irritation tests were negative. Ames mutagenicity test with and without activation was negative. Some animal studies are reported to indicate a probability of increased incidence of mesotheliomas in male rats and adenomas of the lacrimal gland of the eyes in mice of both sexes (NIH 79-1708). NCI concluded that data are insufficient to establish carcinogenicity.

--- REACTIVITY DATA ---

Stability and Conditions to Avoid: None reported by manufacturer other than to minimize skin exposure and inhalation.

Materials to Avoid (Incompatibility): Acids and oxidizing agents.

Hazard Decomposition Products: Highly toxic oxides of lead, nitrogen, and sulfur.

Hazardous Polymerization: Will not occur.

--- SPILL AND EMERGENCY RESPONSE ---

For Spills: Sweep up. Wash residuals with soap and water. Transfer to a closed container.

Waste Disposal Method: According to RCRA 40 CFR Section 261.24..

Empty Containers: The container for this product can present explosion or fire hazards, even when emptied. To avoid the risk of injury, do not cut, puncture, or weld on or near this container.

--- SPECIAL PROTECTION AND PRECAUTIONS ---

Respiratory Protection: Dust mask.

Type of Ventilation Required: General mechanical. Avoid breathing dusts and vapors. Adequate ventilation should be provided to keep mist and dust concentrations below acceptable exposure limits. Discharge from the ventilation system should comply with applicable air-pollution regulations. Local exhaust ventilation is recommended with a capture velocity of 150 to 200 fpm.

Eye Protection Requirements: Chemical goggles. Eyewash fountains and safety showers should be easily accessible.

Protective Gloves and Other: Rubber gloves, goggles.

Storage and Handling Precautions: Containers should be stored in a cool, dry, well-ventilated area. Store away from flammable materials, sources of heat, flame, sparks, and foodstuffs. Exercise due caution to prevent damage to or leakage from the container.

Prevent inhalation of dust. Prevent contact with eyes and skin. Under dusty conditions, static electricity may cause an explosion. Keep containers closed. Store in a cool, dry place.

Use with adequate ventilation. Wash thoroughly after handling before eating and smoking. Work clothing should be frequently laundered and not worn away from work premises.

Other Precautions: Use all standard practices for good personal hygiene. Completely isolate and thoroughly clean all equipment, piping, or vessels before beginning maintenance or repairs. Keep area clean. As with any powder or dust-like material, improper handling can lead to spontaneous combustion (dust explosions).

--- REGULATORY INFORMATION ---

Carcinogenic Ingredients/OSHA/NTP/IARC: None.

SARA Title III Section 313 Ingredients: None.

▼   ▼   ▼   ▼   ▼   ▼   ▼   ▼   ▼   ▼   ▼   ▼   ▼   ▼

--- IDENTIFIERS ---

Name: CADMIUM DIETHYLDITHIOCARBAMATE
Synonyms: Diethyldithiocarbamate De Cadmium
CAS: 14239-68-0; **RTECS:** EU9850000
Formula: $C_{10}H_{20}CdN_2S_4$
WLN: 2N2
Chemical Class: Organic acid salt

See other identifiers listed below under Regulations.

--- PROPERTIES ---

Boiling Point: NA
Melting Point: NA
Flash Point: NA
Autoignition: NA
UEL: NA
LEL: NA
Vapor Density: No data
Specific Gravity: No data

Reactivity with Water: No data on water reactivity
Reactivity with Common Materials: No data
Stability during Transport: No data
Neutralizing Agents: No data
Polymerization Possibilities: No data

Toxic Fire Gases: None reported other than possible unburned vapors
Odor Detected at (ppm): Unknown
Odor Description: No data
100% Odor Detection: No data

--- REGULATIONS ---

DOT hazard class: Not given
Packaging exceptions: 173.
Nonbulk packaging: 173.
Bulk packaging: 173.

STCC Number: Not listed

Clean Water Act Sect. 307: Yes
Clean Water Act Sect. 311: No
Clean Air Act: CAA '90 by category
EPA Waste Number: D006
CERCLA Ref: Not listed
RQ Designation: Not listed
SARA TPQ Value: Not listed
SARA Sect. 312 Categories:
  Chronic toxicity: carcinogen.
  Chronic toxicity: mutagen.

Listed in SARA Sect. 313: Yes
De Minimus Concentration: 0.1%

U.S. Postal Service Mailability: Not given

--- TOXICITY DATA ---

Short-term Toxicity: Unknown

Long-term Toxicity: Unknown

Conc IDLH: Unknown

NIOSH REL: Not given

ACGIH TLV: TLV = dust 0.05 mg/m^3 as cadmium
ACGIH STEL: As cadmium

OSHA PEL: Transitional Limits: PEL = (fume) 0.1 mg/m^3; (dust) 0.2 mg/m^3; Ceiling = (fume) 0; Final Rule Limits: TWA = (fume) 0.1 mg/m^3; (dust) 0.2 mg/m^3; Ceiling = (fume) 0.3 mg/m^3; (dust) 0.6 mg/m^3.

MAK Information: Not listed

Carcinogen: Y; Status: See below

Carcinogen Lists: *IARC*: Not listed; *MAK*: Not listed; *NIOSH*: Not listed; *NTP*: Carcinogen defined by NTP as reasonably anticipated to be carcinogenic, with limited evidence in humans or sufficient evidence in experimental animals; *ACGIH*: Not listed; *OSHA*: Not listed.

LD50 Value: No LD50 in RTECS 1992

Reproductive Toxicity (1992 RTECS): This chemical has no known mammalian reproductive toxicity.

--- INITIAL INCIDENT RESPONSE ---

No DOT Guide information for this product.

▼   ▼   ▼   ▼   ▼   ▼   ▼   ▼   ▼   ▼   ▼   ▼   ▼   ▼

--- IDENTIFIERS ---

Name: ZINC STEARATE
CAS: 557-05-1; **RTECS:** ZH5200000
Formula: $C_{36}H_{70}O_4Zn$
Chemical Class: Organic acid salt

See other identifiers listed below under Regulations.

--- PROPERTIES ---

Boiling Point: NA
Melting Point: 393 K; 119.8°C; 247.7°F
Flash Point: NA
Autoignition: NA
UEL: NA
LEL: NA
Vapor Density: No data
Specific Gravity: No data

Reactivity with Water: No data on water reactivity
Reactivity with Common Materials: No data
Stability during Transport: No data
Neutralizing Agents: No data
Polymerization Possibilities: No data

Toxic Fire Gases: None reported other than possible unburned vapors
Odor Detected at (ppm): Unknown
Odor Description: No data
100% Odor Detection: No data

--- REGULATIONS ---

DOT hazard class: Not given
Packaging exceptions: 173.
Nonbulk packaging: 173.
Bulk packaging: 173.

STCC Number: Not listed

Clean Water Act Sect. 307: Yes
Clean Water Act Sect. 311: No
Clean Air Act: Not listed
EPA Waste Number: None
CERCLA Ref: Not listed
RQ Designation: Not listed
SARA TPQ Value: Not listed
SARA Sect. 312 Categories:
  Acute toxicity: irritant.
Listed in SARA Sect. 313: Yes
De Minimus Concentration: 1.0%

U.S. Postal Service Mailability: Not given

NFPA Codes:
  Health Hazard (blue): Unspecified
  Flammability (red): Unspecified
  Reactivity (yellow): Unspecified
  Special: Unspecified

--- TOXICITY DATA ---

Short-term Toxicity: Unknown

Long-term Toxicity: Unknown

Symptoms: Inhalation has been reported as causing pulmonary fibrosis.
  (Source: SAX)

Conc IDLH: Unknown

ACGIH TLV: Not listed
ACGIH STEL: Not listed

OSHA PEL: Transitional Limits: PEL = total dust 15 mg/m³; respirable fraction 5 mg/m³; Final Rule Limits: TWA = total dust 10 mg/m³; respirable fraction 5 mg/m³.

MAK Information: Not listed

Carcinogen: N; **Status:** See below

Carcinogen Lists: *IARC*: Not listed; *MAK*: Not listed; *NIOSH*: Not listed; *NTP*: Not listed; *ACGIH*: Not listed; *OSHA*: Not listed.

LD50 Value: No LD50 in RTECS 1992

Other Species Toxicity Data: (Source: NIOSH RTECS 1992) itr-rat LDLo:250 mg/kg

Reproductive Toxicity (1992 RTECS): This chemical has no known mammalian reproductive toxicity.

--- INITIAL INCIDENT RESPONSE ---

No DOT Guide information for this product.

▼ ▼ ▼ ▼ ▼ ▼ ▼ ▼ ▼ ▼ ▼ ▼ ▼ ▼

--- IDENTIFIERS ---

Chemical Name and Synonyms: DI-ORTHO-TOLYLGUANIDINE, $C_{15}H_{17}N_3$

Trade Name and Synonyms: DOTG Accelerator

Chemical Family: Guanidine

Manufacturer: American Cyanamid, Bound Brook, New Jersey

--- INGREDIENTS AND COMPOSITION DATA ---

Material	Cas No.	%	TLV
DOTG	NA		Not established

--- PHYSICAL DATA ---

Boiling Point (°F): NA

Specific Gravity (H₂O = 1): 1.20

Vapor Pressure (mm Hg): NA

Percent Volatile by Volume (%): Less than 1

Vapor Density (air = 1): NA

Freezing Point: NA

Evaporation Rate: NA

pH: NA

Solubility in Water: Negligible

Melting Point: NA

Appearance/Color/Odor: White powder, practically odorless.

--- FIRE AND EXPLOSION HAZARDS ---

Flash Point: NA

Flammability Limits: NA

Autoignition Temperature: NA

Extinguishing Media: Foam, dry chemical, carbon dioxide, or water.

Special Fire Fighting Procedures: As in any fire, prevent human exposure to fire, smoke, fumes, or products of combustion. Evacuate nonessential personnel from the fire area. Firefighters should wear full-face, self-contained breathing apparatus and impervious protective clothing such as gloves, hoods, and suits. Manufacturer recommends positive-pressure self-contained breathing apparatus. *Do not* use high-pressure water stream. Airborne dust creates an explosion hazard.

Unusual Fire and Explosion Hazards: Dust may be explosive if mixed with air in critical proportions and in the presence of a source of ignition. The hazard is similar to that of any organic solid including sawdust. Maintain normal good housekeeping for control of dust.

Hazardous Products of Combustion and Decomposition: Thermal decomposition may produce carbon monoxide and/or carbon dioxide.

Hazardous Polymerization: Will not occur.

--- HEALTH HAZARD DATA ---

Effects of Overexposure: Dust causes eye burns.

Emergency Response and First Aid:
Skin: Wash with lots of clean water for at least 15 min.
Eyes: Flush eyes with lots of water. If irritation persists, consult a physician.
Inhalation: Expose to fresh air. Keep warm and quiet. Give artificial respiration if necessary.
Ingestion: None recommended. Consult physician immediately.

Cancer Information: None reported.

--- TOXICITY DATA ---

Animal Effects: None reported.

--- REACTIVITY DATA ---

Stability and Conditions to Avoid: None reported by manufacturer other than to minimize skin exposure and inhalation.

Materials to Avoid (Incompatibility): None.

Hazard Decomposition Products: Carbon dioxide and carbon monoxide.

Hazardous Polymerization: Will not occur.

--- SPILL AND EMERGENCY RESPONSE ---

For Spills: Sweep up. Wash residuals with soap and water. Transfer to a closed container.

Empty Containers: The container for this product can present explosion or fire hazards, even when emptied. To avoid the risk of injury, do not cut, puncture, or weld on or near this container.

--- SPECIAL PROTECTION AND PRECAUTIONS ---

Respiratory Protection: Dust mask.

Type of Ventilation Required: General mechanical. Avoid breathing dusts and vapors. Adequate ventilation should be provided to keep mist and dust concentrations below acceptable exposure limits. Discharge from the ventilation system should comply with applicable air-pollution regulations. Local exhaust ventilation is recommended with a capture velocity of 150 to 200 fpm.

Eye Protection Requirements: Chemical goggles. Eyewash fountains and safety showers should be easily accessible.

Protective Gloves and Other: Rubber gloves, goggles.

Storage and Handling Precautions: Containers should be stored in a cool, dry, well-ventilated area. Store away from flammable materials, sources of heat, flame, sparks, and foodstuffs. Exercise due caution to prevent damage to or leakage from the container.

Prevent inhalation of dust. Prevent contact with eyes and skin. Under dusty conditions, static electricity may cause an explosion. Keep containers closed. Store in a cool, dry place.

Use with adequate ventilation. Wash thoroughly after handling before eating and smoking. Work clothing should be frequently laundered and not worn away from work premises.

Other Precautions: Use all standard practices for good personal hygiene. Completely isolate and thoroughly clean all equipment, piping, or vessels before beginning maintenance or repairs. Keep area clean. As with any powder or dust-like material, improper handling can lead to spontaneous combustion (dust explosions).

--- REGULATORY INFORMATION ---

Carcinogenic Ingredients/OSHA/NTP/IARC: None
SARA Title III Section 313 Ingredients: None

▼ ▼ ▼ ▼ ▼ ▼ ▼ ▼ ▼ ▼ ▼ ▼ ▼ ▼

--- IDENTIFIERS ---

Chemical Name and Synonyms: COPPER MERCAPTOBENZOTHIA-ZOLE

Trade Name and Synonyms: CUPSAC Accelerator

Chemical Family: Thiazole

Manufacturer: American Cyanamid, Bound Brook, New Jersey

UN ID Number: NA

--- INGREDIENTS AND COMPOSITION DATA ---

Material	Cas No.	%	TLV
Mercaptobenzo-thiazole, copper salt	032510-27-3	55.5	1 mg/m³
Mineral oil	0080120-95-1	2.5	5 mg/m³

--- PHYSICAL DATA ---

Boiling Point (°F): NA

Specific Gravity (H₂O = 1): 1.77

Vapor Pressure (mm Hg): NA

Percent Volatile by Volume (%): Negligible

Vapor Density (air = 1): NA

Freezing Point: NA

Evaporation Rate: NA

pH: NA

Solubility in Water: Negligible

Melting Point (°C): 165 to 175

Appearance/Color/Odor: Yellow-orange powder; no odor description given.

--- FIRE AND EXPLOSION HAZARDS ---

Flash Point: NA

Flammability Limits: NA

Autoignition Temperature: NA

Decomposition Temperature: 350°C

Extinguishing Media: Foam, dry chemical, carbon dioxide, or water.

Special Fire Fighting Procedures: As with many solids, any dust that is generated may be explosive if mixed with air in critical proportions and in the presence of a source of ignition. Use water, carbon dioxide, or dry chemical to extinguish fires. Wear self-contained, positive-pressure breathing apparatus.

Unusual Fire and Explosion Hazards: Dust may be explosive if mixed with air in critical proportions and in the presence of a source of ignition. The hazard is similar to that of any organic solid including sawdust. Maintain normal good housekeeping for control of dust.

Hazardous Products of Combustion and Decomposition: Thermal decomposition may produce carbon monoxide and/or carbon dioxide.

Hazardous Polymerization: Will not occur.

--- HEALTH HAZARD DATA ---

Effects of Overexposure: Overexposure to this material is not likely to cause significant acute toxic effects. Inhalation of dusts and mists of copper salts can result in irritation of nasal mucous membranes, sometimes of the pharynx, and, on occasion, ulceration with perforation of the nasal septum. If copper salts in sufficient concentration reach the gastrointestinal tract, they act as irritants, producing salivation, nausea, vomiting, gastric pain, hemorrhagic gastritis, and diarrhea. Chronic exposures may result in an anemia, but chronic poisoning as from lead does not occur. On the skin, copper salts also act as irritants, producing itching eczema, and on the eye, conjunctivitis or even ulceration and turbidity of the cornea.

Overexposure to the mineral oil in this product is not likely to cause significant acute toxic effects.

Emergency Response and First Aid:
Skin: Wash with lots of clean water for at least 15 min.
Eyes: Flush eyes with lots of water. If irritation persists, consult a physician.
Inhalation: Expose to fresh air. Keep warm and quiet. Give artificial respiration if necessary.
Ingestion: None recommended. Consult physician immediately.

Cancer Information: None reported.

--- TOXICITY DATA ---

Animal Effects: The acute oral (rat) and acute dermal (rabbit) LD50 values for this material are greater than 10 g/kg and greater than 5 g/kg, respectively. The material caused no significant skin and no eye irritation in rabbit primary irritation studies.

--- REACTIVITY DATA ---

Stability and Conditions to Avoid: None reported by manufacturer other than to minimize skin exposure and inhalation. This material is considered stable.

Materials to Avoid (Incompatibility): Strong oxidizing agents.

Hazard Decomposition Products: Carbon dioxide, carbon monoxide, and/or oxides of sulfur.

Hazardous Polymerization: Will not occur.

--- SPILL AND EMERGENCY RESPONSE ---

For Spills: Sweep up. Wash residuals with soap and water. Transfer to a closed container.

Empty Containers: The container for this product can present explosion or fire hazards, even when emptied. To avoid the risk of injury, do not cut, puncture, or weld on or near this container.

--- SPECIAL PROTECTION AND PRECAUTIONS ---

Respiratory Protection: Dust mask.

Type of Ventilation Required: General mechanical. Avoid breathing dusts and vapors. Adequate ventilation should be provided to keep mist and dust concentrations below acceptable exposure limits. Discharge from the ventilation system should comply with applicable air-pollution regulations. Local exhaust ventilation is recommended with a capture velocity of 150 to 200 fpm.

Eye Protection Requirements: Chemical goggles. Eyewash fountains and safety showers should be easily accessible.

Protective Gloves and Other: Rubber gloves, goggles.

Storage and Handling Precautions: Containers should be stored in a cool, dry, well-ventilated area. Store away from flammable materials, sources of heat, flame, sparks, and foodstuffs. Exercise due caution to prevent damage to or leakage from the container.

Prevent inhalation of dust. Prevent contact with eyes and skin. Under dusty conditions, static electricity may cause an explosion. Keep containers closed. Store in a cool, dry place.

Use with adequate ventilation. Wash thoroughly after handling before eating and smoking. Work clothing should be frequently laundered and not worn away from work premises.

Other Precautions: Use all standard practices for good personal hygiene. Completely isolate and thoroughly clean all equipment, piping, or vessels before beginning maintenance or repairs. Keep area clean. As with any powder or dust-like material, improper handling can lead to spontaneous combustion (dust explosions).

--- REGULATORY INFORMATION ---

Proper Shipping Name: Not applicable/not regulated

Hazard Class: Not applicable

UN/NA: Not applicable

DOT Hazardous Substances: (Report Quantity of Product) Not applicable

DOT Label Required: Not applicable

TSCA Information: This product is manufactured in compliance with all provisions of the Toxic Substances Control Act, 15 U.S.C.

▼　▼　▼　▼　▼　▼　▼　▼　▼　▼　▼　▼　▼　▼

--- IDENTIFIERS ---

Chemical Name and Synonyms: THIURAM, MIXTURE OF THIURAMAS

Trade Name and Synonyms: MOTS No. 1 Accelerator

Chemical Family: Thiuram

Manufacturer: American Cyanamid, Bound Brook, New Jersey

UN ID Number: NA

--- INGREDIENTS AND COMPOSITION DATA ---

Material	Cas No.	%	TLV
Thiuram	NA		5 mg/m^3

--- PHYSICAL DATA ---

Boiling Point (°F): NA

Specific Gravity (H₂O = 1): 1.44

Vapor Pressure (mm Hg): NA

Percent Volatile by Volume (%): Negligible

Vapor Density (air = 1): NA

Freezing Point: NA

Evaporation Rate: NA

pH: NA

Solubility in Water: Negligible

Melting Point (°C): NA

Appearance/Color/Odor: Grey powder; practically odorless.

--- FIRE AND EXPLOSION HAZARDS ---

Flash Point: NA

Flammability Limits: NA

Autoignition Temperature: NA

Decomposition Temperature: NA

Extinguishing Media: Foam, dry chemical, carbon dioxide, or water.

Special Fire Fighting Procedures: As with many solids, any dust that is generated may be explosive if mixed with air in critical proportions and in the presence of a source of ignition. Use water, carbon dioxide, or dry chemical to extinguish fires. Wear self-contained, positive-pressure breathing apparatus.

Unusual Fire and Explosion Hazards: Dust may be explosive if mixed with air in critical proportions and in the presence of a source of ignition. The hazard is similar to that of any organic solid including sawdust. Maintain normal good housekeeping for control of dust.

Hazardous Products of Combustion and Decomposition: Thermal decomposition may produce carbon monoxide and/or carbon dioxide.

Hazardous Polymerization: Will not occur.

--- HEALTH HAZARD DATA ---

Effects of Overexposure: Dust causes eye irritation.

Emergency Response and First Aid:
Skin: Wash with lots of clean water for at least 15 min.
Eyes: Flush eyes with lots of water. If irritation persists, consult a physician.
Inhalation: Expose to fresh air. Keep warm and quiet. Give artificial respiration if necessary.
Ingestion: None recommended. Consult physician immediately.

Cancer Information: None reported.

--- TOXICITY DATA ---

Animal Effects: None reported by manufacturer.

--- REACTIVITY DATA ---

Stability and Conditions to Avoid: None reported by manufacturer other than to minimize skin exposure and inhalation. This material is considered stable.

Materials to Avoid (Incompatibility): No specific incompatibility.

Hazard Decomposition Products: Carbon dioxide, carbon monoxide, and or oxides of sulfur.

Hazardous Polymerization: Will not occur.

--- SPILL AND EMERGENCY RESPONSE ---

For Spills: Sweep up. Wash residuals with soap and water. Transfer to a closed container.

Empty Containers: The container for this product can present explosion or fire hazards, even when emptied. To avoid the risk of injury, do not cut, puncture, or weld on or near this container.

--- SPECIAL PROTECTION AND PRECAUTIONS ---

Respiratory Protection: Dust mask.

Type of Ventilation Required: General mechanical. Avoid breathing dusts and vapors. Adequate ventilation should be provided to keep mist and dust concentrations below acceptable exposure limits. Discharge from the ventilation system should comply with applicable air-pollution regulations. Local exhaust ventilation is recommended with a capture velocity of 150 to 200 fpm.

Eye Protection Requirements: Chemical goggles. Eyewash fountains and safety showers should be easily accessible.

Protective Gloves and Other: Rubber gloves, goggles.

Storage and Handling Precautions: Containers should be stored in a cool, dry, well-ventilated area. Store away from flammable materials, sources of heat, flame, sparks, and foodstuffs. Exercise due caution to prevent damage to or leakage from the container.

Prevent inhalation of dust. Prevent contact with eyes and skin. Under dusty conditions, static electricity may cause an explosion. Keep containers closed. Store in a cool, dry place.

Use with adequate ventilation. Wash thoroughly after handling before eating and smoking. Work clothing should be frequently laundered and not worn away from work premises.

Other Precautions: Use all standard practices for good personal hygiene. Completely isolate and thoroughly clean all equipment, piping, or vessels before beginning maintenance or repairs. Keep area clean. As with any

powder or dust-like material, improper handling can lead to spontaneous combustion (dust explosions).

--- REGULATORY INFORMATION ---

None reported.

▼   ▼   ▼   ▼   ▼   ▼   ▼   ▼   ▼   ▼   ▼   ▼   ▼   ▼

--- IDENTIFIERS ---

Chemical Name and Synonyms: TETRAMETHYLTHIURAM MONO-SULFIDE

Trade Name and Synonyms: Monex

Chemical Family: Thiuram

Manufacturer: Uniroyal Chemical Co., Middlebury, Connecticut

UN ID Number: NA

--- INGREDIENTS AND COMPOSITION DATA ---

Material	Cas No.	%	TLV
Thiuram	97-74-5		Not established

--- PHYSICAL DATA ---

Boiling Point (°F): NA

Specific Gravity (H₂O = 1): 1.38

Vapor Pressure (mm Hg): NA

Percent Volatile by Volume (%): Low

Vapor Density (air = 1): NA

Freezing Point: NA

Evaporation Rate: NA

pH: NA

Solubility in Water: Insoluble in water. Soluble in acetone, benzene, and ethylene dichloride.

Melting Point (°C): 106

Appearance/Color/Odor: Yellow pellets or powder with characteristic odor.

--- FIRE AND EXPLOSION HAZARDS ---

Flash Point: 355°F (179°C) TCC

Flammability Limits: NA

Autoignition Temperature: NA

Decomposition Temperature: NA

Extinguishing Media: Foam, dry chemical, carbon dioxide, or water.

Special Fire Fighting Procedures: As with many solids, any dust that is generated may be explosive if mixed with air in critical proportions and in the presence of a source of ignition. Use water, carbon dioxide, or dry chemical to extinguish fires. Wear self-contained, positive-pressure breathing apparatus.

Unusual Fire and Explosion Hazards: Dust may be explosive if mixed with air in critical proportions and in the presence of a source of ignition. The hazard is similar to that of any organic solid including sawdust. Maintain normal good housekeeping for control of dust.

Hazardous Products of Combustion and Decomposition: Thermal decomposition may produce carbon monoxide and/or carbon dioxide.

Hazardous Polymerization: Will not occur.

--- HEALTH HAZARD DATA ---

Effects of Overexposure: Dust causes eye irritation. Repeated minimal contact with skin may cause sensitization. Exposure can produce an adverse reaction when alcohol is consumed and can affect the activity of certain prescription drugs; consult physician.

Emergency Response and First Aid:
Skin: Wash with lots of clean water for at least 15 min.
Eyes: Flush eyes with lots of water. If irritation persists, consult a physician.
Inhalation: Expose to fresh air. Keep warm and quiet. Give artificial respiration if necessary.
Ingestion: None recommended. Consult physician immediately.

Cancer Information: None reported.

--- TOXICITY DATA ---

Animal Effects:
Oral toxicity: LD50 (rats), 1.3 g/kg
Dermal toxicity: LD50 (rabbits), >2g/kg
Irritation: eye (rabbits), moderate
          skin (rabbits), negative
Sensitization: skin-positive based on human experience
Genotoxicity: Ames Salmonella, positive
          L5178Y mouse lymphoma, negative
          Balb/3T3, cell transformation, negative
Chronic: The feeding to mice of up to 337 ppm for 18 months did not produce an increased tumor incidence.

--- REACTIVITY DATA ---

Stability and Conditions to Avoid: None reported by manufacturer other than to minimize skin exposure and inhalation. This material is considered stable.

Materials to Avoid (Incompatibility): Strong oxidizing agents.

Hazard Decomposition Products: Carbon dioxide, carbon monoxide, nitrogen, and/or oxides of sulfur.

Hazardous Polymerization: Will not occur.

--- SPILL AND EMERGENCY RESPONSE ---

For Spills: Sweep up. Wash residuals with soap and water. Transfer to a closed container.

Empty Containers: The container for this product can present explosion or fire hazards, even when emptied. To avoid the risk of injury, do not cut, puncture, or weld on or near this container.

--- SPECIAL PROTECTION AND PRECAUTIONS ---

Respiratory Protection: Dust mask.

Type of Ventilation Required: General mechanical. Avoid breathing dusts and vapors. Adequate ventilation should be provided to keep mist and dust concentrations below acceptable exposure limits. Discharge from the ventilation system should comply with applicable air-pollution regulations. Local exhaust ventilation is recommended with a capture velocity of 150 to 200 fpm.

Eye Protection Requirements: Chemical goggles. Eyewash fountains and safety showers should be easily accessible.

Protective Gloves and Other: Rubber gloves, goggles.

Storage and Handling Precautions: Containers should be stored in a cool, dry, well-ventilated area. Store away from flammable materials, sources of heat, flame, sparks, and foodstuffs. Exercise due caution to prevent damage to or leakage from the container.

Prevent inhalation of dust. Prevent contact with eyes and skin. Under dusty conditions, static electricity may cause an explosion. Keep containers closed. Store in a cool, dry place.

Use with adequate ventilation. Wash thoroughly after handling before eating and smoking. Work clothing should be frequently laundered and not worn away from work premises.

Other Precautions: Use all standard practices for good personal hygiene. Completely isolate and thoroughly clean all equipment, piping, or vessels before beginning maintenance or repairs. Keep area clean. As with any

powder or dust-like material, improper handling can lead to spontaneous combustion (dust explosions).

--- REGULATORY INFORMATION ---

None reported.

▼ ▼ ▼ ▼ ▼ ▼ ▼ ▼ ▼ ▼ ▼ ▼ ▼ ▼

--- IDENTIFIERS ---

Name: HEXAMINE
Synonyms: Methanamine;   1,3,5,7-Tetraazatricyclo[3.3.1.1 3,7]-Decane; Hexamethylene Tetramine; HMT; HMTA; 1,3,5,7-Tetraazaadamantane; Hexamethylenamine; Aminoform; Ammoform; Cystamin; Cystogen; Formin; Uritone; Urotropin; Ammonioformaldehyde; Formamine; Hexaform; Hexilmethylenamine; Methenamine; Urotropine
CAS: 100-97-0; **RTECS:** MN4725000
Formula: $C_6H_{12}N_4$; **Mol Wt:** 140.22
WLN: T66 B6 A B- C 1B I BN DN FN HNTJ
Chemical Class: Aliphatic amine

See other identifiers listed below under Regulations.

--- PROPERTIES ---

Boiling Point: 553 K; 279.8°C; 535.7°F
Melting Point: 563 K; 289.8°C; 553.7°F
Flash Point: 523 K; 249.8°C; 481.7°F
Autoignition: NA
UEL: NA
LEL: NA
Vapor Density: No data
Specific Gravity: No data

Reactivity with Water: No data on water reactivity
Reactivity with Common Materials: No data
Stability during Transport: No data
Neutralizing Agents: No data
Polymerization Possibilities: No data

Toxic Fire Gases: None reported other than possible unburned vapors
Odor Detected at (ppm): Not pertinent
Odor Description: Mild, ammonia-like (Source: CHRIS)
100% Odor Detection: No data

--- REGULATIONS ---

DOT hazard class: 4.1 FLAMMABLE SOLID
DOT guide: 32
Identification number: UN1328
DOT shipping name: HEXAMINE
Packing group: III
Label(s) required: FLAMMABLE SOLID
Special Provisions: A1
Packaging exceptions: 173.151
Nonbulk packaging: 173.213
Bulk packaging: 173.240
Quantity limitations:
  Passenger air/rail: 25 kg
  Cargo aircraft only: 100 kg
  Vessel stowage: A

STCC Number: Not listed

Clean Water Act Sect. 307: No
Clean Water Act Sect. 311: No
Clean Air Act: Not listed
EPA Waste Number: None
CERCLA Ref: Not listed
RQ Designation: Not listed
SARA TPQ Value: Not listed
SARA Sect. 312 Categories:
  Acute toxicity: adverse effect to target organs.
  Fire hazard: flammable.

U.S. Postal Service Mailability: Not given

NFPA Codes:
  Health Hazard (blue): Unspecified
  Flammability (red): Unspecified
  Reactivity (yellow): Unspecified
  Special: Unspecified

--- TOXICITY DATA ---

Short-term Toxicity: Unknown

Long-term Toxicity: Unknown

Target Organs: Skin, eyes, mucous membranes

Symptoms: Prolonged and repeated contact may cause skin irritation. (Source: CHRIS)

Conc IDLH: Unknown

NIOSH REL: Not given

ACGIH TLV: Not listed
ACGIH STEL: Not listed

OSHA PEL: Not in Table Z-1-A

MAK Information: Not listed

Carcinogen: N; **Status:** See below

Carcinogen Lists: *IARC*: Not listed; *MAK*: Not listed; *NIOSH*: Not listed; *NTP*: Not listed; *ACGIH*: Not listed; *OSHA*: Not listed.

LD50 Value: No LD50 in RTECS 1992

Other Species Toxicity Data: (Source: NIOSH RTECS 1992)
scu-rat LDLo:200 mg/kg
ivn-rat LD50:9200 mg/kg
par-rat LDLo:200 mg/kg
scu-mus LD50:215 mg/kg
par-mus LDLo:450 mg/kg
scu-cat LDLo:200 mg/kg
par-cat LDLo:200 mg/kg
scu-gpg LDLo:300 mg/kg
par-mam LDLo:300 mg/kg

Reproductive Toxicity (1992 RTECS): This chemical has no known mammalian reproductive toxicity.

--- PROTECTION AND FIRST AID ---

First Aid Source: CT THIC
Eye: Flush with water.
Skin: Wash with soap and water.
Inhalation: None given.
Ingestion: Gastric lavage followed by saline catharsis.

First Aid Source: CHRIS Manual 1991.
Wash skin or eyes thoroughly with water. Call a physician.

First Aid Source: DOT Emergency Response Guide 1993.
Move victim to fresh air; call emergency medical care. In case of contact with material, immediately flush skin or eyes with running water for at least 15 min. Removal of solidified molten material from skin requires medical assistance. Remove and isolate contaminated clothing and shoes at the site.

--- INITIAL INCIDENT RESPONSE ---

Fire Extinguishment: Water, foam, carbon dioxide, dry chemical.
(CHRIS 91)

U.S. Department of Transportation Guide to Hazardous Materials Transport Information - Publication DOT 5800.5 (1990).
DOT Shipping Name: HEXAMINE
DOT ID Number: UN1328

ERG90   GUIDE 32
* POTENTIAL HAZARDS *

*Health Hazards
Fire may produce irritating or poisonous gases.
Contact may cause burns to skin and eyes.
Runoff from fire control or dilution water may cause pollution.

*Fire or Explosion
Flammable/combustible material; may be ignited by heat, sparks, or flames.
May burn rapidly with flare-burning effect.

* EMERGENCY ACTION *

Keep unnecessary people away; isolate hazard area and deny entry.
Stay upwind; keep out of low areas.
Positive-pressure self-contained breathing apparatus (SCBA) and structural
firefighters' protective clothing will provide limited protection.
**CALL CHEMTREC AT 1-800-424-9300 FOR EMERGENCY ASSIS-
TANCE.** If water pollution occurs, notify the appropriate authorities.

*Fire
Small fires: Dry chemical, sand, earth, water spray, or regular foam.
Large fires: Water spray, fog, or regular foam.
Move container from fire area if you can do it without risk.
Apply cooling water to sides of containers that are exposed to flames until
well after fire is out. Stay away from ends of tanks.
For massive fire in cargo area, use unmanned hose holder or monitor nozzles;
if this is impossible, withdraw from area and let fire burn.
Magnesium fires: Use dry sand, Met-L-X R powder, or G-1 graphite powder.

*Spill or Leak
Shut off ignition sources; no flares, smoking, or flames in hazard area.
Do not touch or walk through spilled material.
Small dry spills: With clean shovel, place material into clean, dry container
and cover loosely; move containers from spill area.
Large spills: Wet down with water and dike for later disposal.

*First Aid
Move victim to fresh air; call emergency medical care.
In case of contact with material, immediately flush skin or eyes with running
water for at least 15 min.
Removal of solidified molten material from skin requires medical assistance.
Remove and isolate contaminated clothing and shoes at the site.

▼　▼　▼　▼　▼　▼　▼　▼　▼　▼　▼　▼　▼　▼

--- IDENTIFIERS ---

Name: ZINC ETHYLPHENYLDITHIOCARBAMATE
Synonyms: Accelerator EFK; Fenyl-Ethyldithiokarbaminan Zinfonaty (Czech);
Hermat Fedk; Vulkacit P Extra N; Zinc, Bis(Ethylphenylcarbamodithioato-
S,S')-, (T-4)- (9CI); Zinc Ethylphenyldithiocarbamate; Zinc Ethylphenyl-
thiocarbamate

CAS: 14634-93-6; **RTECS:** ZH0890000
Formula: $C_{18}H_{20}N_2S_4Zn$; **Mol Wt:** 458.01
Chemical Class: Organic acid salt

See other identifiers listed below under Regulations.

--- PROPERTIES ---

Boiling Point: NA
Melting Point: NA
Flash Point: NA
Autoignition: NA
UEL: NA
LEL: NA
Vapor Density: No data
Specific Gravity: No data

Reactivity with Water: No data on water reactivity
Reactivity with Common Materials: No data
Stability during Transport: No data
Neutralizing Agents: No data
Polymerization Possibilities: No data

Toxic Fire Gases: None reported other than possible unburned vapors
Odor Detected at (ppm): Unknown
Odor Description: No data
100% Odor Detection: No data

--- REGULATIONS ---

DOT hazard class: Not given
Packaging exceptions: 173.
Nonbulk packaging: 173.
Bulk packaging: 173.

STCC Number: Not listed

Clean Water Act Sect. 307: Yes
Clean Water Act Sect. 311: No
Clean Air Act: Not listed
EPA Waste Number: None
CERCLA Ref: Not listed

RQ Designation: Not listed
SARA TPQ Value: Not listed
SARA Sect. 312 Categories:
Acute toxicity: irritant.
Chronic toxicity: mutagen.
Listed in SARA Sect. 313: Yes
De Minimus Concentration: 1.0%

U.S. Postal Service Mailability: Not given

--- TOXICITY DATA ---

Short-term Toxicity: Unknown

Long-term Toxicity: Unknown

Conc IDLH: Unknown

NIOSH REL: Not given

ACGIH TLV: Not listed
ACGIH STEL: Not listed

OSHA PEL: Not in Table Z-1-A

MAK Information: Not listed

Carcinogen: N; **Status:** See below

Carcinogen Lists: *IARC*: Not listed; *MAK*: Not listed; *NIOSH*: Not listed; *NTP*: Not listed; *ACGIH*: Not listed; *OSHA*: Not listed.

LD50 Value: orl-rat LD50:10800 mg/kg

Other Species Toxicity Data: (Source: NIOSH RTECS 1992)
orl-rat LD50:10800 mg/kg
ipr-mus LD50:533 mg/kg

Reproductive Toxicity (1992 RTECS): This chemical has no known mammalian reproductive toxicity.

--- PROTECTION AND FIRST AID ---

Recommended Respiration Protection Source: NIOSH Pocket Guide (85-114)
OSHA (Zinc Ethylphenyldithiocarbamate) -
2000 ppm: Any supplied-air respirator. Any self-contained breathing
apparatus.
5000 ppm: Any supplied-air respirator operated in a continuous-flow mode.
10,000 ppm: Any self-contained breathing apparatus with a full facepiece.
Any supplied-air respirator with a full facepiece. Any supplied-air respirator
with a tight-fitting facepiece operated in a continuous-flow mode.
25,000 ppm: Any supplied-air respirator with a full facepiece and operated
in a pressure-demand or other positive-pressure mode.
Emergency or planned entry in unknown concentrations or IDLH conditions:
Any self-contained breathing apparatus with a full facepiece and operated in
a pressure-demand or other positive-pressure mode. Any supplied-air
respirator with a full facepiece and operated in a pressure-demand or other
positive-pressure mode in combination with an auxiliary self-contained
breathing apparatus operated in a pressure-demand or other positive-pressure
mode.
Escape: Any appropriate escape-type self-contained breathing apparatus.

--- INITIAL INCIDENT RESPONSE ---

No DOT Guide information for this product.

▼   ▼   ▼   ▼   ▼   ▼   ▼   ▼   ▼   ▼   ▼   ▼   ▼   ▼

--- IDENTIFIERS ---

Name: CARBAMIC ACID, DIMETHYLDITHIO-, COPPER (II) SALT
Synonyms: Compound-4018; Copper, Bis(Dimethyldithiocarbamato)-; Copper
Dimethyldithiocarbamate; Cumate; Dimethyldithiocarbamic Acid Copper Salt;
Wolfen
CAS: 137-29-1; **RTECS:** FA0175000
Formula: $C_6H_{12}N_2S_4$ .Cu; **Mol Wt:** 303.98
WLN: SUYSHN1
Chemical Class: Organic acid salt

See other identifiers listed below under Regulations.

--- PROPERTIES ---

Boiling Point: NA
Melting Point: NA
Flash Point: NA
Autoignition: NA
UEL: NA
LEL: NA
Vapor Density: No data
Specific Gravity: No data

Reactivity with Water: No data on water reactivity
Reactivity with Common Materials: No data
Stability during Transport: No data
Neutralizing Agents: No data
Polymerization Possibilities: No data

Toxic Fire Gases: Oxides of nitrogen
Odor Detected at (ppm): Unknown
Odor Description: No data
100% Odor Detection: No data

--- REGULATIONS ---

DOT hazard class: Not given
Packaging exceptions: 173.
Nonbulk packaging: 173.
Bulk packaging: 173.

STCC Number: Not listed

Clean Water Act Sect. 307: Yes
Clean Water Act Sect. 311: No
Clean Air Act: Not listed
EPA Waste Number: None
CERCLA Ref: Not listed
RQ Designation: Not listed
SARA TPQ Value: Not listed
SARA Sect. 312 Categories:
   Chronic toxicity: adverse effect to target organs after long periods of
   exposure.
   Chronic toxicity: mutagen.

Listed in SARA Sect. 313: Yes
De Minimus Concentration: 1.0%
U.S. Postal Service Mailability: Not given

--- TOXICITY DATA ---

Short-term Toxicity: Unknown

Long-term Toxicity: Unknown

Conc IDLH: Unknown

NIOSH REL: Not given

ACGIH TLV: TLV = fume, 0.2 mg/m^3; dust and mists, 1 mg/m^3 as copper
ACGIH STEL: As copper

OSHA PEL: Transitional limits: PEL = (fume) 0.1 mg/m^3; (dust and mists) 1 mg/m^3; Final Rule Limits: TWA = (fume) 0.1 mg/m^3; (dust and mists) 1 mg/m^3.

MAK Information: Not listed

Carcinogen: N; **Status:** See below

Carcinogen Lists: *IARC*: Not listed; *MAK*: Not listed; *NIOSH*: Not listed; *NTP*: Not listed; *ACGIH*: Not listed; *OSHA*: Not listed.

LD50 Value: No LD50 in RTECS 1991

Other Species Toxicity Data: (Source: NIOSH RTECS 1992)
ihl-rat LCLo:210 mg/m^3/4H
ipr-rat LDLo:25 mg/kg
ipr-mus LDLo:7800 mg/kg

Reproductive Toxicity (1993 RTECS): This chemical has no mammalian reproductive toxicity.

--- INITIAL INCIDENT RESPONSE ---
No DOT Guide information for this product.

▼   ▼   ▼   ▼   ▼   ▼   ▼   ▼   ▼   ▼   ▼   ▼   ▼   ▼

--- IDENTIFIERS ---

Name: 2-BENZOTHIAZOLETHIOL
Synonyms: Captax; MBT; 2-Mercaptobenzothiazole; NCI-C56519; Rotax; Sul
Fadene; USAF GY-3; USAF XR-29
CAS: 149-30-4; **RTECS:** DL6475000
Formula: $C_7H_5NS_2$; **Mol Wt:** 167.25
WLN: T56 BN DSJ CSH
Chemical Class: Azole

See other identifiers listed below under Regulations.

--- PROPERTIES ---

Boiling Point: NA
Melting Point: 443 K; 169.8°C; 337.7°F
Flash Point: NA
Autoignition: NA
UEL: NA
LEL: NA
Vapor Density: No data
Specific Gravity: No data
Density: 1.42 at 25°C

Reactivity with Water: No data on water reactivity
Reactivity with Common Materials: No data
Stability during Transport: No data
Neutralizing Agents: No data
Polymerization Possibilities: No data

Toxic Fire Gases: None reported other than possible unburned vapors
Odor Detected at (ppm): Unknown

Odor Description: No data
100% Odor Detection: No data

--- REGULATIONS ---

DOT hazard class: 6.1 POISON
DOT guide: 53
Identification number: UN2811

DOT shipping name: POISONOUS SOLIDS, N.O.S.
Packing group: III
Label(s) required: KEEP AWAY FROM FOOD
Packaging exceptions: 173.153
Nonbulk packaging: 173.213
Bulk packaging: 173.240
Quantity limitations:
  Passenger air/rail: 100 kg
  Cargo aircraft only: 200 kg
  Vessel stowage: A

STCC Number: Not listed

Clean Water Act Sect. 307: No
Clean Water Act Sect. 311: No
Clean Air Act: Not listed
EPA Waste Number: None
CERCLA Ref: Not listed
RQ Designation: Not listed
SARA TPQ Value: Not listed
SARA Sect. 312 Categories:
  Acute toxicity: toxic. LD50 > 50 and < = 500 mg/kg (oral rat).
  Chronic toxicity: mutagen.
  Chronic toxicity: reproductive toxin.

U.S. Postal Service Mailability: Not given

--- TOXICITY DATA ---

Short-term Toxicity: Unknown

Long-term Toxicity: Unknown

Conc IDLH: Unknown

NIOSH REL: Not given.

ACGIH TLV: Not listed
ACGIH STEL: Not listed

OSHA PEL: Not in Table Z-1-A

MAK Information: Not listed

Carcinogen: N; **Status:** See below

Carcinogen Lists: *IARC*: Not listed; *MAK*: Not listed; *NIOSH*: Not listed; *NTP*: Not listed; *ACGIH*: Not listed; *OSHA*: Not listed.

LD50 Value: orl-rat LD50:100 mg/kg

Other Species Toxicity Data: (Source: NIOSH RTECS 1992)
orl-rat LD50:100 mg/kg
ihl-rat LC50: > 1270 mg/m^3
ipr-rat LD50:300 mg/kg
orl-mus LD50:1851 mg/kg
ipr-mus LD50:100 mg/kg
scu-mus LD50:1158 mg/kg
skn-rbt LD50: > 7940 mg/kg

Reproductive Toxicity (1992 RTECS): This chemical is a mammalian reproductive toxin.

Reproductive Toxicity Data (1992 RTECS):
par-rat TDLo:800 mg/kg (2D male/2D pre) BEXBAN 93, 107, 82
    Effects on Fertility
        Post-implantation mortality
    Effects on Embryo or Fetus
        Fetotoxicity (except death, e.g., stunted fetus)
        Fetal death

par-rat TDLo:400 mg/kg (4-11D preg) BEXBAN 93, 107, 82
    Effects on Embryo or Fetus
        Fetotoxicity (except death, e.g., stunted fetus)
        Fetal death

--- PROTECTION AND FIRST AID ---

First Aid Source: DOT Emergency Response Guide 1993.
Move victim to fresh air; call emergency medical care. In case of contact with material, immediately flush skin or eyes with running water for at least 15 min. Remove and isolate contaminated clothing and shoes at the site.

--- INITIAL INCIDENT RESPONSE ---

U.S. Department of Transportation Guide to Hazardous Materials Transport
Information - Publication DOT 5800.5 (1990).
DOT Shipping Name: POISONOUS SOLIDS, N.O.S.
DOT ID Number: UN2811

ERG90   GUIDE 53
                        * POTENTIAL HAZARDS *

*Health Hazards
Poisonous if swallowed.
Inhalation of dust poisonous.
Fire may produce irritating or poisonous gases.
Runoff from fire control or dilution water may cause pollution.

*Fire or Explosion
Some of these materials may burn, but none of them ignites readily.

                        * EMERGENCY ACTION *

Keep unnecessary people away; isolate hazard area and deny entry.
Stay upwind; keep out of low areas.
Positive-pressure self-contained breathing apparatus (SCBA) and structural
   firefighters' protective clothing will provide limited protection.
**CALL CHEMTREC AT 1-800-424-9300 FOR EMERGENCY ASSIS-
TANCE.** If water pollution occurs, notify the appropriate authorities.

*Fire
Small fires: Dry chemical, CO_2, water spray, or regular foam.
Large fires: Water spray, fog, or regular foam.
Move container from fire area if you can do it without risk.

*Spill or Leak
Do not touch or walk through spilled material; stop leak if you can do it
   without risk.
Small spills: Take up with sand or other noncombustible absorbent material
   and place into containers for later disposal.
Small dry spills: With clean shovel, place material into clean, dry container
   and cover; move containers from spill area.
Large spills: Dike far ahead of liquid spill for later disposal.

*First Aid
Move victim to fresh air; call emergency medical care.
In case of contact with material, immediately flush skin or eyes with running water for at least 15 min.
Remove and isolate contaminated clothing and shoes at the site.

▼   ▼   ▼   ▼   ▼   ▼   ▼   ▼   ▼   ▼   ▼   ▼   ▼   ▼

--- IDENTIFIERS ---

Name: 2-BENZOTHIAZOLETHIOL, ZINC SALT
Synonyms: Bis(2-Benzothiazolylthio)Zinc; Bis(Mercaptobenzothiazolato)Zinc; Hermat Zn-MBT; 2-Mercaptobenzothiazole Zinc Salt; OXAF; Pennac ZT; Tisperse MB-58; USAF GY-7; Vulkacit ZM; Zenite; Zenite Special; Zetax; Zinc 2-Benzothiazolethiolate; Zinc Benzothiazolyl Mercaptide; Zinc Benzothiazol-2-Ylthiolate; Zinc Benzothiazyl-2-Mercaptide; Zinc, Bis(2-Benzothiazolethiolato)-; Zinc Mercaptobenzothiazolate; Zinc 2-Mercaptobenzothiazole; Zinc Mercaptobenzothiazole Salt; ZMBT; ZnMB
CAS: 155-04-4; **RTECS:** DL7000000
Formula: $C_{14}H_8N_2S_4$. Zn; **Mol Wt:** 397.85
WLN: T56 BN DSJ CSH
Chemical Class: Organic acid salt

See other identifiers listed below under Regulations.

--- PROPERTIES ---

Boiling Point: NA
Melting Point: NA
Flash Point: NA
Autoignition: NA
UEL: NA
LEL: NA
Vapor Density: No data
Specific Gravity: No data

Reactivity with Water: No data on water reactivity
Reactivity with Common Materials: No data
Stability during Transport: No data
Neutralizing Agents: No data
Polymerization Possibilities: No data

Toxic Fire Gases: None reported other than possible unburned vapors
Odor Detected at (ppm): Unknown
Odor Description: No data
100% Odor Detection: No data

--- REGULATIONS ---

DOT hazard class: Not given
Packaging exceptions: 173.
Nonbulk packaging: 173.
Bulk packaging: 173.

STCC Number: Not listed

Clean Water Act Sect. 307: Yes
Clean Water Act Sect. 311: No
Clean Air Act: Not listed
EPA Waste Number: None
CERCLA Ref: Not listed
RQ Designation: Not listed
SARA TPQ Value: Not listed
SARA Sect. 312 Categories: No category
Listed in SARA Sect. 313: Yes
De Minimus Concentration: 1.0%

U.S. Postal Service Mailability: Not given

--- TOXICITY DATA ---

Short-term Toxicity: Unknown

Long-term Toxicity: Unknown

Conc IDLH: Unknown

NIOSH REL: Not given

ACGIH TLV: Not listed
ACGIH STEL: Not listed

OSHA PEL: Not in Table Z-1-A

MAK Information: Not listed

Carcinogen: N; **Status:** See below

Carcinogen Lists: *IARC*: Not listed; *MAK*: Not listed; *NIOSH*: Not listed; *NTP*: Not listed; *ACGIH*: Not listed; *OSHA*: Not listed.

LD50 Value: orl-rat LD50:540 mg/kg

Other Species Toxicity Data: (Source: NIOSH RTECS 1992)
orl-rat LD50:540 mg/kg
ipr-mus LD50:200 mg/kg

Reproductive Toxicity (1992 RTECS): This chemical has no known mammalian reproductive toxicity.

--- INITIAL INCIDENT RESPONSE ---

No DOT Guide information for this product.

▼   ▼   ▼   ▼   ▼   ▼   ▼   ▼   ▼   ▼   ▼   ▼   ▼   ▼

--- IDENTIFIERS ---

Name: UREA, 1,3-DIETHYL-2-THIO-
Synonyms: *N,N'*-Diethylthiocarbamide; *N,N'*-Diethylthiourea; 1,3-Diethylthiourea; 1,3-Diethyl-2-Thiourea; NCI-C03816; Pennzone E; Thiate H; Thiourea, *N,N'*-Diethyl-; U 15030; USAF EK-1803
CAS: 105-55-5; **RTECS:** YS9800000
Formula: $C_5H_{12}N_2S$; **Mol Wt:** 132.25
WLN: SUYM1

See other identifiers listed below under Regulations.

--- PROPERTIES ---

Boiling Point: NA
Melting Point: 349.16 TO 351.16 K; 76 to 78°C; 168.8 to 172.4°F
Flash Point: NA
Autoignition: NA
UEL: NA
LEL: NA

Vapor Density: No data
Specific Gravity: No data

Reactivity with Water: No data on water reactivity
Reactivity with Common Materials: No data
Stability during Transport: No data
Neutralizing Agents: No data
Polymerization Possibilities: No data

Toxic Fire Gases: None reported other than possible unburned vapors
Odor Detected at (ppm): Unknown
Odor Description: No data
100% Odor Detection: No data

--- REGULATIONS ---

DOT hazard class: Not given
Packaging exceptions: 173.
Nonbulk packaging: 173.
Bulk packaging: 173.

STCC Number: Not listed

Clean Water Act Sect. 307: No
Clean Water Act Sect. 311: No
Clean Air Act: Not listed
EPA Waste Number: None
CERCLA Ref: Not listed
RQ Designation: Not listed
SARA TPQ Value: Not listed
SARA Sect. 312 Categories:
  Acute toxicity: toxic. LD50 > 50 and < = 500 mg/kg (oral rat).
  Chronic toxicity: mutagen.

U.S. Postal Service Mailability: Not given

--- TOXICITY DATA ---

Short-term Toxicity: Unknown

Long-term Toxicity: Unknown

Conc IDLH: Unknown

NIOSH REL: Not given

ACGIH TLV: Not listed
ACGIH STEL: Not listed

OSHA PEL: Not in Table Z-1-A

MAK Information: Not listed

Carcinogen: N; **Status:** See below

Carcinogen Lists: *IARC*: Not listed; *MAK*: Not listed; *NIOSH*: Not listed; *NTP*: Not listed; *ACGIH*: Not listed; *OSHA*: Not listed.

LD50 Value: orl-rat LD50:316 mg/kg

Other Species Toxicity Data: (Source: NIOSH RTECS 1992)
orl-rat LD50:316 mg/kg
orl-mus LDLo:62 mg/kg
ipr-mus LD50:500 mg/kg

Reproductive Toxicity (1992 RTECS): This chemical has no known mammalian reproductive toxicity.

--- INITIAL INCIDENT RESPONSE ---

No DOT Guide information for this product.

▼   ▼   ▼   ▼   ▼   ▼   ▼   ▼   ▼   ▼   ▼   ▼   ▼   ▼

--- IDENTIFIERS ---

Name: BIS(DIMETHYLTHIOCARBAMYL) DISULFIDE
Synonyms: Thiram; Bis((Dimethylamino)Carbonothioyl)Disulphide; Disolfuro Di Tetrametiltiourame (Italian); Disulfure De Tetramethylthiourame; Alpha, Alpha'-Dithiobis(Dimethylthio) Formamide; *N,N'*-(Dithiodicarborothioyl)-Bis(*N*-Methyl-Methanamine); Methyl Thiuramdisulfide; Tetramethyl Thiuram Disulfide; Tetramethyl Diurane Sulphite; Tetramethylene Thiuram Disulphide; Tetramethylthiocarbamoyl Disulphide; Tetramethylthioperoxydicarbonic Diamide; Tetramethyl Thioram Disulfide (Dutch); Tetramethyl-Thiram

Disulfid (German); Tetramethylthiuram Bisulfide; Tetramethylthiuram Disulfide; *N,N,N',N'*-Tetramethyl Thiuram Disulfide; Tetramethyl Thiurane Disulfide; Tetramethylthiuram Disulfide; Thiram; Tiuram (Polish); USAF B-30; USAF EK-2089; USAF P-5; TMTD; ENT 987; SQ 1489; NSC 1771; Thiurad; Thiosan; Thylate; Tiuramyl; Thiuramyl; Duralyn; Fernasan; Nowergan; Pezifilm; Pomarsol; Tersan; Tuads; Tulisan; Arasan; HCDB
CAS: 137-26-8; **RTECS:** JO1400000
Formula: $C_6H_{12}N_2S_4$; **Mol Wt:** 240.44
WLN: 1N1
Chemical Class: Organic sulfide

See other identifiers listed below under Regulations.

--- PROPERTIES ---

Physical Description: Colorless to cream solid (some commercial products dyed blue)
Boiling Point: Decomposes
Melting Point: 413.15 K; 140°C; 284°F
Flash Point: 362 K; 88.8°C; 191.9°F
Autoignition: NA
Vapor Pressure: About 0 mm
UEL: NA
LEL: NA
Vapor Density: No data
Specific Gravity: 1.43 at 20°C
Density: 1.43 g/cc or 13.299 lb/gal
Water Solubility: INSOL
Incompatibilities: Strong oxidizers, strong acids, oxidizable materials

Reactivity with Water: No data on water reactivity
Reactivity with Common Materials: No data
Stability during Transport: No data
Neutralizing Agents: No data
Polymerization Possibilities: No data

Toxic Fire Gases: Oxides of nitrogen, oxides of sulfur
Odor Detected at (ppm): Data not available
Odor Description: Data not available (Source: CHRIS)
100% Odor Detection: No data

--- REGULATIONS ---

DOT hazard class: CLASS 9
DOT guide: 31
Identification number: UN3077
DOT shipping name: ENVIRONMENTALLY HAZARDOUS
SUBSTANCES, SOLID, N.O.S.
Packing group: III
Label(s) required: CLASS 9
Special Provisions: 8, B54
Packaging exceptions: 173.155
Nonbulk packaging: 173.213
Bulk packaging: 173.240
Quantity limitations:
Passenger air/rail: None
Cargo aircraft only: None
Vessel stowage: A

STCC Number: 4921632, 4921631, 4921634, 492133

Clean Water Act Sect. 307: No
Clean Water Act Sect. 311: No
Clean Air Act: Not listed
EPA Waste Number: U244
CERCLA Ref: Y
RQ Designation: A, 10 lb (4.54 kg) CERCLA
SARA TPQ Value: Not listed
SARA Sect. 312 Categories:
Acute toxicity: toxic. LD50 > 50 and < = 500 mg/kg (oral rat).
Acute toxicity: irritant.
Acute toxicity: adverse effect to target organs.
Chronic toxicity: adverse effect to target organs after long periods of exposure.
Chronic toxicity: reproductive toxin.

U.S. Postal Service Mailability:
Hazard class: ORM-A
Mailability: Domestic service and air transportation; shipper's declaration
Max per parcel: 70 lb; 5 lb (air)

NFPA Codes:
Health Hazard (blue): Unspecified
Flammability (red): Unspecified
Reactivity (yellow): Unspecified
Special: Unspecified

--- TOXICITY DATA ---

Short-term Toxicity: *Inhalation:* No information on human exposure is available. Animal studies indicate that irritation of the nose and throat may occur at levels above 5 mg/m^3. *Eyes:* May cause irritation, tearing, and sensitivity to light. *Skin:* Exposure to spray containing 45% thiram resulted in irritation and skin sensitization. *Ingestion:* No information available on human exposure. In animal studies, 38 ppm in food caused nausea, vomiting, diarrhea, hyperexcitability, weakness, and loss of muscle control. Death may occur from ingestion of approximately one teaspoonful. (NYDH)

Long-term Toxicity: Occupational exposure to 0.03 mg/m^3 over a 5-yr period has caused mild irritation of the nose and throat. Prolonged contact has caused eye irritation, tearing, increased sensitivity to light, reduced night vision, and blurred vision. Thiram has caused birth defects in laboratory animals. Whether it has this effect in humans is not known. (NYDH)

Target Organs: Skin, mucous membranes, kidneys, respiratory system, lungs.

Symptoms: Inhalation of dust may cause respiratory irritation. Liquid irritates eyes and skin and may cause allergic eczema in sensitive individuals. Ingestion causes nausea, vomiting, and diarrhea, all of which may be persistent; paralysis may develop. (Source: CHRIS)

Conc IDLH: 1500 mg/m^3

ACGIH TLV: TLV = 1 mg/m^3
ACGIH STEL: Not listed

OSHA PEL: Transitional Limits: PEL = 5 mg/m^3; Final Rule Limits: TWA = 5 mg/m^3.

MAK Information: 5 calculated as total dust, mg/m^3; Substance with systemic effects; onset of effect less than or equal to 2 hr: Peak = 5xMAK for 30 min, 2 times per shift of 8 hr.

Carcinogen: N; **Status:** See below
References: Animal indefinite IARC** 12, 225, 76; Human indefinite IARC** 12, 225, 76

Carcinogen Lists: *IARC:* Not classified as to human carcinogenicity or probably not carcinogenic to humans; *MAK:* Not listed; *NIOSH:* Not listed; *NTP:* Not listed; *ACGIH:* Not listed; *OSHA:* Not listed.

Human Toxicity Data: (Source: NIOSH RTECS)
ihl-hmn TCLo:30 mg/m^3/5Y-I VRDEA5 (10), 136, 71
　Sense Organs
　　Eye
　　　Conjunctive irritation
　Cardiac
　　Other changes
　Lungs, Thorax, or Respiration
　　Structural or functional changes in trachea or bronchi

LD50 Value: orl-rat LD50:560 mg/kg

Other Species Toxicity Data: (Source: NIOSH RTECS 1992)
orl-rat LD50:560 mg/kg
ihl-rat LC50:500 mg/m^3/4H
skn-rat LD50: >5 g/kg
ipr-rat LD50:138 mg/kg
scu-rat LD50:646 mg/kg
unr-rat LD50:740 mg/kg
orl-mus LD50:1350 mg/kg
ipr-mus LD50:70 mg/kg
scu-mus LD50:1109 mg/kg
unr-mus LD50:1150 mg/kg
orl-cat LDLo:230 mg/kg
orl-rbt LD50:210 mg/kg
skn-rbt LDLo:1 g/kg
unr-rbt LD50:210 mg/kg
unr-ckn LD50:840 mg/kg
unr-dom LD50:225 mg/kg
unr-mam LD50:400 mg/kg
orl-bwd LD50:300 mg/kg

Reproductive Toxicity (1992 RTECS): This chemical is a mammalian reproductive toxin.

Reproductive Toxicity Data (1992 RTECS):
orl-rat TDLo:1200 mg/kg (7-12D preg) TXAPA9 35, 83, 76
  Effects on Fertility
    Preimplantation mortality
    Postimplantation mortality
    Litter size (number of fetuses per litter; measured before birth)

orl-rat TDLo:300 mg/kg (15D preg) GISAAA 43(6), 37, 78
  Effects on Embryo or Fetus
    Fetotoxicity (except death, e.g., stunted fetus)
    Fetal death
  Specific Developmental Abnormalities
    Other developmental abnormalities

orl-rat TDLo:1190 mg/kg (16-22D preg/21D post) TXAPA9 35, 83, 76
  Effects on Newborn
    Growth statistics (e.g., reduced weight gain)

orl-rat TDLo:550 mg/kg (1-22D preg) AEEDDS 2, 215, 76
  Effects on Newborn
    Behavioral

orl-rat TDLo:420 mg/kg (1-20D preg) GISAAA 51(6), 23, 86
  Specific Developmental Abnormalities
    Cardiovascular (circulatory) system

par-rat TDLo:400 mg/kg (4-11D preg) BEXBAN 93, 107, 82
  Effects on Embryo or Fetus
    Fetotoxicity (except death, e.g., stunted fetus)
    Fetal death

par-rat TDLo:800 mg/kg (2D male/2D pre) BEXBAN 93, 107, 82
  Effects on Embryo or Fetus
    Fetotoxicity (except death, e.g., stunted fetus)
    Fetal death

orl-mus TDLo:100 mg/kg (6-15D preg) ATXKA8 30, 251, 73
  Specific Developmental Abnormalities
    Craniofacial (including nose and tongue)
    Musculoskeletal system

orl-mus TDLo:300 mg/kg (6-15D preg) ATXKA8 30, 251, 73
Effects on Fertility
Postimplantation mortality

orl-mus TDLo:80 mg/kg (3D male) FCTOD7 25, 709, 87
Paternal Effects
Spermatogenesis

scu-mus TDLo:90 mg/kg (6-14D preg) NTIS** PB223-160
Effects on Fertility
Preimplantation mortality
Litter size (number of fetuses per litter; measured before birth)
Effects on Embryo or Fetus
Extra embryonic features (e.g., placenta, umbilical cord)

scu-mus TDLo:1035 mg/kg (6-14D preg) NTIS** PB223-160
Effects on Embryo or Fetus
Fetotoxicity (except death, e.g., stunted fetus)

orl-ham TDLo:125 mg/kg (7D preg) TXAPA9 15, 152, 69
Specific Developmental Abnormalities
Musculoskeletal system

orl-ham TDLo:250 mg/kg (8D preg) TXAPA9 15, 152, 69
Specific Developmental Abnormalities
Craniofacial (including nose and tongue)

orl-ham TDLo:250 mg/kg (7D preg) TXAPA9 15, 152, 69
Specific Developmental Abnormalities
Body wall

orl-ham TDLo:300 mg/kg (7D preg) TXAPA9 15, 152, 69
Specific Developmental Abnormalities
Central nervous system

--- PROTECTION AND FIRST AID ---

NIOSH Pocket Guide to Chemical Hazards:
**Wear Appropriate Equipment to Prevent: Reasonable probability of skin
contact.

**Wear Eye Protection to Prevent: Reasonable probability of eye contact.

**Exposed Personnel Should Wash: Promptly when skin becomes contaminated.

**Work Clothing Should be Changed Daily: If there is any reasonable possibility that the clothing may be contaminated.

**Remove Clothing: Promptly remove nonimpervious clothing that becomes contaminated.

**Reference: NIOSH

Recommended Respiration Protection Source: NIOSH Pocket Guide (85-114) OSHA (Bis(Dimethylthiocarbamyl) Disulfide)
50 mg/m³: Any chemical cartridge respirator with organic vapor cartridge(s) in combination with a dust, mist, and fume filter. Substance reported to cause eye irritation or damage; may require eye protection. Any supplied-air respirator. Any self-contained breathing apparatus.
125 mg/m³: Any powered air-purifying respirator with organic vapor cartridge(s) in combination with a dust, mist, and fume filter. Any supplied-air respirator operated in a continuous-flow mode. Substance reported to cause eye irritation or damage; may require eye protection.
250 mg/m³: Any self-contained breathing apparatus with a full facepiece. Any chemical cartridge respirator with a full facepiece and organic vapor cartridge(s) in combination with a high-efficiency particulate filter. Any supplied-air respirator with a full facepiece. Any powered air-purifying respirator with a tight-fitting facepiece and organic vapor cartridge(s) in combination with a high-efficiency particulate filter. Substance reported to cause eye irritation or damage; may require eye protection. Any air-purifying full facepiece respirator (gas mask) with a chin-style or front- or back-mounted organic vapor canister having a high-efficiency particulate filter.
1500 mg/m³: Any supplied-air respirator with a full facepiece and operated in a pressure-demand or other positive-pressure mode.
Emergency or planned entry in unknown concentrations or IDLH conditions: Any self-contained breathing apparatus with a full facepiece and operated in a pressure-demand or other positive-pressure mode. Any supplied-air respirator with a full facepiece and operated in a pressure-demand or other positive-pressure mode in combination with an auxiliary self-contained breathing apparatus operated in a pressure-demand or other positive-pressure mode.

Escape: Any air-purifying full facepiece respirator (gas mask) with a chin-style or front- or back-mounted organic vapor canister having a high-efficiency particulate filter. Any appropriate escape-type self-contained breathing apparatus.

First Aid Source: HCDB
Eye: Wash with water; if irritation persists, consult a physician.
Skin: Wash with water; if irritation persists, consult a physician.
Inhalation: Remove victim from exposure. If breathing has stopped or is difficult, give artificial respiration and call physician.
Ingestion: Call physician; induce vomiting and flow with gastric lavage; treatment thereafter is symptomatic and supportive; avoid fats, oils, and lipid solvents, which enhance absorption; rigorously prohibit ethyl alcohol in all forms for at least 10 days; inform doctor if patient has used alcohol within 48 hr.

First Aid Source: CHRIS Manual 1991.
Inhalation: Remove victim from exposure; if breathing has stopped or is difficult, give artificial respiration and call physician.
Eye and skin: Wash with water; if irritation persists, consult a physician.
Ingestion: Call physician; induce vomiting and follow with gastric lavage; treatment thereafter is symptomatic and supportive; avoid fats, oils, and lipid solvents, which enhance absorption; rigorously prohibit ethyl alcohol in all forms for at least 10 days; inform doctor if patient has used alcohol within 48 hr.

First Aid Source: DOT Emergency Response Guide 1993.
In case of contact with material, immediately flush eyes with running water for at least 15 min. Wash skin with soap and water. Remove and isolate contaminated clothing and shoes at the site.

--- INITIAL INCIDENT RESPONSE ---

Fire Extinguishment: Water, dry chemical, carbon dioxide. (CHRIS 91)

U.S. Department of Transportation Guide to Hazardous Materials Transport Information - Publication DOT 5800.5 (1990).
DOT Shipping Name: ENVIRONMENTALLY HAZARDOUS
SUBSTANCES, SOLID, N.O.S.
DOT ID Number: UN3077

ERG90   GUIDE 31
* POTENTIAL HAZARDS *

*Health Hazards
Contact may cause burns to skin and eyes.
Fire may produce irritating or poisonous gases.
Runoff from fire control or dilution water may cause pollution.

*Fire or Explosion
Some of these materials may burn, but none of them ignites readily.

* EMERGENCY ACTION *

Keep unnecessary people away; isolate hazard area and deny entry.
Positive-pressure self-contained breathing apparatus (SCBA) and structural
firefighters' protective clothing will provide limited protection.
**CALL CHEMTREC AT 1-800-424-9300 FOR EMERGENCY ASSIS-
TANCE.** If water pollution occurs, notify the appropriate authorities.

*Fire
Small fires: Dry chemical, CO_2, water spray, or regular foam.
Large fires: Water spray, fog, or regular foam.
Move container from fire area if you can do it without risk.
Do not scatter spilled material with high-pressure water streams.
Dike fire-control water for later disposal.

*Spill or Leak
Stop leak if you can do it without risk.
Small spills: Take up with sand or other noncombustible absorbent material
and place into containers for later disposal.
Small dry spills: With clean shovel, place material into clean, dry container
and cover loosely; move containers from spill area.
Large spills: Dike far ahead of liquid spill for later disposal.
Cover powder spill with plastic sheet or tarp to minimize spreading.

*First Aid
In case of contact with material, immediately flush skin or eyes with running
water for at least 15 min. Wash skin with soap and water.
Remove and isolate contaminated clothing and shoes at the site.

▼   ▼   ▼   ▼   ▼   ▼   ▼   ▼   ▼   ▼   ▼   ▼   ▼   ▼

--- IDENTIFIERS ---

Chemical Name and Synonyms: 1,3-DIPHENYL-2-THIOUREA

Trade Name and Synonyms: A-1 Thiocarbanilide

Chemical Family: Rubber Accelerator

Manufacturer: Monsanto Co., Akron, Ohio

UN ID Number: NA

--- INGREDIENTS AND COMPOSITION DATA ---

Material	Cas No.	%	TLV
Product	NA		Not established

--- PHYSICAL DATA ---

Boiling Point (°F): NA

Specific Gravity (H₂O = 1): 1.32

Vapor Pressure (mm Hg): NA

Percent Volatile by Volume (%): Low

Vapor Density (air = 1): NA

Freezing Point: NA

Evaporation Rate: NA

pH: NA

Solubility in Water: Negligible

Melting Point (°C): NA

Appearance/Color/Odor: Grey powder, aromatic odor.

--- FIRE AND EXPLOSION HAZARDS ---

Flash Point: NA

Flammability Limits: NA

Autoignition Temperature: NA

Decomposition Temperature: NA

Extinguishing Media: Foam, dry chemical, carbon dioxide, or water.

Special Fire Fighting Procedures: As with many solids, any dust that is generated may be explosive if mixed with air in critical proportions and in the presence of a source of ignition. Use water, carbon dioxide, or dry chemical to extinguish fires. Wear self-contained, positive-pressure breathing apparatus.

Unusual Fire and Explosion Hazards: Dust may be explosive if mixed with air in critical proportions and in the presence of a source of ignition. The hazard is similar to that of any organic solid including sawdust. Maintain normal good housekeeping for control of dust.

Hazardous Products of Combustion and Decomposition: Thermal decomposition may produce carbon monoxide and/or carbon dioxide.

Hazardous Polymerization: Will not occur.

--- HEALTH HAZARD DATA ---

Effects of Overexposure: Dust causes eye irritation. Repeated minimal contact with skin may cause sensitization. Exposure can produce an adverse reaction when alcohol is consumed and can affect the activity of certain prescription drugs; consult physician.

Emergency Response and First Aid:
Skin: Wash with lots of clean water for at least 15 min.
Eyes: Flush eyes with lots of water. If irritation persists, consult a physician.
Inhalation: Expose to fresh air. Keep warm and quiet. Give artificial respiration if necessary.
Ingestion: None recommended. Consult physician immediately.

Cancer Information: None reported.

--- TOXICITY DATA ---

Animal Effects: None reported by manufacturer.

--- REACTIVITY DATA ---

Stability and Conditions to Avoid: None reported by manufacturer other than to minimize skin exposure and inhalation. This material is considered stable.

Materials to Avoid (Incompatibility): None reported by manufacturer.

Hazard Decomposition Products: Carbon dioxide, carbon monoxide, nitrogen, and/or oxides of sulfur.

Hazardous Polymerization: Will not occur.

--- SPILL AND EMERGENCY RESPONSE ---

For Spills: Sweep up. Wash residuals with soap and water. Transfer to a closed container.

Empty Containers: The container for this product can present explosion or fire hazards, even when emptied. To avoid the risk of injury, do not cut, puncture, or weld on or near this container.

--- SPECIAL PROTECTION AND PRECAUTIONS ---

Respiratory Protection: Dust mask.

Type of Ventilation Required: General mechanical. Avoid breathing dusts and vapors. Adequate ventilation should be provided to keep mist and dust concentrations below acceptable exposure limits. Discharge from the ventilation system should comply with applicable air-pollution regulations. Local exhaust ventilation is recommended with a capture velocity of 150 to 200 fpm.

Eye Protection Requirements: Chemical goggles. Eyewash fountains and safety showers should be easily accessible.

Protective Gloves and Other: Rubber gloves, goggles.

Storage and Handling Precautions: Containers should be stored in a cool, dry, well-ventilated area. Store away from flammable materials, sources of heat, flame, sparks, and foodstuffs. Exercise due caution to prevent damage to or leakage from the container.

Prevent inhalation of dust. Prevent contact with eyes and skin. Under dusty conditions, static electricity may cause an explosion. Keep containers closed. Store in a cool, dry place.

Use with adequate ventilation. Wash thoroughly after handling before eating and smoking. Work clothing should be frequently laundered and not worn away from work premises.

Other Precautions: Use all standard practices for good personal hygiene. Completely isolate and thoroughly clean all equipment, piping, or vessels before beginning maintenance or repairs. Keep area clean. As with any powder or dust-like material, improper handling can lead to spontaneous combustion (dust explosions).

--- REGULATORY INFORMATION ---

None reported.

▼   ▼   ▼   ▼   ▼   ▼   ▼   ▼   ▼   ▼   ▼   ▼   ▼   ▼

--- IDENTIFIERS ---

Chemical Name and Synonyms: CARBAMIDE

Trade Name and Synonyms: BIK

Chemical Family: Rubber Accelerator

Manufacturer: Uniroyal Chemical Co., Naugatuck, Connecticut

UN ID Number: NA

--- INGREDIENTS AND COMPOSITION DATA ---

Material	Cas No.	%	TLV
Product	NA		Not established

--- PHYSICAL DATA ---

Boiling Point (°F): Boils at 310°F (no flash)

Specific Gravity (H₂O = 1): 1.30

Vapor Pressure (mm Hg): NA

Percent Volatile by Volume (%): NA

Vapor Density (air = 1): NA

Freezing Point: NA

Evaporation Rate: NA

pH: NA

Solubility in Water: Partially

Melting Point (°C): 130

Appearance/Color/Odor: White, free-flowing powder.

--- FIRE AND EXPLOSION HAZARDS ---

Flash Point: NA

Flammability Limits: NA

Autoignition Temperature: NA

Decomposition Temperature: NA

Extinguishing Media: Foam, dry chemical, carbon dioxide, or water.

Special Fire Fighting Procedures: As with many solids, any dust that is generated may be explosive if mixed with air in critical proportions and in the presence of a source of ignition. Use water, carbon dioxide, or dry chemical to extinguish fires. Wear self-contained, positive-pressure breathing apparatus.

Unusual Fire and Explosion Hazards: Dust may be explosive if mixed with air in critical proportions and in the presence of a source of ignition. The hazard is similar to that of any organic solid including sawdust. Maintain normal good housekeeping for control of dust.

Hazardous Products of Combustion and Decomposition: Thermal decomposition may produce carbon monoxide and/or carbon dioxide.

Hazardous Polymerization: Will not occur.

--- HEALTH HAZARD DATA ---

Effects of Overexposure: Dust causes eye irritation.

Emergency Response and First Aid:
Skin: Wash with lots of clean water for at least 15 min.
Eyes: Flush eyes with lots of water. If irritation persists, consult a physician.
Inhalation: Expose to fresh air. Keep warm and quiet. Give artificial respiration if necessary.
Ingestion: None recommended. Consult physician immediately.

Cancer Information: None reported.

--- TOXICITY DATA ---

Animal Effects: None reported by manufacturer.

--- REACTIVITY DATA ---

Stability and Conditions to Avoid: None reported by manufacturer other than to minimize skin exposure and inhalation. This material is considered stable.

Materials to Avoid (Incompatibility): None reported by manufacturer.

Hazard Decomposition Products: Carbon dioxide, carbon monoxide, nitrogen, and/or oxides of sulfur.

Hazardous Polymerization: Will not occur.

--- SPILL AND EMERGENCY RESPONSE ---

For Spills: Sweep up. Wash residuals with soap and water. Transfer to a closed container.

Empty Containers: The container for this product can present explosion or fire hazards, even when emptied. To avoid the risk of injury, do not cut, puncture, or weld on or near this container.

--- SPECIAL PROTECTION AND PRECAUTIONS ---

Respiratory Protection: Dust mask.

Type of Ventilation Required: General mechanical. Avoid breathing dusts and vapors. Adequate ventilation should be provided to keep mist and dust concentrations below acceptable exposure limits. Discharge from the ventilation system should comply with applicable air-pollution regulations. Local exhaust ventilation is recommended with a capture velocity of 150 to 200 fpm.

Eye Protection Requirements: Chemical goggles. Eyewash fountains and safety showers should be easily accessible.

Protective Gloves and Other: Rubber gloves, goggles.

Storage and Handling Precautions: Containers should be stored in a cool, dry, well-ventilated area. Store away from flammable materials, sources of heat, flame, sparks, and foodstuffs. Exercise due caution to prevent damage to or leakage from the container.

Prevent inhalation of dust. Prevent contact with eyes and skin. Under dusty conditions, static electricity may cause an explosion. Keep containers closed. Store in a cool, dry place.

Use with adequate ventilation. Wash thoroughly after handling before eating and smoking. Work clothing should be frequently laundered and not worn away from work premises.

Other Precautions: Use all standard practices for good personal hygiene. Completely isolate and thoroughly clean all equipment, piping, or vessels before beginning maintenance or repairs. Keep area clean. As with any

powder or dust-like material, improper handling can lead to spontaneous combustion (dust explosions).

--- REGULATORY INFORMATION ---

None reported.

▼   ▼   ▼   ▼   ▼   ▼   ▼   ▼   ▼   ▼   ▼   ▼   ▼

--- IDENTIFIERS ---

Chemical Name and Synonyms: 1,3-DIPHENYL GUANIDINE; N,N'-DIPHENYLGUANIDINE

Trade Name and Synonyms: DPG

Chemical Family: Rubber Accelerator

Manufacturer: Monsanto Chemical Co., St. Louis, Missouri

UN ID Number: NA

--- INGREDIENTS AND COMPOSITION DATA ---

Material	Cas No.	%	TLV
Product	102-06-7		Not established

--- PHYSICAL DATA ---

Boiling Point (°F): NA

Specific Gravity (H_2O = 1): 1.2

Vapor Pressure (mm Hg): NA

Percent Volatile by Volume (%): NA

Vapor Density (air = 1): NA

Freezing Point: NA

Evaporation Rate: NA

pH: NA

Solubility in Water: Insoluble in water; soluble in acetone

Melting Point (°C): 143

Appearance/Color/Odor: Off-white powder and pellets; oiled and nondusting powder. Has an aromatic odor.

--- FIRE AND EXPLOSION HAZARDS ---

Flash Point: 315°F (Cleveland closed cup)

Flammability Limits: NA

Autoignition Temperature: NA

Decomposition Temperature: NA

Extinguishing Media: Water spray or other Class A extinguishing media.

Special Fire Fighting Procedures: As with many solids, any dust that is generated may be explosive if mixed with air in critical proportions and in the presence of a source of ignition. Use water, carbon dioxide, or dry chemical to extinguish fires. Wear self-contained, positive-pressure breathing apparatus.

Unusual Fire and Explosion Hazards: Dust may be explosive if mixed with air in critical proportions and in the presence of a source of ignition. The hazard is similar to that of any organic solid including sawdust. Maintain normal good housekeeping for control of dust.

Hazardous Products of Combustion and Decomposition: Thermal decomposition may produce carbon monoxide and/or carbon dioxide and nitrogen oxides. Tetraphenylmelamine, aniline, triphenyldicarbimide and triphenyl-melamine may also be liberated during a fire.

Hazardous Polymerization: Will not occur.

--- HEALTH HAZARD DATA ---

Effects of Overexposure: Dermal contact and inhalation are the primary routes of occupational exposure. Occupational exposure to this material has not been reported to cause significant adverse human health effects. However, repeated or prolonged exposure to diphenylguanidine may cause an allergic skin response in susceptible individuals.

Emergency Response and First Aid:
Skin: Wash with lots of clean water for at least 15 min.
Eyes: Flush eyes with lots of water. If irritation persists, consult a physician.
Inhalation: Expose to fresh air. Keep warm and quiet. Give artificial respiration if necessary.
Ingestion: None recommended. Consult physician immediately.

Cancer Information: None reported.

--- TOXICITY DATA ---

Animal Effects:
Data from Monsanto studies and from the available scientific literature indicate the following:
Oral LD50 (rat): 350 mg/kg, moderately toxic
Dermal LD50 (rabbit): >794 mg/kg, moderately toxic
Eye Irritation (rabbit): (FHSA) 20.2 on a scale of 110.0, moderately
  irritating
Skin Irritation (rabbit): (FHSA) 0.0 on a scale of 8.0, nonirritating

Patch testing of 49 human volunteers with 70% DPG accelerator in petrolatum produced no positive reactions following initial application; 19 of the 49 subjects displayed positive reactions during subsequent exposures. Four subjects exhibited positive reactions upon rechallenge 2 weeks later. Under the conditions of this test, diphenylguanidine was considered to be a potential skin sensitizer, and a weak cumulative irritant, but was not considered a primary irritant.

Rats were given DPG accelerator at concentrations of 0, 300, 500, 800, 1500 and 3000 ppm in the diet for 2 weeks. Increases in mortality, reductions in body weight gain, piloerection, and ataxia were exhibited by the high-dose-group animals. A dose-related reduction in body weight gain and food consumption was also observed; the reductions were significant at dosages of 500 ppm and above.

DPG accelerator was administered in the diet at concentrations of 0, 50, 150 or 500 ppm to groups of rats for 90 days. A reduction in body weight gain, along with reduced food consumption, and changes in organ weight and organ/body weight ratios were observed in the high-dose animals. No increases in mortality or abnormal clinical signs were observed. No increases in the incidences of gross or microscopic lesions were observed in treated animals. The no-effect level in this study was considered to be 150 ppm.

A teratology study was conducted in which mice were given diphenylguanidine by oral intubation at concentrations of 0.25, 1.0, 4.0 or 10.0 mg/kg on days 0 through 18 of gestation. A significant reduction in the mean number of implants in the high-dose group was observed. No adverse effects on fetuses in the lower-dose groups were observed, and no teratogenic effects were noted in the study.

Decreased sperm counts, alterations in sperm morphology, microscopic changes in testicular tissue, and reduced fertility (in mice) were reported for mice and hamsters following continuous (ad libitum) administration of diphenylguanidine at dietary concentrations equivalent to 4 or 8 mg/kg body weight. Testicular weights were decreased after 5 weeks of treatment. Changes in sperm were noted after 7 weeks of treatment.

DPG accelerator was evaluated in the L5178Y TK mouse lymphoma mutation assay and in microbial mutagenicity assays. The microbial assays, both with and without microsomal activation, used five strains of *Salmonella* and one yeast strain. No mutagenic effects were observed.

Environmental Toxicity Information:

96-hr LC50 Bluegill: 9.6 mg/l, moderately toxic
96-hr LC50 Trout: 11 mg/l, slightly toxic
96-hr LC50 Fathead Minnow: 4.2 mg/l, moderately toxic
96-hr EC50 Algae, cell count: 1.7 mg/l, moderately toxic
48-hr LC50 *Daphnia*: 17 mg/l, slightly toxic

--- REACTIVITY DATA ---

Stability and Conditions to Avoid: None reported by manufacturer other than to minimize skin exposure and inhalation. This material is considered stable. If strongly heated, may catch fire and give off toxic fumes. The material is stable to 160°F.

Materials to Avoid (Incompatibility): None reported by manufacturer.

Hazard Decomposition Products: Carbon dioxide, carbon monoxide, nitrogen, and/or oxides of sulfur.

Hazardous Polymerization: Will not occur.

--- SPILL AND EMERGENCY RESPONSE ---

For Spills: Sweep up. Wash residuals with soap and water. Transfer to a closed container.

Empty Containers: The container for this product can present explosion or fire hazards, even when emptied. To avoid the risk of injury, do not cut, puncture, or weld on or near this container.

--- SPECIAL PROTECTION AND PRECAUTIONS ---

Respiratory Protection: Avoid breathing dust. Use NIOSH/OSHA approved equipment when airborne exposure is excessive. Full facepiece equipment is recommended and, if used, replaces the need for chemical splash goggles. Consult respirator manufacturer to determine appropriate type of equipment for given application. The respirator use limitations specified by NIOSH/MSHA or the manufacturer must be observed. High airborne concentrations may require the use of self-contained breathing apparatus or a supplied-air respirator. Respiratory protection must be in compliance with 29 CFR 1910.134.

Although OSHA and ACGIH have not established specific exposure limits for this material, they have established the following limits for nuisance dusts:

OSHA PEL/8-hr TWA: Total 15 mg/m^3, respirable 5 mg/m^3.
ACGIH TLV/8-hr TWA: Total 10 mg/m^3, respirable 5 mg/m^3.
These limits are stated only to indicate the least stringent airborne dust exposure levels applicable to nuisance dusts.

Type of Ventilation Required: General mechanical. Avoid breathing dusts and vapors. Adequate ventilation should be provided to keep mist and dust concentrations below acceptable exposure limits. Discharge from the ventilation system should comply with applicable air-pollution regulations. Local exhaust ventilation is recommended with a capture velocity of 150 to 200 fpm.

Eye Protection Requirements: Chemical goggles. Eyewash fountains and safety showers should be easily accessible.

Protective Gloves and Other: Rubber gloves, goggles.

Storage and Handling Precautions: Containers should be stored in a cool, dry, well-ventilated area. Store away from flammable materials, sources of heat, flame, sparks, and foodstuffs. Exercise due caution to prevent damage to or leakage from the container.

Prevent inhalation of dust. Prevent contact with eyes and skin. Under dusty conditions, static electricity may cause an explosion. Keep containers closed. Store in a cool, dry place.

Use with adequate ventilation. Wash thoroughly after handling before eating and smoking. Work clothing should be frequently laundered and not worn away from work premises.

Other Precautions: Use all standard practices for good personal hygiene. Completely isolate and thoroughly clean all equipment, piping, or vessels before beginning maintenance or repairs. Keep area clean. As with any powder or dust-like material, improper handling can lead to spontaneous combustion (dust explosions).

--- REGULATORY INFORMATION ---

FDA Information: Diphenylguanidine is regulated under the Federal Food, Drug and Cosmetic Act, and applicable regulations issued thereunder for use as a substance in the preparation of Rubber Articles Intended for Repeated Use, under 21 CFR 177.2600, subject to the provisions, conditions and limitations in that regulation.

DOT Proper Shipping Name: ORM-A, N.O.S. (via air on oiled product)

DOT Hazard Class/I.D. No.: ORM-A/ NA1693 (via air on oiled product)

DOT Label: Not applicable

U.S. Surface Freight Classification: Rubber accelerator, N.O.I.B.N.

Reportable Quantity (RQ)
(40 CFR Part 117)
Under Clean Water Act Regulations: Not applicable

This substance is identified as a hazardous chemical under the criteria of the OSHA Hazard Communication Standard (29 CFR 1910.1200).

▼ ▼ ▼ ▼ ▼ ▼ ▼ ▼ ▼ ▼ ▼ ▼ ▼ ▼

--- IDENTIFIERS ---

Chemical Name and Synonyms: POLYETHYLENE GLYCOL, HO-$(CH_2CH_2O)_n$-H

Trade Name and Synonyms: Carbowax Polyethylene Glycol 3350

Chemical Family: Oxyalkylene Polymer

Manufacturer: Union Carbide Corp., Akron, Ohio

UN ID Number: NA

--- INGREDIENTS AND COMPOSITION DATA ---

Material	Cas No.	%	TLV
Poly(oxy-1,2,-ethanedyl), a-hydro-w-hydroxy	25322-68-3	100	See toxicity section

--- PHYSICAL DATA ---

Boiling Point (°F): 392, decomposes

Specific Gravity (H_2O = 1): 1.072

Vapor Pressure (mm Hg): Nil

Percent Volatile by Volume (%): NA

Vapor Density (air = 1): >1

Freezing Point: NA

Evaporation Rate: Nil

pH: NA

Solubility in Water: 62 at 20°C

Melting Point (°C): 54 to 58

Appearance/Color/Odor: White solid (flake, powder, or molten); mild odor.

--- FIRE AND EXPLOSION HAZARDS ---

Flash Point: >350°F, Pensky-Martens closed cup ASTM D93

Flammability Limits: NA

Autoignition Temperature: NA

Decomposition Temperature: NA

Extinguishing Media: Apply alcohol-type or all-purpose-type foams by manufacturers' recommended techniques for large fires. Use carbon dioxide or dry chemical media for small fires.

Special Fire Fighting Procedures: Do not direct a solid stream of water or foam into hot, burning pools; this may cause frothing and increase fire intensity. Use self-contained breathing apparatus and protective clothing.

Unusual Fire and Explosion Hazards: Avoid dispersion of dust in air to reduce potential for dust ignition and explosion.

Hazardous Products of Combustion and Decomposition: Carbon monoxide and/or carbon dioxide.

Hazardous Polymerization: Will not occur.

--- HEALTH HAZARD DATA ---

Effects of Overexposure: This material may contribute to nuisance dusts and possibly to respirable dusts; avoid breathing dusts. Overexposure to vapors generated at high temperatures may result in eye and respiratory tract irritation and in the inhalation of harmful amounts of material.

Emergency Response and First Aid:
Skin: Wash with lots of clean water for at least 15 min.
Eyes: Flush eyes with lots of water. If irritation persists, consult a physician.
Inhalation: Expose to fresh air. Keep warm and quiet. Give artificial respiration if necessary.
Ingestion: None recommended. Consult physician immediately.

Cancer Information: This product may contain trace amounts of ethylene oxide, which is considered by OSHA, IARC, and NTP as a potential carcinogen for humans. See toxicity section also.

--- TOXICITY DATA ---

Animal Effects: Manufacturer reports that toxicology studies have shown this material to be very low acute toxicity and nonirritating. There is no specific antidote. Treatment of overexposure should be directed at the control of symptoms and the clinical condition.

This product may contain trace amounts of ethylene oxide (CAS No. 75-21-8), a condition that creates the potential for accumulation of ethylene oxide in the head space of shipping and storage containers and in enclosed areas where the product is being handled or used. Ethylene oxide is considered as a potential carcinogen for humans. Ethylene oxide may also present reproductive, mutagenic, genotoxic, neurologic, and sensitization hazards in humans. If this product is handled with adequate ventilation, the presence of these trace amounts is not expected to result in any short- or long-term hazards.

This product may not be exempt from OSHA's ethylene oxide standard 29 CFR 1910.1017. Users should comply with all applicable provisions. Personnel should be monitored to determine levels of exposure to ethylene oxide. If necessary, protective measures should be taken. The OSHA PEL for ethylene oxide is 1 ppm TWA, 8 hr. The action level is 0.5 ppm TWA, 8 hr. The ACGIH TLV is 1 ppm TWA, 8 hr, and OSHA has proposed an excursion limit of 5 ppm (15-min average).

--- REACTIVITY DATA ---

Stability and Conditions to Avoid: None reported by manufacturer other than to minimize skin exposure and inhalation. This material is considered stable. If strongly heated, may catch fire and give off toxic fumes.

Materials to Avoid (Incompatibility): None reported by manufacturer.

Hazard Decomposition Products: Carbon dioxide and carbon monoxide.

Hazardous Polymerization: Will not occur.

--- SPILL AND EMERGENCY RESPONSE ---

For Spills: Sweep up. Wash residuals with soap and water. Transfer to a closed container.

Waste Disposal Method: Potential for a dust explosion exists if attempt is made to incinerate organic powders. If incineration is desired, dissolve in a suitable solvent and incinerate as a solution.

Empty Containers: The container for this product can present explosion or fire hazards, even when emptied. To avoid the risk of injury, do not cut, puncture, or weld on or near this container.

--- SPECIAL PROTECTION AND PRECAUTIONS ---

Respiratory Protection: Avoid breathing dust. Use NIOSH/OSHA-approved equipment when airborne exposure is excessive. Full facepiece equipment is recommended and, if used, replaces the need for chemical splash goggles. Consult respirator manufacturer to determine appropriate type of equipment for given application. The respirator use limitations specified by NIOSH/MSHA or the manufacturer must be observed. High airborne concentrations may require use of self-contained breathing apparatus or supplied-air respirator. Respiratory protection must be in compliance with 29 CFR 1910.134.

Although OSHA and ACGIH have not established specific exposure limits for this material, they have established the following limits for nuisance dusts:

OSHA PEL/8-hr TWA: Total 15 mg/m^3, respirable 5 mg/m^3.
ACGIH TLV/8-hr TWA: Total 10 mg/m^3, respirable 5 mg/m^3.

These limits are stated only to indicate the least stringent airborne dust exposure levels applicable to nuisance dusts.

Type of Ventilation Required: General mechanical. Avoid breathing dusts and vapors. Adequate ventilation should be provided to keep mist and dust concentrations below acceptable exposure limits. Discharge from the ventilation system should comply with applicable air-pollution regulations. Local exhaust ventilation is recommended with a capture velocity of 150 to 200 fpm.

Eye Protection Requirements: Chemical goggles. Eyewash fountains and safety showers should be easily accessible.

Protective Gloves and Other: Rubber gloves, goggles.

Storage and Handling Precautions: Containers should be stored in a cool, dry, well-ventilated area. Store away from flammable materials, sources of heat, flame, sparks, and foodstuffs. Exercise due caution to prevent damage to or leakage from the container.

Prevent inhalation of dust. Prevent contact with eyes and skin. Under dusty conditions, static electricity may cause an explosion. Keep containers closed. Store in a cool, dry place.

Use with adequate ventilation. Wash thoroughly after handling before eating and smoking. Work clothing should be frequently laundered and not worn away from work premises.

Other Precautions: Use all standard practices for good personal hygiene. Completely isolate and thoroughly clean all equipment, piping, or vessels before beginning maintenance or repairs. Keep area clean. As with any powder or dust-like material, improper handling can lead to spontaneous combustion (dust explosions).

--- REGULATORY INFORMATION ---

Status on Substance Lists: The concentrations shown are maximum or ceiling levels (weight %) to be used for calculations for regulations. Trade Secrets are indicated by "TS".

Federal EPA: Comprehensive Environmental Response, Compensation, and Liability Act of 1980 (CERCLA) requires notification of the National

Response Center of release of quantities of Hazardous Substances equal to or greater than the reportable quantities (RQs) in 40 CFR 302.4.

Components present in this product at a level which could require reporting under the statute are:

CHEMICAL	CAS NUMBER	UPPER BOUND CONCENTRATION %
Ethylene Oxide	75-21-8	.0005
Dioxane	123-91-1	.0005

Superfund Amendments and Reauthorization Act of 1986 (SARA) Title III: Requires emergency planning based on Threshold Planning Quantities (TPQs) and release reporting based on Reportable Quantities (RQs) in 40 CFR 355 (used for SARA 302, 304, 311 and 312).

Component present in this product at a level which could require reporting under the statute are: *** NONE ***.

Superfund Amendments and Reauthorization Act of 1986 (SARA) Title III: Requires submission of annual reports of release of toxic chemicals that appear in 40 CFR 372 (for SARA 313). This information must be included in all MSDSs that are copied and distributed for this material.

Components present in this product at a level which could require reporting under the statute are: *** NONE ***.

State Right-to-Know

CALIFORNIA Proposition 65: This product contains trace levels of ACETALDEHYDE, DIOXANE, ETHYLENE OXIDE AND FORMALDE-HYDE which the state of California has found to cause cancer, birth defects or other reproductive harm.

MASSACHUSETTS Right-to-Know, Substance List (MSL) Hazardous Substances and Extraordinarily Hazardous Substances on the MSL must be identified when present in products.

Components present in this product at a level which could require reporting under the statute are:

EXTRAORDINARILY HAZARDOUS SUBSTANCES (= > 0.0001%)

CHEMICAL	CAS NUMBER	UPPER BOUND CONCENTRATION %
Acetaldehyde	75-07-0	.0006
Dioxane	123-91-1	.0005
Ethylene Oxide	75-21-8	.0005
Formaldehyde	50-00-0	.0004

PENNSYLVANIA Right-to-Know, Hazardous Substance List Hazardous Substances and Special Hazardous Substances on the List must be identified when present in products.

Components present in this product at a level which could require reporting under the statute are: *** NONE ***.

Toxic Substances Control Act (TSCA) Status: The ingredients of this product are on the TSCA inventory.

California SCAQMD Rule 443.1 VOC'S: Not presently available.

5 MINERAL FILLERS, ACTIVATORS, AND DUSTING AGENTS

The following chemicals are covered in this section:

CHEMICAL NAME OR TYPE	TRADE NAME
Hydrated Alumina or Aluminum Hydroxide	Alcoa Aluminas: Hydrals, Hydrated Alumina
Carbon Black	Black Pearls, Elftex, Mogul, Monarch, Regal, Sterling, Vulcan, Acetylene Black
Calcium Carbonate, Whiting, Chalk, Calcite (Limestone)	Omyalite 95T
Calcium Stearate	Calcium Stearate
Kaolin Clays	Catalpo
Hydrous Magnesium Silicate, Talc, Steatite, Soapstone	Mistron Vapor
Magnesium Stearate	Magnesium Stearate
Organophilic Clay	Bentone
Hydrated Amorphous Silicon Dioxide	Hi-Sil

--- IDENTIFIERS ---

Chemical Name and Synonyms: HYDRATED ALUMINA, $Al_2O_3 \cdot 3H_2O$

Trade Name and Synonyms: Alcoa Aluminas, Hydral 705, 710, 710S, C31, C33, C331, C333

Chemical Family: Metal Salt

Manufacturer: Aluminum Company of America, Pittsburgh, Pennsylvania

UN ID Number: NA

--- INGREDIENTS AND COMPOSITION DATA ---

Material	Cas No.	Composition (ppm)
Antimony	NA	1
Arsenic	NA	1
Bismuth	NA	1
Boron	NA	2
Cadmium	NA	1
Chromium	NA	5
Copper	NA	5
Lead	NA	10
Manganese	NA	5
Mercury	NA	20 ppb

No detectable levels of the following materials:

Amines (low molecular weight)

Cyanides (inorganic)

Phenols (monomeric)

--- PHYSICAL DATA ---

Boiling Point (°F): NA

Specific Gravity (H₂O = 1): 0.8 to 0.22 for hydrals
0.7 to 1.3 for C-products

Vapor Pressure (mm Hg): Nil

Percent Volatile by Volume (%): NA

Vapor Density (air = 1): NA

Freezing Point: NA

Evaporation Rate: Nil

pH: Noncorrosive

Solubility in Water: Insoluble in water. Soluble in acids and alkalies.

Melting Point (°F): 3700

Appearance/Color/Odor: White crystalline powder; no odor.

--- FIRE AND EXPLOSION HAZARDS ---

Flash Point: NA

Flammability Limits: NA

Autoignition Temperature: NA

Decomposition Temperature: NA

Extinguishing Media: Apply alcohol-type or all-purpose-type foams by manufacturers' recommended techniques for large fires. Use carbon dioxide or dry chemical media for small fires.

Special Fire Fighting Procedures: Do not direct a solid stream of water or foam into hot, burning pools; this may cause frothing and increase fire intensity. Use self-contained breathing apparatus and protective clothing.

Unusual Fire and Explosion Hazards: Avoid dispersion of dust in air to reduce potential for dust ignition and explosions.

Hazardous Products of Combustion and Decomposition: Carbon monoxide and/or carbon dioxide.

Hazardous Polymerization: Will not occur.

--- HEALTH HAZARD DATA ---

Effects of Overexposure: "By ingestion negligible. By inhalation low," according to the "Hygienic Guide Series--Aluminum and Aluminum Oxides," issued by the American Industrial Hygiene Association.

Emergency Response and First Aid:
Skin: Wash with lots of clean water for at least 15 min.
Eyes: Flush eyes with lots of water. If irritation persists, consult a physician.
Inhalation: Expose to fresh air. Keep warm and quiet. Give artificial respiration if necessary.
Ingestion: None recommended. Consult physician immediately.

Cancer Information: None reported.

--- TOXICITY DATA ---

Animal Effects: Threshold limit value for "inert" or nuisance particulates is 10 mg/m^3 or 30 mppcf, whichever is lower, per ACGIH Threshold Limit Values of Airborne Contaminants, Notice of Intended Changes, 1971.

Food Additives Amendment Status Alcoa Hydral 705, Hydral 710, C31, C33, C331, and C333 considered "Generally Recognized as Safe" (GRAS) for use in foods or food packaging materials as defined in the Food additives Amendment of 1958 to the Federal Food, Drug, and Cosmetic Act.

--- REACTIVITY DATA ---

Stability and Conditions to Avoid: None reported by manufacturer other than to minimize skin exposure and inhalation. This material is considered stable. If strongly heated, may catch fire and give off toxic fumes.

Materials to Avoid (Incompatibility):   None reported by manufacturer.

Hazard Decomposition Products: Carbon dioxide and carbon monoxide.

Hazardous Polymerization: Will not occur.

--- SPILL AND EMERGENCY RESPONSE ---

For Spills: Sweep up. Wash residuals with soap and water. Transfer to a closed container.

Waste Disposal Method: No specific recommendations.

Empty Containers: The container for this product can present explosion or fire hazards, even when emptied. To avoid the risk of injury, do not cut, puncture, or weld on or near this container.

--- SPECIAL PROTECTION AND PRECAUTIONS ---

Respiratory Protection: Avoid breathing dust. Use NIOSH/OSHA-approved equipment when airborne exposure is excessive. Full facepiece equipment is recommended and, if used, replaces the need for chemical splash goggles. Consult respirator manufacturer to determine appropriate type of equipment for given application. The respirator use limitations specified by NIOSH/MSHA or the manufacturer must be observed. High airborne concentrations may require use of self-contained breathing apparatus or supplied-air respirator. Respiratory protection must be in compliance with 29 CFR 1910.134.

Although OSHA and ACGIH have not established specific exposure limits for this material, they have established the following limits for nuisance dusts:

OSHA PEL/8-hr TWA: Total 15 mg/m^3, respirable 5 mg/m^3.
ACGIH TLV/8-hr TWA: Total 10 mg/m^3, respirable 5 mg/m^3.
These limits are stated only to indicate the least stringent airborne dust exposure levels applicable to nuisance dusts.

Type of Ventilation Required: General mechanical. Avoid breathing dusts and vapors. Adequate ventilation should be provided to keep mist and dust concentrations below acceptable exposure limits. Discharge from the ventilation system should comply with applicable air-pollution regulations. Local exhaust ventilation is recommended with a capture velocity of 150 to 200 fpm.

Eye Protection Requirements: Chemical goggles. Eyewash fountains and safety showers should be easily accessible.

Protective Gloves and Other: None recommended.

Storage and Handling Precautions: Containers should be stored in a cool, dry, well-ventilated area. Store away from flammable materials, sources of heat, flame, sparks, and foodstuffs. Exercise due caution to prevent damage to or leakage from the container.

Prevent inhalation of dust. Prevent contact with eyes and skin. Under dusty conditions, static electricity may cause an explosion. Keep containers closed. Store in a cool, dry place.

Use with adequate ventilation. Wash thoroughly after handling before eating and smoking. Work clothing should be frequently laundered and not worn away from work premises.

Other Precautions: Use all standard practices for good personal hygiene. Completely isolate and thoroughly clean all equipment, piping, or vessels before beginning maintenance or repairs. Keep area clean. As with any powder or dust-like material, improper handling can lead to spontaneous combustion (dust explosions).

--- REGULATORY INFORMATION ---

None reported.

▼   ▼   ▼   ▼   ▼   ▼   ▼   ▼   ▼   ▼   ▼   ▼   ▼   ▼

--- IDENTIFIERS ---

Name: ALUMINUM HYDROXIDE
Synonyms: AF 260; Alcoa 331; Alcoa C 30BF; Alumigel; Alumina Hydrate; Alumina Hydrated; Alumina Trihydrate; alpha-Alumina Trihydrate; Aluminic Acid; Aluminium Hydroxide; Aluminum Hydrate; Aluminum (III) Hydroxide; Aluminum Hydroxide Gel; Aluminum Oxide-3H$_2$O; Aluminum Oxide Hydrate; Aluminum Oxide Trihydrate; Aluminum Trihydrat; Aluminum Trihydroxide; Alusal; Amberol ST 140F; Amphojel; Baco AF 260; British Aluminum AF 260; C 31; C 33; C 31C; C 4D; C 31F; C-31-F; C.I. 77002; GHA 331; GHA 332; H 46; Higilite; Higilite H 32; Higilite H 42; Higilite

H 31S; Hychol 705; Hydral 705; Hydral 710; Hydrated Alumina; Martinal;
P 30BF; Trihydrated Alumina; Trihydroxyaluminum
CAS: 21645-51-2; **RTECS:** BD0940000
Formula: AlH_3O_3; **Mol Wt:** 78.01
WLN: .AL..H3.O3
Chemical Class: Metal oxide

See other identifiers listed below under Regulations.

--- PROPERTIES ---

Boiling Point: NA
Melting Point: NA
Flash Point: NA
Autoignition: NA
UEL: NA
LEL: NA
Vapor Density: No data
Specific Gravity: No data

Reactivity with Water: No data on water reactivity
Reactivity with Common Materials: No data
Stability during Transport: No data
Neutralizing Agents: No data
Polymerization Possibilities: No data

Toxic Fire Gases: None reported other than possible unburned vapors
Odor Detected at (ppm): Unknown
Odor Description: No data
100% Odor Detection: No data

--- REGULATIONS ---

DOT hazard class: Not given
Packaging exceptions: 173.
Nonbulk packaging: 173.
Bulk packaging: 173.

STCC Number: Not listed

Clean Water Act Sect. 307: No
Clean Water Act Sect. 311: No

Clean Air Act: Not listed
EPA Waste Number: None
CERCLA Ref: Not listed
RQ Designation: Not listed
SARA TPQ Value: Not listed
SARA Sect. 312 Categories:
Acute toxicity: irritant.

U.S. Postal Service Mailability: Not given

--- TOXICITY DATA ---

Short-term Toxicity: Unknown

Long-term Toxicity: Unknown

Conc IDLH: Unknown

NIOSH REL: Not given

ACGIH TLV: TLV = 10 mg/m^3 metal dust
ACGIH STEL: Metal dust

OSHA PEL: Transitional Limits: PEL = (total dust) 15 mg/m^3; (respirable fraction) 5 mg/m^3; Final Rule Limits: TWA = (total dust) 10 mg/m^3; (respirable fraction) 5 mg/m^3.

MAK Information: 6F ppm

Carcinogen: N; **Status:** See below

Carcinogen Lists: *IARC*: Not listed; *MAK*: Not listed; *NIOSH*: Not listed; *NTP*: Not listed; *ACGIH*: Not listed; *OSHA*: Not listed.

Human Toxicity Data: (Source: NIOSH RTECS)
orl-chd TDLo:79 g/kg/2Y-I PEDIAU 71, 56, 83
    Behavioral
        Changes in motor activity (specific assay)
        Muscle contraction or spasticity
    Musculoskeletal
        Osteomalacia

orl-chd TDLo:122 g/kg/4D JOPDAB 92, 592, 78
Gastrointestinal
Other changes
Nutritional and Gross Metabolic
Changes in: body temperature increase

LD50 Value: No LD50 in RTECS 1992

Other Species Toxicity Data: (Source: NIOSH RTECS 1992)
ipr-rat LDLo:150 mg/kg

Reproductive Toxicity (1992 RTECS): This chemical has no known mammalian reproductive toxicity.

▼ ▼ ▼ ▼ ▼ ▼ ▼ ▼ ▼ ▼ ▼ ▼ ▼ ▼

--- IDENTIFIERS ---

Name: KAOLIN
Synonyms: Argilla; Bolusalba; China Clay; Porcelain Clay; White Bole
CAS: 1332-58-7; **RTECS:** GF1670500
Formula: $AlHO_8Si_3.K$

See other identifiers listed below under Regulations.

--- PROPERTIES ---

Physical Description: White or yellowish-white, earthy mass or white powder.
Boiling Point: NA
Melting Point: NA
Flash Point: NA
Autoignition: NA
UEL: NA
LEL: NA
Vapor Density: No data
Specific Gravity: No data
Water Solubility: INSOL

Reactivity with Water: No data on water reactivity
Reactivity with Common Materials: No data
Stability during Transport: No data

Neutralizing Agents: No data
Polymerization Possibilities: No data

Toxic Fire Gases: None reported other than possible unburned vapors
Odor Detected at (ppm): Unknown
Odor Description: No data
100% Odor Detection: No data

--- REGULATIONS ---

DOT hazard class: Not given
Packaging exceptions: 173.
Nonbulk packaging: 173.
Bulk packaging: 173.

STCC Number: Not listed

Clean Water Act Sect. 307: No
Clean Water Act Sect. 311: No
Clean Air Act: Not listed
EPA Waste Number: None
CERCLA Ref: Not listed
RQ Designation: Not listed
SARA TPQ Value: Not listed
SARA Sect. 312 Categories:
    Chronic toxicity: reproductive toxin.

U.S. Postal Service Mailability: Not given

NFPA Codes:
    Health Hazard (blue): Unspecified
    Flammability (red): Unspecified
    Reactivity (yellow): Unspecified
    Special: Unspecified

--- TOXICITY DATA ---

Short-term Toxicity: Unknown

Long-term Toxicity: Unknown

Conc IDLH: Unknown

NIOSH REL: Not given

ACGIH TLV: TLV = 10 mg/m^3
ACGIH STEL: Not listed

OSHA PEL: Transitional Limits: PEL = (total dust) 15 mg/m^3; (respirable fraction) 5 mg/m^3; Final Rule Limits: TWA = (total dust) 10 mg/m^3; (respirable fraction) 5 mg/m^3.

MAK Information: Not listed

Carcinogen: N; **Status:** See below

Carcinogen Lists: *IARC*: Not listed; *MAK*: Not listed; *NIOSH*: Not listed; *NTP*: Not listed; *ACGIH*: Not listed; *OSHA*: Not listed.

LD50 Value: No LD50 in RTECS 1992

Reproductive Toxicity (1992 RTECS): This chemical is a mammalian reproductive toxin.

Reproductive Toxicity Data (1992 RTECS):
orl-rat TDLo:590 g/kg (37D pre/1-22D preg) JONUAI 107, 2020, 77
  Effects on Newborn
    Growth statistics (e.g., reduced weight gain)

--- **INITIAL INCIDENT RESPONSE** ---

No DOT Guide information for this product.

▼ ▼ ▼ ▼ ▼ ▼ ▼ ▼ ▼ ▼ ▼ ▼ ▼ ▼

--- **IDENTIFIERS** ---

Name: CARBON BLACK
Synonyms: Channel Black; Lamp Black; Furnace Black; Thermal Black; Acetylene Black
CAS: 1333-86-4; **RTECS:** FF5800000
Formula: C; **Mol Wt:** 12
Chemical Class: Nonmetal

See other identifiers listed below under Regulations.

--- PROPERTIES ---

Physical Description: Dark brown to black powder, pellets or paste. (NYDH)
Boiling Point: NA
Melting Point: NA
Flash Point: NA
Autoignition: NA
Vapor Pressure: About 0 mm
UEL: NA
LEL: NA
Vapor Density: No data
Specific Gravity: No data
Water Solubility: INSOL
Incompatibilities: Strong oxidizers such as chlorates, bromates, nitrates

Reactivity with Water: No data on water reactivity
Reactivity with Common Materials: No data
Stability during Transport: No data
Neutralizing Agents: No data
Polymerization Possibilities: No data

Toxic Fire Gases: None reported other than possible unburned vapors
Odor Detected at (ppm): Unknown
Odor Description: No data
100% Odor Detection: No data

--- REGULATIONS ---

DOT hazard class: Not given
Packaging exceptions: 173.
Nonbulk packaging: 173.
Bulk packaging: 173.

STCC Number: Not listed

Clean Water Act Sect. 307: No
Clean Water Act Sect. 311: No
Clean Air Act: Not listed
EPA Waste Number: None
CERCLA Ref: Not listed
RQ Designation: Not listed
SARA TPQ Value: Not listed

SARA Sect. 312 Categories:
Acute toxicity: irritant.
Chronic toxicity: mutagen.

U.S. Postal Service Mailability: Not given

NFPA Codes:
Health Hazard (blue): Unspecified
Flammability (red): Unspecified
Reactivity (yellow): Unspecified
Special: Unspecified

--- TOXICITY DATA ---

Short-term Toxicity: *Inhalation:* May cause irritation to nose, mouth, and throat. *Skin:* May cause irritation. *Eyes:* May cause irritation. *Ingestion:* Animal studies show that toxic effects are unlikely. (NYDH)

Long-term Toxicity: Exposure to levels well above 3.5 mg/m^3 for several months may result in damage to the skin and nails and temporary or permanent damage to the lungs and breathing passages and may adversely affect the heart. Carbon black containing PAH greater than 0.1% should be considered a suspect carcinogen. (NYDH)

Symptoms: Conjunctivitis, corneal hypoplasia, eczema, bronchitis, pneumatotumor. (Source: THIC)

Conc IDLH: Unknown

NIOSH REL: Potential occupational carcinogen 3.5 mg/m^3. Time-weighted averages for 8-hr exposure 0.1 in presence mg/m^3 of PAHS.

ACGIH TLV: TLV = 3.5 mg/m^3
ACGIH STEL: Not listed

OSHA PEL: Transitional Limits: PEL = 3.5 mg/m^3; Final Rule Limits: TWA = 3.5 mg/m^3.

MAK Information: Not listed

Carcinogen: N; **Status:** See below

Carcinogen Lists: *IARC*: Not classified as to human carcinogenicity or probably not carcinogenic to humans; *MAK*: Not listed; *NIOSH*: Carcinogen defined by NIOSH with no further categorization; *NTP*: Not listed; *ACGIH*: Not listed; *OSHA*: Not listed.

LD50 Value: No LD50 in RTECS 1992

Reproductive Toxicity (1992 RTECS): This chemical has no known mammalian reproductive toxicity.

--- PROTECTION AND FIRST AID ---

NIOSH Pocket Guide to Chemical Hazards:
**Wear Eye Protection to Prevent: Reasonable probability of eye contact.

**Exposed Personnel Should Wash: At the end of each work shift.

**Reference: NIOSH

First Aid Source: NIOSHP FAM
Inhalation: Make victim blow nose but discourage sniffling.
Ingestion: Induce vomiting; then administer 1 tbsp. mineral oil (nujol) or half-glass peanut oil.
Skin: Remove contaminated clothes; wash skin with soap and water.
Eyes: Irrigate promptly.

--- INITIAL INCIDENT RESPONSE ---

No DOT Guide information for this product.

▼ ▼ ▼ ▼ ▼ ▼ ▼ ▼ ▼ ▼ ▼ ▼ ▼ ▼

--- IDENTIFIERS ---

Name: CARBONIC ACID, CALCIUM SALT (1:1)
Synonyms: Atomit; Calcium Carbonate
CAS: 471-34-1; **RTECS:** FF9335000
Formula: $CO_3.Ca$; **Mol Wt:** 100.09
Chemical Class: Mineral acid salt

See other identifiers listed below under Regulations.

--- PROPERTIES ---

Boiling Point: NA
Melting Point: NA
Flash Point: NA
Autoignition: NA
UEL: NA
LEL: NA
Vapor Density: No data
Specific Gravity: No data

Reactivity with Water: No data on water reactivity
Reactivity with Common Materials: No data
Stability during Transport: No data
Neutralizing Agents: No data
Polymerization Possibilities: No data

Toxic Fire Gases: None reported other than possible unburned vapors
Odor Detected at (ppm): Unknown
Odor Description: No data
100% Odor Detection: No data

--- REGULATIONS ---

DOT hazard class: Not given
Packaging exceptions: 173.
Nonbulk packaging: 173.
Bulk packaging: 173.

STCC Number: Not listed

Clean Water Act Sect. 307: No
Clean Water Act Sect. 311: No
Clean Air Act: Not listed
EPA Waste Number: None
CERCLA Ref: Not listed
RQ Designation: Not listed
SARA TPQ Value: Not listed
SARA Sect. 312 Categories:
  Acute toxicity: irritant.

U.S. Postal Service Mailability: Not given

--- TOXICITY DATA ---

Short-term Toxicity: Unknown

Long-term Toxicity: Unknown

Conc IDLH: Unknown

NIOSH REL: Not given

ACGIH TLV: Not listed
ACGIH STEL: Not listed

OSHA PEL: Not in Table Z-1-A

MAK Information: Not listed

Carcinogen: N; **Status:** See below

Carcinogen Lists: *IARC*: Not listed; *MAK*: Not listed; *NIOSH*: Not listed; *NTP*: Not listed; *ACGIH*: Not listed; *OSHA*: Not listed.

LD50 Value: orl-rat LD50:6450 mg/kg

Other Species Toxicity Data: (Source: NIOSH RTECS 1992) orl-rat LD50:6450 mg/kg

Reproductive Toxicity (1992 RTECS): This chemical has no known mammalian reproductive toxicity.

--- INITIAL INCIDENT RESPONSE ---

No DOT Guide information for this product.

▼   ▼   ▼   ▼   ▼   ▼   ▼   ▼   ▼   ▼   ▼   ▼   ▼

--- IDENTIFIERS ---

Chemical Name and Synonyms: CALCIUM STEARATE,
$Ca(CH_3(CH_2)16...)_2$

Trade Name and Synonyms: Calcium Stearate

Chemical Family: Metal Soap

Manufacturer: Tenneco Chemicals, Inc., Piscataway, New Jersey

UN ID Number: NA

--- INGREDIENTS AND COMPOSITION DATA ---

Material	Cas No.	%	TLV
Calcium stearate	NA		Not established

--- PHYSICAL DATA ---

Boiling Point (°F): NA

Specific Gravity (H₂O = 1): 1.03

Vapor Pressure (mm Hg): NA

Percent Volatile by Volume (%): 2

Vapor Density (air = 1): NA

Freezing Point: NA

Evaporation Rate: Nil

pH: Noncorrosive

Solubility: Negligible

Melting Point (°F): NA

Appearance/Color/Odor: White, fluffy, fine powder, slight fatty acid odor.

--- FIRE AND EXPLOSION HAZARDS ---

Flash Point: NA

Flammability Limits: NA

Autoignition Temperature: NA

Decomposition Temperature: NA

Extinguishing Media: Water spray, foam, carbon dioxide, or dry chemical.

Special Fire Fighting Procedures: Water or foam may cause frothing.

Unusual Fire and Explosion Hazards: Avoid dispersion of dust in air to reduce potential for dust ignition and explosion.

Hazardous Products of Combustion and Decomposition: Carbon monoxide and/or carbon dioxide.

Hazardous Polymerization: Will not occur.

--- HEALTH HAZARD DATA ---

Effects of Overexposure: None reported

Emergency Response and First Aid:
Skin: Wash with lots of clean water for at least 15 min.
Eyes: Flush eyes with lots of water. If irritation persists, consult a physician.
Inhalation: Expose to fresh air. Keep warm and quiet. Give artificial respiration if necessary.
Ingestion: None recommended. Consult physician immediately.

Cancer Information: None reported.

--- TOXICITY DATA ---

Animal Effects: None reported

--- REACTIVITY DATA ---

Stability and Conditions to Avoid: None reported by manufacturer other than to minimize skin exposure and inhalation. This material is considered stable.

Materials to Avoid (Incompatibility): Strong oxidizing materials.

Hazard Decomposition Products: Carbon dioxide and carbon monoxide.

Hazardous Polymerization: Will not occur.

--- SPILL AND EMERGENCY RESPONSE ---

For Spills: Sweep up. Wash residuals with soap and water. Transfer to a closed container.

Waste Disposal Method: No specific recommendations.

Empty Containers: The container for this product can present explosion or fire hazards, even when emptied. To avoid the risk of injury, do not cut, puncture, or weld on or near this container.

--- SPECIAL PROTECTION AND PRECAUTIONS ---

Respiratory Protection: Avoid breathing dust. Dust respirator recommended. Although OSHA and ACGIH have not established specific exposure limits for this material, they have established the following limits for nuisance dusts:

OSHA PEL/8-hr TWA: Total 15 mg/m^3, respirable 5 mg/m^3.
ACGIH TLV/8-hr TWA: Total 10 mg/m^3, respirable 5 mg/m^3.
These limits are stated only to indicate the least stringent airborne dust exposure levels applicable to nuisance dusts.

Type of Ventilation Required: General mechanical. Avoid breathing dusts and vapors. Adequate ventilation should be provided to keep mist and dust concentrations below acceptable exposure limits. Discharge from the ventilation system should comply with applicable air-pollution regulations. Local exhaust ventilation is recommended with a capture velocity of 150 to 200 fpm.

Eye Protection Requirements: Chemical goggles. Eyewash fountains and safety showers should be easily accessible.

Protective Gloves and Other: None recommended.

Storage and Handling Precautions: Containers should be stored in a cool, dry, well-ventilated area. Store away from flammable materials, sources of heat, flame, sparks, and foodstuffs. Exercise due caution to prevent damage to or leakage from the container.

Prevent inhalation of dust. Prevent contact with eyes and skin. Under dusty conditions, static electricity may cause an explosion. Keep containers closed. Store in a cool, dry place.

Use with adequate ventilation. Wash thoroughly after handling before eating and smoking. Work clothing should be frequently laundered and not worn away from work premises.

Other Precautions: Use all standard practices for good personal hygiene. Completely isolate and thoroughly clean all equipment, piping, or vessels before beginning maintenance or repairs. Keep area clean. As with any powder or dust-like material, improper handling can lead to spontaneous combustion (dust explosions).

--- REGULATORY INFORMATION ---

None reported.

▼ ▼ ▼ ▼ ▼ ▼ ▼ ▼ ▼ ▼ ▼ ▼ ▼ ▼

--- IDENTIFIERS ---

Chemical Name and Synonyms: MAGNESIUM STEARATE, $Mg(C_{18}H_{35}O_2)2$

Trade Name and Synonyms: Magnesium Stearate

Chemical Family: Metal Soap

Manufacturer: Diamond Shamrock Chemical Co., Cleveland, Ohio

UN ID Number: NA

--- INGREDIENTS AND COMPOSITION DATA ---

Material	Cas No.	%	TLV
Magnesium stearate	NA		Not established

--- PHYSICAL DATA ---

Boiling Point (°F): NA

Specific Gravity (H₂O = 1): NA

Vapor Pressure (mm Hg): NA

Percent Volatile by Volume (%): NA

Vapor Density (air = 1): NA

Freezing Point: NA

Evaporation Rate: Nil

pH: Noncorrosive

Solubility: Negligible

Melting Point (°C): 145

Appearance/Color/Odor: Odorless white powder.

--- FIRE AND EXPLOSION HAZARDS ---

Flash Point: NA

Flammability Limits: NA

Autoignition Temperature: NA

Decomposition Temperature: NA

Extinguishing Media: Foam, carbon dioxide, or dry chemical.

Special Fire Fighting Procedures: None reported.

Unusual Fire and Explosion Hazards: Avoid dispersion of dust in air to reduce potential for dust ignition and explosion.

Hazardous Products of Combustion and Decomposition: Carbon monoxide and/or carbon dioxide.

Hazardous Polymerization: Will not occur.

--- HEALTH HAZARD DATA ---

Effects of Overexposure: None reported

Emergency Response and First Aid:
Skin: Wash with lots of clean water for at least 15 min.
Eyes: Flush eyes with lots of water. If irritation persists, consult a physician.
Inhalation: Expose to fresh air. Keep warm and quiet. Give artificial respiration if necessary.
Ingestion: None recommended. Consult physician immediately.

Cancer Information: None reported.

--- TOXICITY DATA ---

Animal Effects: None reported

--- REACTIVITY DATA ---

Stability and Conditions to Avoid: None reported by manufacturer other than to minimize skin exposure and inhalation. This material is considered stable.

Materials to Avoid (Incompatibility): Strong oxidizing materials.

Hazard Decomposition Products: Carbon dioxide and carbon monoxide.

Hazardous Polymerization: Will not occur.

--- SPILL AND EMERGENCY RESPONSE ---

For Spills: Sweep up. Wash residuals with soap and water. Transfer to a closed container.

Waste Disposal Method: No specific recommendations.

Empty Containers: The container for this product can present explosion or fire hazards, even when emptied. To avoid the risk of injury, do not cut, puncture, or weld on or near this container.

--- SPECIAL PROTECTION AND PRECAUTIONS ---

Respiratory Protection: Avoid breathing dust. Dust respirator recommended. Although OSHA and ACGIH have not established specific exposure limits for this material, they have established the following limits for nuisance dusts:

OSHA PEL/8-hr TWA: Total 15 mg/m^3, respirable 5 mg/m^3.
ACGIH TLV/8-hr TWA: Total 10 mg/m^3, respirable 5 mg/m^3.
These limits are stated only to indicate the least stringent airborne dust exposure levels applicable to nuisance dusts.

Type of Ventilation Required: General mechanical. Avoid breathing dusts and vapors. Adequate ventilation should be provided to keep mist and dust concentrations below acceptable exposure limits. Discharge from the ventilation system should comply with applicable air-pollution regulations. Local exhaust ventilation is recommended with a capture velocity of 150 to 200 fpm.

Eye Protection Requirements: Chemical goggles. Eyewash fountains and safety showers should be easily accessible.

Protective Gloves and Other: Wear rubber gloves and goggles.

Storage and Handling Precautions: Containers should be stored in a cool, dry, well-ventilated area. Store away from flammable materials, sources of heat, flame, sparks, and foodstuffs. Exercise due caution to prevent damage to or leakage from the container.

Prevent inhalation of dust. Prevent contact with eyes and skin. Under dusty conditions, static electricity may cause an explosion. Keep containers closed. Store in a cool, dry place.

Use with adequate ventilation. Wash thoroughly after handling before eating and smoking. Work clothing should be frequently laundered and not worn away from work premises.

Other Precautions: Use all standard practices for good personal hygiene. Completely isolate and thoroughly clean all equipment, piping, or vessels

before beginning maintenance or repairs. Keep area clean. As with any powder or dust-like material, improper handling can lead to spontaneous combustion (dust explosions).

--- REGULATORY INFORMATION ---

None reported.

▼ ▼ ▼ ▼ ▼ ▼ ▼ ▼ ▼ ▼ ▼ ▼ ▼ ▼

--- IDENTIFIERS ---

Chemical Name and Synonyms: ORGANOPHILIC CLAY

Trade Name and Synonyms: Bentone 34

Chemical Family: Clay Product

Manufacturer: NL Industries, Inc., Hightstown, New Jersey

UN ID Number: NA

--- INGREDIENTS AND COMPOSITION DATA ---

Material	Cas No.	%	TLV
Cristobalite	NA	< 1	Not established

--- PHYSICAL DATA ---

Boiling Point (°F): NA

Specific Gravity (H₂O = 1): 1.7

Vapor Pressure (mm Hg): NA

Percent Volatile by Volume (%): NA

Vapor Density (air = 1): NA

Freezing Point: NA

Evaporation Rate: Nil

pH: Noncorrosive

Solubility: Negligible

Melting Point (°C): NA

Appearance/Color/Odor: Odorless white powder.

--- FIRE AND EXPLOSION HAZARDS ---

Flash Point: NA

Flammability Limits: NA

Autoignition Temperature: NA

Decomposition Temperature: NA

Extinguishing Media: NFPA Class A agents acceptable: dry chemical, foam CO_2, water (fog).

Special Fire Fighting Procedures: None reported.

Unusual Fire and Explosion Hazards: Avoid dispersion of dust in air to reduce potential for dust ignition and explosion.

Hazardous Products of Combustion and Decomposition: Carbon monoxide and/or carbon dioxide and traces of ammonia.

Hazardous Polymerization: Will not occur.

--- HEALTH HAZARD DATA ---

Effects of Overexposure: Threshold limit value, 13.5 mppcf. The prolonged inhalation of dusts may result in shortness of breath and decreased chest expansion, but with absence of fever.

Emergency Response and First Aid:
Skin: Wash with lots of clean water for at least 15 min.
Eyes: Flush eyes with lots of water. If irritation persists, consult a physician.

Inhalation: Expose to fresh air. Keep warm and quiet. Give artificial respiration if necessary.
Ingestion: None recommended. Consult physician immediately.

Cancer Information: None reported.

--- TOXICITY DATA ---

Animal Effects: None reported

--- REACTIVITY DATA ---

Stability and Conditions to Avoid: None reported by manufacturer other than to minimize skin exposure and inhalation. This material is considered stable.

Materials to Avoid (Incompatibility): Strong oxidizing materials.

Hazard Decomposition Products: Carbon dioxide and carbon monoxide.

Hazardous Polymerization: Will not occur.

--- SPILL AND EMERGENCY RESPONSE ---

For Spills: Sweep up. Wash residuals with soap and water. Transfer to a closed container.

Waste Disposal Method: No specific recommendations.

Empty Containers: The container for this product can present explosion or fire hazards, even when emptied. To avoid the risk of injury, do not cut, puncture, or weld on or near this container.

--- SPECIAL PROTECTION AND PRECAUTIONS ---

Respiratory Protection: Avoid breathing dust. Dust respirator recommended. Although OSHA and ACGIH have not established specific exposure limits for this material, they have established the following limits for nuisance dusts:

OSHA PEL/8-hr TWA: Total 15 mg/m^3, respirable 5 mg/m^3.
ACGIH TLV/8-hr TWA: Total 10 mg/m^3, respirable 5 mg/m^3.
These limits are stated only to indicate the least stringent airborne dust exposure levels applicable to nuisance dusts.

Type of Ventilation Required: General mechanical. Avoid breathing dusts and vapors. Adequate ventilation should be provided to keep mist and dust concentrations below acceptable exposure limits. Discharge from the ventilation system should comply with applicable air-pollution regulations. Local exhaust ventilation is recommended with a capture velocity of 150 to 200 fpm.

Eye Protection Requirements: Chemical goggles. Eyewash fountains and safety showers should be easily accessible.

Protective Gloves and Other: Wear rubber gloves and goggles.

Storage and Handling Precautions: Containers should be stored in a cool, dry, well-ventilated area. Store away from flammable materials, sources of heat, flame, sparks, and foodstuffs. Exercise due caution to prevent damage to or leakage from the container.

Prevent inhalation of dust. Prevent contact with eyes and skin. Under dusty conditions, static electricity may cause an explosion. Keep containers closed. Store in a cool, dry place.

Use with adequate ventilation. Wash thoroughly after handling before eating and smoking. Work clothing should be frequently laundered and not worn away from work premises.

Other Precautions: Use all standard practices for good personal hygiene. Completely isolate and thoroughly clean all equipment, piping, or vessels before beginning maintenance or repairs. Keep area clean. As with any powder or dust-like material, improper handling can lead to spontaneous combustion (dust explosions).

--- REGULATORY INFORMATION ---

None reported.

▼ ▼ ▼ ▼ ▼ ▼ ▼ ▼ ▼ ▼ ▼ ▼ ▼ ▼

--- IDENTIFIERS ---

Name: SILICA, AMORPHOUS
Synonyms: Amorphous Fused Silica; Fused Quartz; Fused Silica; Quartz Glass; SG-67; Silica, Fused; Silicon Dioxide; Silicone Dioxide; Suprasil;

Vitreous Quartz; Diatomite; Silicon Dioxide (Amorphous); Diatomaceous Earth; Diatomaceous Silica; Colloidal Silica
CAS: 60676-86-0; **RTECS:** VV7320000
Formula: SiO_2; **Mol Wt:** 60.09
WLN: SI O2
Chemical Class: Nonmetal oxide

See other identifiers listed below under Regulations.

--- PROPERTIES ---

Physical Description: Colorless to gray, odorless powder
Boiling Point: NA
Melting Point: 1977.58 K; 1704.4°C; 3099.9°F
Flash Point: NA
Autoignition: NA
Vapor Pressure: About 0 mm
UEL: NA
LEL: NA
Vapor Density: No data
Specific Gravity: No data
Density: 2.600
Water Solubility: INSOL
Incompatibilities: Fluorine, oxygen difluoride, chlorine trifluoride

Reactivity with Water: No data on water reactivity
Reactivity with Common Materials: No data
Stability during Transport: No data
Neutralizing Agents: No data
Polymerization Possibilities: No data

Toxic Fire Gases: None reported other than possible unburned vapors
Odor Detected at (ppm): Unknown
Odor Description: No data
100% Odor Detection: No data

--- REGULATIONS ---

DOT hazard class: Not given
Packaging exceptions: 173.
Nonbulk packaging: 173.
Bulk packaging: 173.

STCC Number: Not listed

Clean Water Act Sect. 307: No
Clean Water Act Sect. 311: No
Clean Air Act: CAA '90 by category
EPA Waste Number: None
CERCLA Ref: N
RQ Designation: Not listed
SARA TPQ Value: Not listed
SARA Sect. 312 Categories:
Acute toxicity: adverse effect to target organs.

U.S. Postal Service Mailability: Not given

NFPA Codes:
Health Hazard (blue): Unspecified
Flammability (red): Unspecified
Reactivity (yellow): Unspecified
Special: Unspecified

--- TOXICITY DATA ---

Short-term Toxicity: Unknown

Long-term Toxicity: Unknown

Target Organs: Respiratory system, lungs

Symptoms: Pneumoconiosis (Source: NIOSHP)

Conc IDLH: Unknown

NIOSH REL: Potential occupational carcinogen 0.05 mg/m^3. Time-weighted averages for 8-hr exposure.

ACGIH TLV: TLV = 0.1 mg/m^3 respirable dust
ACGIH STEL: Respirable dust

OSHA PEL: Transitional Limits: PEL = see table Z-3 ppm; Final Rule Limits: TWA = respirable dust, 0.1 mg/m^3.

MAK Information: 0.3 mg/m^3

Carcinogen: N; **Status:** See below

Carcinogen Lists: *IARC:* Not classified as to human carcinogenicity or probably not carcinogenic to humans; *MAK:* Not listed; *NIOSH:* Carcinogen defined by NIOSH with no further categorization; *NTP:* Not listed; *ACGIH:* Not listed; *OSHA:* Not listed.

LD50 Value: No LD50 in RTECS 1992

Other Species Toxicity Data: (Source: NIOSH RTECS 1992)
ipr-rat LDLo:400 mg/kg
itr-rat LDLo:120 mg/kg
ipr-mus LDLo:40 mg/kg
ivn-cat LDLo:5 mg/kg
ivn-rbt LDLo:35 mg/kg

Reproductive Toxicity (1992 RTECS): This chemical has no known mammalian reproductive toxicity.

--- PROTECTION AND FIRST AID ---

First Aid Source: NIOSH
Inhalation: None given
Ingestion: None given
Skin: None given
Eyes: Irrigate immediately

6 ANTIOXIDANTS, STABILIZERS, AND FIRE RETARDANTS

Three classes of chemicals are covered in this section:

	CHEMICAL FAMILY/NAME	TRADE NAMES
Antioxidants	Phenolic Antioxidants	Good-Rite
		TMQ
		Antioxidant 2246
		Aminox
		Dilaury/Thiodipropionate
		Flexzone
		Irganox
		Agerite DPPD and HP-S, MA, Resin D
		Vanox
		Akrochem Antioxidant 33
Stabilizers	Alkylated Cresol	BHT
Flame Retardants	Brominated Organic Compounds	Citex BN-451
	Chlorinated Organics	Dechlorane 602
	Metal Oxide, Antimony	Fire Shield
	Trioxide	Antimony Oxide
	Antimony Trioxide	Thermoguard S

329

--- IDENTIFIERS ---

Chemical Name and Synonyms: TRIS(3,5-di-t-BUTYL-4-HYDROXY-BENZYL)ISOCYANURATE, $C_{48}H_{69}N_3O_6$

Trade Name and Synonyms: Good-Rite 3114

Chemical Family: Phenolic Antioxidant

Manufacturer: BF Goodrich Chemical Co., Cleveland, Ohio

UN ID Number: NA

--- INGREDIENTS AND COMPOSITION DATA ---

Material	Cas No.	%	TLV
Product	NA		Not established

Note: Good-rite 3114 has FDA approval under Regulation 178.2010* for use in polyolefins in contact with fatty and nonfatty foods.

--- PHYSICAL DATA ---

Boiling Point (°F): NA

Specific Gravity (H_2O = 1): 1.03

Vapor Pressure (mm Hg): NA

Percent Volatile by Volume (%): NA

Vapor Density (air = 1): NA

Freezing Point: NA

Evaporation Rate: Nil

pH: Noncorrosive

Solubility in Water: 0.002 g/100 g at 25°C

Melting Point (°C): NA

Appearance/Color/Odor: White powder, negligible odor.

--- FIRE AND EXPLOSION HAZARDS ---

Flash Point: 553°F, C.O.C.

Flammability Limits: NA

Autoignition Temperature: NA

Decomposition Temperature: NA

Extinguishing Media: NFPA Class A agents acceptable: dry chemical, foam CO_2, water (fog).

Special Fire Fighting Procedures: None reported.

Unusual Fire and Explosion Hazards: Avoid dispersion of dust in air to reduce potential for dust ignition and explosion.

Hazardous Products of Combustion and Decomposition: Carbon monoxide and/or carbon dioxide and traces of ammonia.

Hazardous Polymerization: Will not occur.

--- HEALTH HAZARD DATA ---

Effects of Overexposure: None reported; see toxicity data below.

Emergency Response and First Aid:
Skin: Wash with lots of clean water for at least 15 min.
Eyes: Flush eyes with lots of water. If irritation persists, consult a physician.
Inhalation: Expose to fresh air. Keep warm and quiet. Give artificial respiration if necessary.
Ingestion: None recommended. Consult physician immediately.

Cancer Information: None reported.

--- TOXICITY DATA ---

Animal Effects: Threshold limit value, LD50 single oral dose in rats is greater than 6800 mg/kg. Acute dermal LD50 for rabbits is greater than 10,000 mg/kg of body weight. Slight eye irritation without corneal damage produced in rabbits with direct application. No deaths among rats following exposure to dust at a concentration of 66.5 mg/liter of air.

--- REACTIVITY DATA ---

Stability and Conditions to Avoid: None reported by manufacturer other than to minimize skin exposure and inhalation. This material is considered stable at temperatures up to 500°F.

Materials to Avoid (Incompatibility): Strong oxidizing materials.

Hazard Decomposition Products: Carbon dioxide and carbon monoxide.

Hazardous Polymerization: Will not occur.

--- SPILL AND EMERGENCY RESPONSE ---

For Spills: Sweep up. Wash residuals with soap and water. Transfer to a closed container.

Waste Disposal Method: No specific recommendations.

Empty Containers: The container for this product can present explosion or fire hazards, even when emptied. To avoid the risk of injury, do not cut, puncture, or weld on or near this container.

--- SPECIAL PROTECTION AND PRECAUTIONS ---

Respiratory Protection: Avoid breathing dust. Dust respirator recommended. Although OSHA and ACGIH have not established specific exposure limits for this material, they have established the following limits for nuisance dusts:

OSHA PEL/8-hr TWA: Total 15 mg/m³, respirable 5 mg/m³.
ACGIH TLV/8-hr TWA: Total 10 mg/m³, respirable 5 mg/m³.
These limits are stated only to indicate the least stringent airborne dust exposure levels applicable to nuisance dusts.

Type of Ventilation Required: General mechanical. Avoid breathing dusts and vapors. Adequate ventilation should be provided to keep mist and dust concentrations below acceptable exposure limits. Discharge from the ventilation system should comply with applicable air-pollution regulations. Local exhaust ventilation is recommended with a capture velocity of 150 to 200 fpm.

Eye Protection Requirements: Chemical goggles. Eyewash fountains and safety showers should be easily accessible.

Protective Gloves and Other: Wear rubber gloves and goggles.

Storage and Handling Precautions: Containers should be stored in a cool, dry, well-ventilated area. Store away from flammable materials, sources of heat, flame, sparks, and foodstuffs. Exercise due caution to prevent damage to or leakage from the container.

Prevent inhalation of dust. Prevent contact with eyes and skin. Under dusty conditions, static electricity may cause an explosion. Keep containers closed. Store in a cool, dry place.

Use with adequate ventilation. Wash thoroughly after handling before eating and smoking. Work clothing should be frequently laundered and not worn away from work premises.

Other Precautions: Use all standard practices for good personal hygiene. Completely isolate and thoroughly clean all equipment, piping, or vessels before beginning maintenance or repairs. Keep area clean. As with any powder or dustlike material, improper handling can lead to spontaneous combustion (dust explosions).

--- REGULATORY INFORMATION ---

None reported.

▼ ▼ ▼ ▼ ▼ ▼ ▼ ▼ ▼ ▼ ▼ ▼ ▼ ▼

--- IDENTIFIERS ---

Chemical Name and Synonyms: 2,2'-METHYLENEBIS(4-METHYL-6-TERTIARYBUTYL PHENOL, $C_{23}H_{32}O_2$

Trade Name and Synonyms: Antioxidant 2246

Chemical Family: Phenol

Manufacturer: American Cyanamid Co., Bound Brook, New Jersey

UN ID Number: NA

--- INGREDIENTS AND COMPOSITION DATA ---

Material	Cas No.	%	TLV
Product	NA		Not established

--- PHYSICAL DATA ---

Boiling Point (°F): NA

Specific Gravity (H₂O = 1): 1.08

Vapor Pressure (mm Hg): NA

Percent Volatile by Volume (%): Negligible

Vapor Density (air = 1): NA

Freezing Point: NA

Evaporation Rate: NA

pH: NA

Solubility in Water: Negligible

Melting Point (°C): NA

Appearance/Color/Odor: White powder; mild phenolic odor.

--- FIRE AND EXPLOSION HAZARDS ---

Flash Point: NA

Flammability Limits: NA

Autoignition Temperature: NA

Decomposition Temperature: NA

Extinguishing Media: Carbon dioxide, dry chemical, or water.

Special Fire Fighting Procedures: Do not use high-pressure water stream. Airborne dust creates an explosion hazard.

Unusual Fire and Explosion Hazards: Dust may be explosive if mixed with air in critical proportions and in the presence of a source of ignition. The hazard is similar to that of any organic solid including sawdust. Maintain normal good housekeeping for control of dust.

Hazardous Products of Combustion and Decomposition: This material, if ignited, can give off nitrogen oxides and carbon monoxide gases.

Hazardous Polymerization: Will not occur.

--- HEALTH HAZARD DATA ---

Effects of Overexposure: None expected.

Emergency Response and First Aid:
Skin: Wash with lots of clean water for at least 15 min.
Eyes: Flush eyes with lots of water. If irritation persists, consult a physician.
Inhalation: Expose to fresh air. Keep warm and quiet. Give artificial respiration if necessary.
Ingestion: None recommended. Consult physician immediately.

Cancer Information: None reported.

--- TOXICITY DATA ---

Animal Effects: None reported.

--- REACTIVITY DATA ---

Stability and Conditions to Avoid: No specific incompatibility.

Materials to Avoid (Incompatibility): None.

Hazard Decomposition Products: Thermal decomposition may produce carbon monoxide and/or carbon dioxide.

Hazardous Polymerization: Will not occur.

--- SPILL AND EMERGENCY RESPONSE ---

For Spills: Vacuum or sweep up and place into dry, clean, covered container for disposal. Keep this material out of watersheds and waterways.

Waste Disposal Method: No specific recommendations made by manufacturer.

Empty Containers: Burn containers in an approved incinerator or dispose of in an approved chemical landfill in accordance with all applicable local, state, and federal laws and regulations.

--- SPECIAL PROTECTION AND PRECAUTIONS ---

Respiratory Protection: Avoid breathing dust. Dust respirator recommended. Although OSHA and ACGIH have not established specific exposure limits for this material, they have established the following limits for nuisance dusts:

OSHA PEL/8-hr TWA: Total 15 mg/m^3, respirable 5 mg/m^3.
ACGIH TLV/8-hr TWA: Total 10 mg/m^3, respirable 5 mg/m^3.
These limits are stated only to indicate the least stringent airborne dust exposure levels applicable to nuisance dusts.

Type of Ventilation Required: General mechanical. Avoid breathing dusts and vapors. Adequate ventilation should be provided to keep mist and dust concentrations below acceptable exposure limits. Discharge from the ventilation system should comply with applicable air-pollution regulations. Local exhaust ventilation is recommended with a capture velocity of 150 to 200 fpm.

Eye Protection Requirements: Chemical goggles. Eyewash fountains and safety showers should be easily accessible.

Protective Gloves and Other: Wear rubber gloves and goggles.

Storage and Handling Precautions: Containers should be stored in a cool, dry, well-ventilated area. Store away from flammable materials, sources of heat, flame, sparks, and foodstuffs. Exercise due caution to prevent damage to or leakage from the container.

Prevent inhalation of dust. Prevent contact with eyes and skin. Under dusty conditions, static electricity may cause an explosion. Keep containers closed. Store in a cool, dry place.

Use with adequate ventilation. Wash thoroughly after handling before eating and smoking. Work clothing should be frequently laundered and not worn away from work premises.

Other Precautions: Use all standard practices for good personal hygiene. Completely isolate and thoroughly clean all equipment, piping, or vessels before beginning maintenance or repairs. Keep area clean. As with any powder or dustlike material, improper handling can lead to spontaneous combustion (dust explosions).

--- REGULATORY INFORMATION ---

None reported.

▼ ▼ ▼ ▼ ▼ ▼ ▼ ▼ ▼ ▼ ▼ ▼ ▼ ▼

--- IDENTIFIERS ---

Chemical Name and Synonyms: POLYMERIZED 1,2-DIHYDRO-2,2,4-TRIMETHYLQUINOLINE, AND 1,2-DIHYDRO-2,2,4-TRIMETHYL-QUINOLINE

Trade Name and Synonyms: TMQ, Flectol H Antioxidant

Chemical Family: Rubber Preservative

Manufacturer: Monsanto Chemical Co., St. Louis, Missouri

UN ID Number: NA

--- INGREDIENTS AND COMPOSITION DATA ---

Material	Cas No.	%	TLV
Product	NA		Not established

--- PHYSICAL DATA ---

Boiling Point (°F): NA

Specific Gravity (H_2O = 1): 1.01 to 1.08

Vapor Pressure (mm Hg): NA

Percent Volatile by Volume (%): NA

Vapor Density (air = 1): NA

Freezing Point: NA

Evaporation Rate: NA

pH: NA

Solubility in Water: 0.005 g/100 ml at 25°C

Melting Point (°C): 85 to 130

Appearance/Color/Odor: Light brown powder or flakes.

--- FIRE AND EXPLOSION HAZARDS ---

Flash Point: 239 to 302°F; Cleveland open cup

Flammability Limits: (Dust cloud) LEL: 20 to 200 mg/l

Autoignition Temperature: NA

Decomposition Temperature: NA

Extinguishing Media: Water spray, foam, dry chemical, carbon dioxide, or any Class B extinguishing agent.

Special Fire Fighting Procedures: Firefighters or others exposed to products of combustion (see Hazardous Decomposition Products below) should wear full protective clothing including self-contained breathing apparatus. Equipment should be thoroughly decontaminated after use.

Unusual Fire and Explosion Hazards: Products of decomposition include hazardous carbon monoxide and nitrogen oxides. Powder or dust of this material when mixed in sufficient quantities in air can form explosive mixtures.

Hazardous Products of Combustion and Decomposition: This material, if ignited, can give off nitrogen oxides and carbon monoxide gases.

Hazardous Polymerization: Will not occur.

--- HEALTH HAZARD DATA ---

Effects of Overexposure: Dermal contact and inhalation are expected to be the primary routes of occupational exposure to Flectol H antioxidant. Occupational exposure to this material has not been reported to cause significant adverse human health effects. Testing in human subjects with a closely related polymeric, 1,2-dihydro-2,2,4-trimethylquinoline, indicated a low potential for dermal sensitization. Following exposure to Flectol H antioxidant, workers have reported staining and discoloration of garments contacting the skin.

Emergency Response and First Aid:
Skin: Wash with lots of clean water for at least 15 min.
Eyes: Flush eyes with lots of water. If irritation persists, consult a physician.
Inhalation: Expose to fresh air. Keep warm and quiet. Give artificial respiration if necessary.
Ingestion: None recommended. Consult physician immediately.

Cancer Information: None reported.

--- TOXICITY DATA ---

Animal Effects:
Data from Monsanto studies and the available scientific literature indicate the following:

Oral LD50 (rat): >7940 mg/kg, practically nontoxic
Dermal LD50 (rabbit): >7940 mg/kg, practically nontoxic

Eye Irritation (rabbit): (FHSA) 1.3 on a scale of 110.0, practically
    nonirritating
Skin Irritation (rabbit): (FHSA) 0.0 on a scale of 8.0, nonirritating

A polymerized 1,2-dihydro-2,2,4-trimethylquinoline similar to Flectol H antioxidant was tested on human volunteers. Tested as a 75% preparation in petrolatum, this material produced irritation responses in 9 of the 51 test subjects after initial application; 6 of the 51 volunteers exhibited mild to moderate cumulative irritation. During the challenge phase of the study, sensitization was observed in one subject, with a possible response in two others. This material was considered to be a potential skin sensitizer.

Flectol H antioxidant was evaluated in a 4-week dust inhalation study in which groups of 5 female and 5 male rats were exposed to atmospheric concentrations of 0, 15, 40, 90 or 150 mg/m^3 6 hr per day, 5 days per week. Significant elevations in liver weights were noted for animals of the two highest exposure level groups, with lesser increases in liver weights seen for males at lower exposure levels.

In a 13-week dust inhalation study, rats were exposed to Flectol H antioxidant at concentrations of 15, 49 and 148 mg/m^3 for 6 hr per day, 5 days per week. Treatment-related effects in certain hematological and clinical chemistry parameters were found at all dose levels. Dose-related increases in liver weights were noted. Microscopic examination of livers indicated an increased incidence of cloudy swelling and vacuolation of the liver cell cytoplasm in high-dose animals. Based on the findings of this study, a no-effect level was not determined.

Polymeric 1,2-dihydro-2,2,4-trimethylquinoline has been evaluated in a series of chronic toxicity studies:

Rats were given 0.01, 0.1, or 1.5% polymeric 1,2-dihydro-2,2,4-trimethylquinoline (reported as Flectol H, but later determined to be of non-Monsanto origin) in the diet for 2 years. Slight decreases in growth rate and food consumption and an increase in the occurrence of cholangiofibrosis were observed in the high-dose group. Increases in lymphomas and adenomas were observed. Based on the results of this study, the FDA classified this quinoline polymer as "at least a probable carcinogen" and removed it from the list of rubber additives approved for indirect food contact.

Polymeric 1,2-dihydro-2,2,4-trimethylquinoline was administered to groups of rats at concentrations of 0.01, 0.1, 1.5, or 3.0% in the diet for 2 years.

Animals administered 3% in the test did not survive beyond 3 weeks; only 20% of the animals administered 1.5% survived to the end of the study. Liver enlargement was present in high-dose animals. Alterations of hepatic cells, liver necrosia, and biliary duct hyperplasia were observed in the high-dose group. Only minimal effects were reported for those animals receiving 0.01 and 0.1% of the test material in the diet.

Mice were treated with either a single subcutaneous injection of 10-mg polymeric 1,2-dihydro-2,2,4-trimethylquinoline, or weekly skin application of 0.1 mg or 10 mg of the test material. All animals were observed for their lifetimes. No tumors were observed at the sites of application and no effects on survival times were noted. However, focal hepatic necrosis was observed in two female mice that received weekly dermal applications of 10 mg of the test material.

Groups of three dogs were given polymeric 1,2-dihydro-2,2,4-trimethylquinoline (reported as Flectol H, but later determined to be of non-Monsanto origin) in the diet at concentrations of 0.008, 0.03, or 0.15% for a 1 year period. Secondary hepatic cell degeneration was observed in two of the high-dose animals. No tumors were observed.

Flectol H antioxidant was evaluated for mutagenic or genotoxic potential in the following systems: microbial assays with five S. typhimurium strains and one yeast strain, with and without activation; in vitro Chinese hamster ovary (CHO) cell point mutation assays; and in vivo/in vitro rat hepatocyte/DNA repair assays. No evidence of mutagenicity or genotoxicity was observed in any of the assays.

Environmental Toxicity Information:

96-hr LC50 Trout: 50 mg/l, slightly toxic
96-hr LC50 Bluegill: 54 mg/l, slightly toxic
96-hr LC50 Fathead Minnow: 64 mg/l, slightly toxic
48-hr LC50 Daphnia: 5.8 mg/l, moderately toxic
96-hr EC50 Algae, Cell Count: > 1000 mg/l, nontoxic

--- REACTIVITY DATA ---

Stability and Conditions to Avoid: Thermally stable.

Materials to Avoid (Incompatibility): None

Hazard Decomposition Products: This material, if ignited, can give off nitrogen oxides and carbon monoxide gases.

Hazardous Polymerization: Will not occur.

--- SPILL AND EMERGENCY RESPONSE ---

For Spills: Vacuum or sweep up and place into dry, clean, covered container for disposal. Keep this material out of watersheds and waterways.

Waste Disposal Method: "Waste disposal" as that term is defined in 40 CFR 261, "Identification and Listing of Hazardous Waste." Burn in an approved incinerator or dispose of in an approved chemical landfill in accordance with all applicable local, state, and federal laws and regulations. Consult your attorney or appropriate regulatory officials for information on such disposal.

Empty Containers: Burn containers in an approved incinerator or dispose of in an approved chemical landfill in accordance with all applicable local, state, and federal laws and regulations.

--- SPECIAL PROTECTION AND PRECAUTIONS ---

Respiratory Protection: Avoid breathing dust. Dust respirator recommended. Although OSHA and ACGIH have not established specific exposure limits for this material, they have established the following limits for nuisance dusts:

OSHA PEL/8-hr TWA: Total 15 mg/m^3, respirable 5 mg/m^3.
ACGIH TLV/8-hr TWA: Total 10 mg/m^3, respirable 5 mg/m^3.
These limits are stated only to indicate the least stringent airborne dust exposure levels applicable to nuisance dusts.

Type of Ventilation Required: General mechanical. Avoid breathing dusts and vapors. Adequate ventilation should be provided to keep mist and dust concentrations below acceptable exposure limits. Discharge from the ventilation system should comply with applicable air-pollution regulations. Local exhaust ventilation is recommended with a capture velocity of 150 to 200 fpm.

Eye Protection Requirements: Chemical goggles. Eyewash fountains and safety showers should be easily accessible.

Protective Gloves and Other: Wear rubber gloves and goggles.

Storage and Handling Precautions: Containers should be stored in a cool, dry, well-ventilated area. Store away from flammable materials, sources of heat, flame, sparks, and foodstuffs. Exercise due caution to prevent damage to or leakage from the container.

Prevent inhalation of dust. Prevent contact with eyes and skin. Under dusty conditions, static electricity may cause an explosion. Keep containers closed. Store in a cool, dry place.

Use with adequate ventilation. Wash thoroughly after handling before eating and smoking. Work clothing should be frequently laundered and not worn away from work premises.

Other Precautions: Use all standard practices for good personal hygiene. Completely isolate and thoroughly clean all equipment, piping, or vessels before beginning maintenance or repairs. Keep area clean. As with any powder or dustlike material, improper handling can lead to spontaneous combustion (dust explosions).

--- REGULATORY INFORMATION ---

DOT Proper Shipping Name: NA

DOT Hazard Class/I.D. No: NA

DOT Label: Not applicable

U.S. Surface Freight Classification: Rubber Preservative, N.O.I.B.N.

Reportable Quantity (RQ) under DOT (49 CFR) and CERCLA Regulations: NA

SARA Hazard Notification:
Hazard Categories under criteria of SARA Title III rules (40 CFR Part 370):
Delayed
Section 313 Hazardous Chemical(s): NA

Hazardous Chemical(s) under OSHA Hazard Communication Standard:
This substance is identified as a hazardous chemical under the criteria of the OSHA Hazard Communication Standard (29 CFR 1910.1200).

▼ ▼ ▼ ▼ ▼ ▼ ▼ ▼ ▼ ▼ ▼ ▼ ▼ ▼

--- IDENTIFIERS ---

Chemical Name and Synonyms: DIPHENYLAMINE - Acetone low-temperature reaction product

Trade Name and Synonyms: Aminox

Chemical Family: Antioxidant for rubber

Manufacturer: Uniroyal Chemical Co., Naugatuck, Connecticut

UN ID Number: NA

--- INGREDIENTS AND COMPOSITION DATA ---

Material	Cas No.	%	TLV
Product	NA		Not established

--- PHYSICAL DATA ---

Boiling Point (°F): NA

Specific Gravity (H₂O = 1): 1.16

Vapor Pressure (mm Hg): NA

Percent Volatile by Volume (%): Nonvolatile

Vapor Density (air = 1): NA

Freezing Point: NA

Evaporation Rate: NA

pH: Noncorrosive

Solubility: Insoluble in water; soluble in acetone and benzene

Melting Point (°C): 85 min.

Appearance/Color/Odor: Greenish brown powder.

--- FIRE AND EXPLOSION HAZARDS ---

Flash Point: 355°F (TCC)

Flammability Limits: NA

Autoignition Temperature: In excess of normal processing temperatures

Decomposition Temperature: NA

Extinguishing Media: Carbon dioxide, dry chemical, or water.

Special Fire Fighting Procedures: Do not use high-pressure water stream. Airborne dust creates an explosion hazard.

Unusual Fire and Explosion Hazards: Dust may be explosive if mixed with air in critical proportions and in the presence of a source of ignition. The hazard is similar to that of any organic solid including sawdust. Maintain normal good housekeeping for control of dust.

Hazardous Products of Combustion and Decomposition: Unknown.

Hazardous Polymerization: Will not occur.

--- HEALTH HAZARD DATA ---

Effects of Overexposure: None expected; however, manufacturer recommends that user avoid excessive skin contact and observe good personal hygiene.

Emergency Response and First Aid:
Skin: Wash with lots of clean water for at least 15 min.
Eyes: Flush eyes with lots of water. If irritation persists, consult a physician.
Inhalation: Expose to fresh air. Keep warm and quiet. Give artificial respiration if necessary.
Ingestion: None recommended. Consult physician immediately.

Cancer Information: None reported.

--- TOXICITY DATA ---

Animal Effects: No adverse medical history based on over 20 years of production and use. Oral LD50 (rats): >5 g/kg.

--- REACTIVITY DATA ---

Stability and Conditions to Avoid: No specific incompatibility.

Materials to Avoid (Incompatibility): None.

Hazard Decomposition Products: Thermal decomposition may produce carbon monoxide and/or carbon dioxide.

Hazardous Polymerization: Will not occur.

--- SPILL AND EMERGENCY RESPONSE ---

For Spills: Vacuum or sweep up and place into dry, clean, covered container for disposal. Keep this material out of watersheds and waterways.

Waste Disposal Method: No specific recommendations made by manufacturer.

Empty Containers: Burn containers in an approved incinerator or dispose of in an approved chemical landfill in accordance with all applicable local, state, and federal laws and regulations.

--- SPECIAL PROTECTION AND PRECAUTIONS ---

Respiratory Protection: Avoid breathing dust. Dust respirator recommended. Although OSHA and ACGIH have not established specific exposure limits for this material, they have established the following limits for nuisance dusts:

OSHA PEL/8-hr TWA: Total 15 mg/m^3, respirable 5 mg/m^3.
ACGIH TLV/8-hr TWA: Total 10 mg/m^3, respirable 5 mg/m^3.
These limits are stated only to indicate the least stringent airborne dust exposure levels applicable to nuisance dusts.

Type of Ventilation Required: General mechanical. Avoid breathing dusts and vapors. Adequate ventilation should be provided to keep mist and dust concentrations below acceptable exposure limits. Discharge from the ventilation system should comply with applicable air-pollution regulations. Local exhaust ventilation is recommended with a capture velocity of 150 to 200 fpm.

Eye Protection Requirements: Chemical goggles. Eyewash fountains and safety showers should be easily accessible.

Protective Gloves and Other: Wear rubber gloves and goggles.

Storage and Handling Precautions: Containers should be stored in a cool, dry, well-ventilated area. Store away from flammable materials, sources of heat, flame, sparks, and foodstuffs. Exercise due caution to prevent damage to or leakage from the container.

Prevent inhalation of dust. Prevent contact with eyes and skin. Under dusty conditions, static electricity may cause an explosion. Keep containers closed. Store in a cool, dry place.

Use with adequate ventilation. Wash thoroughly after handling before eating and smoking. Work clothing should be frequently laundered and not worn away from work premises.

Other Precautions: Use all standard practices for good personal hygiene. Completely isolate and thoroughly clean all equipment, piping, or vessels before beginning maintenance or repairs. Keep area clean. As with any powder or dustlike material, improper handling can lead to spontaneous combustion (dust explosions).

--- REGULATORY INFORMATION ---

None reported.

▼ ▼ ▼ ▼ ▼ ▼ ▼ ▼ ▼ ▼ ▼ ▼ ▼ ▼

--- IDENTIFIERS ---

Name: DILAURYL BETA-THIODIPROPIONATE
Synonyms: Advastab 800; Antioxidant AS; Antioxidant LTDP; Bis(Dodecyl-oxycarbonylethyl) Sulfide; Carstab DLTDP; Cyanox LTDP; Didodecyl 3,3'-Thiodipropionate; Dilaurylester Kyseliny beta',beta'-Thiodipropionove (Czech); Dilauryl Thiodipropionate; Dilauryl beta-Thiodipropionate; Dilauryl beta',beta'-Thiodipropionate; Dilauryl 3,3'-Thiodipropionate; DLT, DLTDP; DLTP; DMPTP; Ipognox 89; Irganox PS 800; Lauryl 3,3'-Thiodipropionate; Lusmit; Milban F; Neganox DLTP; Plastanox LTDP; Plastanox LTDP Antioxidant; Propanoic Acid, 3,3'-Thio-bis-, Didodecyl Ester; Stabilizer DLT; Thiobis (Dodecyl Proprinate); Tyox B
CAS: 123-28-4; **RTECS:** UF8000000
Formula: $C_{30}H_{58}O_4S$; **Mol Wt:** 514.94

WLN: 12OV2S2VO12
Chemical Class: Ester

See other identifiers listed below under Regulations.

--- PROPERTIES ---

Boiling Point: NA
Melting Point: NA
Flash Point: NA
Autoignition: NA
UEL: NA
LEL: NA
Vapor Density: No data
Specific Gravity: No data

Reactivity with Water: No data on water reactivity
Reactivity with Common Materials: No data
Stability during Transport: No data
Neutralizing Agents: No data
Polymerization Possibilities: No data

Toxic Fire Gases: None reported other than possible unburned vapors
Odor Detected at (ppm): Unknown
Odor Description: No data
100% Odor Detection: No data

--- REGULATIONS ---

DOT hazard class: Not given
Packaging exceptions: 173.
Nonbulk packaging: 173.
Bulk packaging: 173.

STCC Number: Not listed

Clean Water Act Sect. 307: No
Clean Water Act Sect. 311: No
Clean Air Act: Not listed
EPA Waste Number: None
CERCLA Ref: Not listed
RQ Designation: Not listed

SARA TPQ Value: Not listed
SARA Sect. 312 Categories:
Acute toxicity: irritant.

U.S. Postal Service Mailability: Not given

--- TOXICITY DATA ---

Short-term Toxicity: Unknown

Long-term Toxicity: Unknown

Conc IDLH: Unknown

NIOSH REL: Not given

ACGIH TLV: Not listed
ACGIH STEL: Not listed

OSHA PEL: Not in Table Z-1-A

MAK Information: Not listed

Carcinogen: N; **Status:** See below

Carcinogen Lists: *IARC*: Not listed; *MAK*: Not listed; *NIOSH*: Not listed; *NTP*: Not listed; *ACGIH*: Not listed; *OSHA*: Not listed.

LD50 Value: No LD50 in RTECS 1992

Other Species Toxicity Data: (Source: NIOSH RTECS 1992)
unr-mam LD50: >15 g/kg

Reproductive Toxicity (1992 RTECS): This chemical has no known mammalian reproductive toxicity.

--- INITIAL INCIDENT RESPONSE ---

No DOT Guide information for this product.

▼   ▼   ▼   ▼   ▼   ▼   ▼   ▼   ▼   ▼   ▼   ▼   ▼   ▼

--- IDENTIFIERS ---

Chemical Name and Synonyms: *N*-PHENYL-*N*'-CYCLOHEXYL-*p*-PHENYLENEDIAMINE

Trade Name and Synonyms: Flexzone 6H

Chemical Family: Antiozonant/Antioxidant

Manufacturer: Uniroyal Chemical Co., Naugatuck, Connecticut

UN ID Number: NA

--- INGREDIENTS AND COMPOSITION DATA ---

Material	Cas No.	%	TLV
Product	NA		Not established

--- PHYSICAL DATA ---

Boiling Point (°F): NA

Specific Gravity (H_2O = 1): 1.18

Vapor Pressure (mm Hg): NA

Percent Volatile by Volume (%): Nonvolatile

Vapor Density (air = 1): NA

Freezing Point: NA

Evaporation Rate: NA

pH: Noncorrosive

Solubility: Slightly soluble in water and hydrocarbons; soluble in acetone.

Melting Point (°C): 100 min.

Appearance/Color/Odor: Purple powder.

--- FIRE AND EXPLOSION HAZARDS ---

Flash Point: NA

Flammability Limits: NA

Autoignition Temperature: Above normal processing temperatures

Decomposition Temperature: NA

Extinguishing Media: Carbon dioxide, dry chemical, or water.

Special Fire Fighting Procedures: Do not use high-pressure water stream. Airborne dust creates an explosion hazard.

Unusual Fire and Explosion Hazards: Dust may be explosive if mixed with air in critical proportions and in the presence of a source of ignition. The hazard is similar to that of any organic solid including sawdust. Maintain normal good housekeeping for control of dust.

Hazardous Products of Combustion and Decomposition: Unknown

Hazardous Polymerization: Will not occur.

--- HEALTH HAZARD DATA ---

Effects of Overexposure: Some skin irritation potential. Avoid breathing dust or vapors. Avoid skin and eye contact. Wash thoroughly after use.

Emergency Response and First Aid:
Skin: Wash with lots of clean water for at least 15 min.
Eyes: Flush eyes with lots of water. If irritation persists, consult a physician.
Inhalation: Expose to fresh air. Keep warm and quiet. Give artificial respiration if necessary.
Ingestion: None recommended. Consult physician immediately.

Cancer Information: None reported.

--- TOXICITY DATA ---

Animal Effects: Oral LD50 (rats): over 2 g/kg.
Human patch testing and experience have not indicated the material to cause irritation or sensitization. However, as with any phenylenediamine derivative, care should be used in handling. People sensitized to other *p*-phenyldiamines may encounter dermatitis problems.

--- REACTIVITY DATA ---

Stability and Conditions to Avoid: No specific incompatibility.

Materials to Avoid (Incompatibility): None.

Hazard Decomposition Products: Thermal decomposition may produce carbon monoxide and/or carbon dioxide.

Hazardous Polymerization: Will not occur.

--- SPILL AND EMERGENCY RESPONSE ---

For Spills: Vacuum or sweep up and place into dry, clean, covered container for disposal. Keep this material out of watersheds and waterways.

Waste Disposal Method: No specific recommendations made by manufacturer.

Empty Containers: Burn containers in an approved incinerator or dispose of in an approved chemical landfill in accordance with all applicable local, state, and federal laws and regulations.

--- SPECIAL PROTECTION AND PRECAUTIONS ---

Respiratory Protection: Avoid breathing dust. Dust respirator recommended. Although OSHA and ACGIH have not established specific exposure limits for this material, they have established the following limits for nuisance dusts:

OSHA PEL/8-hr TWA: Total 15 mg/m^3, respirable 5 mg/m^3.
ACGIH TLV/8-hr TWA: Total 10 mg/m^3, respirable 5 mg/m^3.
These limits are stated only to indicate the least stringent airborne dust exposure levels applicable to nuisance dusts.

Type of Ventilation Required: General mechanical. Avoid breathing dusts and vapors. Adequate ventilation should be provided to keep mist and dust concentrations below acceptable exposure limits. Discharge from the ventilation system should comply with applicable air-pollution regulations. Local exhaust ventilation is recommended with a capture velocity of 150 to 200 fpm.

Eye Protection Requirements: Chemical goggles. Eyewash fountains and safety showers should be easily accessible.

Protective Gloves and Other: Wear rubber gloves and goggles.

Storage and Handling Precautions: Containers should be stored in a cool, dry, well-ventilated area. Store away from flammable materials, sources of heat, flame, sparks, and foodstuffs. Exercise due caution to prevent damage to or leakage from the container.

Prevent inhalation of dust. Prevent contact with eyes and skin. Under dusty conditions, static electricity may cause an explosion. Keep containers closed. Store in a cool, dry place.

Use with adequate ventilation. Wash thoroughly after handling before eating and smoking. Work clothing should be frequently laundered and not worn away from work premises.

Other Precautions: Use all standard practices for good personal hygiene. Completely isolate and thoroughly clean all equipment, piping, or vessels before beginning maintenance or repairs. Keep area clean. As with any powder or dustlike material, improper handling can lead to spontaneous combustion (dust explosions).

--- REGULATORY INFORMATION ---

None reported.

▼ ▼ ▼ ▼ ▼ ▼ ▼ ▼ ▼ ▼ ▼ ▼ ▼ ▼

--- IDENTIFIERS ---

Chemical Name and Synonyms: TETRAKIS(METHYLENE(3,5-DI-TERT-BUTYL-4-HYDROXYHYDROCINNAMATE)) METHANE

Trade Name and Synonyms: Irganox 1010

Chemical Family: Antioxidant/Stabilizer

Manufacturer: Ciba-Geigy Corp., Hawthorne, New York

UN ID Number: NA

--- INGREDIENTS AND COMPOSITION DATA ---

Material	Cas No.	%	TLV
Product	6683-19-8		Not established

--- PHYSICAL DATA ---

Boiling Point (°F): NA

Specific Gravity (H₂O = 1): 1.15

Vapor Pressure (mm Hg): 1×10^{-12} mm Hg

Percent Volatile by Volume (%): <0.5

Vapor Density (air = 1): NA

Freezing Point: NA

Evaporation Rate: NA

pH: NA

Solubility in Water: 1 ppm

Melting Point (°C): 110 to 125

Appearance/Color/Odor: White to off-white, crystalline powder.

--- FIRE AND EXPLOSION HAZARDS ---

Flash Point: 567°F (Marcusson)

Flammability Limits: NA

Autoignition Temperature: Above normal processing temperatures

Decomposition Temperature: NA

Extinguishing Media: Carbon dioxide, dry chemical, or water.

Special Fire Fighting Procedures: Use self-contained breathing apparatus.

Unusual Fire and Explosion Hazards: The product can form an explosive dust/air mixture. Avoid dust formation and control ignition sources; employ grounding, venting, and explosion-relief provisions in accord with accepted engineering practices in process operations capable of generating dust and/or static electricity.

Hazardous Products of Combustion and Decomposition: Thermal decomposition and burning may produce carbon monoxide and carbon dioxide.

Hazardous Polymerization: Will not occur.

--- HEALTH HAZARD DATA ---

Effects of Overexposure: This material is not intended for use in products for which prolonged contact with mucous membranes or abraded skin or implantation within the human body is specifically intended, unless the finished product has been tested in accordance with the Food and Drug Administration and/or other applicable safety testing requirements.

Primary routes of exposure:
Dermal, ingestion and inhalation
Threshold Limit Value: none established.

Emergency Response and First Aid:
Skin: Wash with lots of clean water for at least 15 min.
Eyes: Flush eyes with lots of water. If irritation persists, consult a physician.
Inhalation: Expose to fresh air. Keep warm and quiet. Give artificial respiration if necessary.
Ingestion: None recommended. Consult physician immediately.

Cancer Information: None reported.

--- TOXICITY DATA ---

Animal Effects:
Oral LD50: (Rats) >5000 mg/kg
Dermal LD50: (Rabbits) >3160 mg/kg
Skin Irritation: (Rabbits) None (Draize score 0/8)
Eye Irritation: (Rabbits) None (Draize score 0/110)
Sensitization:
RIPT (humans): not a primary irritant and no evidence of sensitization under the conditions of the study, which involved testing of a 0.5% W/V solution in dimethyl phthalate.
Inhalation LC50:
(Rats) > 46 mg/l air for a 1-hr dust exposure.
(Rats) exposure for 4 hr to the fumes and vapors emitted when product was heated to 316°C resulted in no deaths or untoward behavioral reactions for an average nominal concentration of 0.11 mg/l air.
Teratogenicity: (Rats and mice) No teratogenic effects.
Mutagenicity:
Ames Test: negative.
Dominant lethal, mice: no evidence of a dominant lethal effect.
Nucleus anomaly test, chinese hamster: negative.
Chromosome studies in somatic cells, Chinese hamster: negative.
Signs and Symptoms of Exposure (Acute): no identified health effects.
Signs and Symptoms of Exposure (Chronic): no identified health effects.
Medical Conditions Aggravated by Exposure: none known.

--- REACTIVITY DATA ---

Stability and Conditions to Avoid: No specific incompatibility.

Materials to Avoid (Incompatibility): Strong oxidizing agents.

Hazard Decomposition Products: Thermal decomposition may produce carbon monoxide and/or carbon dioxide.

Hazardous Polymerization: Will not occur.

--- SPILL AND EMERGENCY RESPONSE ---

For Spills: Sweep or vacuum and place into closable container for disposal. Wear protective equipment specified below. Thoroughly flush residue with water.

Waste Disposal Method: Incinerate in chemical incinerator equipped with an afterburner and scrubber. Follow all federal, state, and local regulations. *Effluent Data-Bod:* Aerobic sewage OECD coupled units test No. 303A: Average elimination measured by specific analysis was 45.2% of the initially measured concentration. *Effluent Data-Cod:* 2.38 G COD/G Irganox 1010. *Sewage Bacterial Toxicity:* Inhibitory concentration on respiration of aerobic waste water bacteria: IC20, IC50, IC80 >100 ppm. *Fish toxicity:* Zebra Fish, LC50 96 H: >100 ppm. *Invertebrate Toxicity:* Daphnia Magna, EC50 24 H: >86 ppm.

Empty Containers: Burn containers in an approved incinerator or dispose of in an approved chemical landfill in accordance with all applicable local, state, and federal laws and regulations.

--- SPECIAL PROTECTION AND PRECAUTIONS ---

Respiratory Protection: Avoid breathing dust. Dust respirator recommended. Although OSHA and ACGIH have not established specific exposure limits for this material, they have established the following limits for nuisance dusts:

OSHA PEL/8-hr TWA: Total 15 mg/m^3, respirable 5 mg/m^3.
ACGIH TLV/8-hr TWA: Total 10 mg/m^3, respirable 5 mg/m^3.
These limits are stated only to indicate the least stringent airborne dust exposure levels applicable to nuisance dusts.

Type of Ventilation Required: General mechanical. Avoid breathing dusts and vapors. Adequate ventilation should be provided to keep mist and dust concentrations below acceptable exposure limits. Discharge from the ventilation system should comply with applicable air-pollution regulations. Local exhaust ventilation is recommended with a capture velocity of 150 to 200 fpm.

Eye Protection Requirements: Chemical goggles. Eyewash fountains and safety showers should be easily accessible.

Protective Gloves and Other: Wear rubber gloves and goggles.

Storage and Handling Precautions: Containers should be stored in a cool, dry, well-ventilated area. Store away from flammable materials, sources of heat, flame, sparks, and foodstuffs. Exercise due caution to prevent damage to or leakage from the container.

Prevent inhalation of dust. Prevent contact with eyes and skin. Under dusty conditions, static electricity may cause an explosion. Keep containers closed. Store in a cool, dry place.

Use with adequate ventilation. Wash thoroughly after handling before eating and smoking. Work clothing should be frequently laundered and not worn away from work premises.

Other Precautions: Use all standard practices for good personal hygiene. Completely isolate and thoroughly clean all equipment, piping, or vessels before beginning maintenance or repairs. Keep area clean. As with any powder or dustlike material, improper handling can lead to spontaneous combustion (dust explosions).

--- REGULATORY INFORMATION ---

DOT Proper Shipping Name: Not regulated as a hazardous material by the U.S. Department of Transportation (DOT) 49 CFR 172.101 Hazardous Materials Table.
DOT Class: None.
DOT Number: None.
RCRA Status: Not a hazardous waste under RCRA (40 CFR 261).
CERCLA Status: Not listed.
SARA/Title III--Toxic Chemicals List: This product does not contain a toxic chemical for routine annual 'Toxic Chemical Reporting' under Sec. 313 (40 CFR 372).
TSCA Inventory Status: Chemical components listed on TSCA inventory.
California Proposition 65:   This product does not contain any chemicals currently on the California List of known carcinogens and reproductive toxins.
New Jersey Right-to-Know Labeling Information: This product contains the following:

Chemical name: Tetrakis(Methylene(3,5-Di-Tert-Butyl-4-Hydroxyhydro-cinnamate)) Methane
Cas number: 6683-19-8

Chemical name: Pentaerythritol Tris Ester with 3-(3,5-Di-Tert-Butyl-4-Hydroxyphenyl)Propionic Acid
Cas number: 84633-54-5
Percent: Impurity

Pennsylvania Right-to-Know Act: The following is required composition information.

Chemical name: Benzenepropanoic Acid, 3,5-Bis(1,1-Dimethylethyl)-4-Hydroxy-, 2,2-Bis((3-(3,5-Bis(1,1-Dimethylethyl)-4-Hydroxyphenyl)-1-Oxopropoxy)Methyl)-1,3-Propanediyl Ester
Cas number: 6683-19-8
Common name: Tetrakis(Methylene(3,5-Di-Tert-Butyl-4-Hydroxyhydrocin-namate))Methane; 3,5-Di-Tert-Butyl-4-Hydroxyhydrocinnamic Acid, Neopentanetetrayl Ester; P Entaerythritol Tetrakis(3,5-Di-Tert-Butyl-4-Hydroxyhydrocinnamate); Irganox 1010
Comments: Not on Pennsylvania Hazardous Substance List.

▼ ▼ ▼ ▼ ▼ ▼ ▼ ▼ ▼ ▼ ▼ ▼ ▼ ▼

--- IDENTIFIERS ---

Chemical Name and Synonyms: HINDERED PHENOLIC ANTIOXIDANT

Trade Name and Synonyms: Irganox 1035

Chemical Family: Hindered Phenolic Antioxidant

Manufacturer: Ciba-Geigy Corp., Hawthorne, New York

UN ID Number: NA

--- INGREDIENTS AND COMPOSITION DATA ---

Material	Cas No.	%	TLV
Product	NA		Not established

--- PHYSICAL DATA ---

Boiling Point (°F): 147 min.

Specific Gravity (H$_2$O = 1): NA

Vapor Pressure (mm Hg): NA

Percent Volatile by Volume (%): NA

Vapor Density (air = 1): NA

Freezing Point: NA

Evaporation Rate: NA

pH: NA

Solubility: 0.001% at 22°C

Melting Point (°C): NA

Appearance/Color/Odor: Crystalline, free-flowing powder; odorless.

--- FIRE AND EXPLOSION HAZARDS ---

Flash Point: NA

Flammability Limits: NA

Autoignition Temperature: Above normal processing temperatures

Decomposition Temperature: NA

Extinguishing Media: Carbon dioxide, dry chemical, or water.

Special Fire Fighting Procedures: None recommended by manufacturer.

Unusual Fire and Explosion Hazards: None recommended by manufacturer.

Hazardous Products of Combustion and Decomposition: Thermal decomposition and burning may produce carbon monoxide and carbon dioxide.

Hazardous Polymerization: Will not occur.

--- HEALTH HAZARD DATA ---

Effects of Overexposure: No effects reported by manufacturer; however, should be treated as any nuisance dust.

Primary routes of exposure:
Dermal, ingestion, and inhalation
Threshold Limit Value: none established.

Emergency Response and First Aid:
Skin: Wash with lots of clean water for at least 15 min.
Eyes: Flush eyes with lots of water. If irritation persists, consult a physician.
Inhalation: Expose to fresh air. Keep warm and quiet. Give artificial respiration if necessary.
Ingestion: None recommended. Consult physician immediately.

Cancer Information: None reported.

--- TOXICITY DATA ---

Animal Effects: Acute oral LD50 in (male and female) mice 4556 mg/kg; not a primary irritant, fatiguing agent, or sensitizer in human test subjects.

--- REACTIVITY DATA ---

Stability and Conditions to Avoid: No specific incompatibility.

Materials to Avoid (Incompatibility): Strong oxidizing agents.

Hazard Decomposition Products: Thermal decomposition may produce carbon monoxide and/or carbon dioxide.

Hazardous Polymerization: Will not occur.

--- SPILL AND EMERGENCY RESPONSE ---

For Spills: Sweep or vacuum and place into closable container for disposal. Wear protective equipment specified below. Thoroughly flush residue with water.

Waste Disposal Method: Normal plant water, bury, or incinerate.

Empty Containers: Burn containers in an approved incinerator or dispose of in an approved chemical landfill in accordance with all applicable local, state, and federal laws and regulations.

--- SPECIAL PROTECTION AND PRECAUTIONS ---

Respiratory Protection: Avoid breathing dust. Dust respirator recommended. Although OSHA and ACGIH have not established specific exposure limits for this material, they have established the following limits for nuisance dusts:

OSHA PEL/8-hr TWA: Total 15 mg/m³, respirable 5 mg/m³.
ACGIH TLV/8-hr TWA: Total 10 mg/m³, respirable 5 mg/m³.
These limits are stated only to indicate the least stringent airborne dust exposure levels applicable to nuisance dusts.

Type of Ventilation Required: General mechanical. Avoid breathing dusts and vapors. Adequate ventilation should be provided to keep mist and dust concentrations below acceptable exposure limits. Discharge from the ventilation system should comply with applicable air-pollution regulations. Local exhaust ventilation is recommended with a capture velocity of 150 to 200 fpm.

Eye Protection Requirements: Chemical goggles. Eyewash fountains and safety showers should be easily accessible.

Protective Gloves and Other: Wear rubber gloves and goggles.

Storage and Handling Precautions: Containers should be stored in a cool, dry, well-ventilated area. Store away from flammable materials, sources of heat, flame, sparks, and foodstuffs. Exercise due caution to prevent damage to or leakage from the container.

Prevent inhalation of dust. Prevent contact with eyes and skin. Under dusty conditions, static electricity may cause an explosion. Keep containers closed. Store in a cool, dry place.

Use with adequate ventilation. Wash thoroughly after handling before eating and smoking. Work clothing should be frequently laundered and not worn away from work premises.

Other Precautions: Use all standard practices for good personal hygiene. Completely isolate and thoroughly clean all equipment, piping, or vessels before beginning maintenance or repairs. Keep area clean. As with any powder or dustlike material, improper handling can lead to spontaneous combustion (dust explosions).

--- REGULATORY INFORMATION ---

None reported.

▼   ▼   ▼   ▼   ▼   ▼   ▼   ▼   ▼   ▼   ▼   ▼   ▼   ▼

--- IDENTIFIERS ---

Chemical Name and Synonyms: OCTADECYL 3,5-DI-TERT-BUTYL-4-HYDROXYHYDROCINNAMATE

Trade Name and Synonyms: Irganox 1076

Chemical Family: Rubber Antioxidant

Manufacturer: Ciba-Geigy Corp., Hawthorne, New York

UN ID Number: NA

--- INGREDIENTS AND COMPOSITION DATA ---

Material	Cas No.	%	TLV
Product	2082-79-3		Not established

--- PHYSICAL DATA ---

Boiling Point (°F): NA

Specific Gravity (H$_2$O = 1): 1.02

Vapor Pressure (mm Hg): 2 x 10^{-9} mm Hg at 20°C

Percent Volatile by Volume (%): <1

Vapor Density (air = 1): NA

Freezing Point: NA

Evaporation Rate: NA

pH: NA

Solubility: Insoluble

Melting Point (°C): 50 to 55

Appearance/Color/Odor: White to off-white crystalline powder.

--- FIRE AND EXPLOSION HAZARDS ---

Flash Point: 523°F (Marcusson)

Flammability Limits: NA

Autoignition Temperature: Above normal processing temperatures

Decomposition Temperature: NA

Extinguishing Media: Carbon dioxide, dry chemical, or water spray.

Special Fire Fighting Procedures: Use self-contained breathing apparatus.

Unusual Fire and Explosion Hazards: Decomposition and combustion products may be toxic.

Hazardous Products of Combustion and Decomposition: Thermal decomposition and burning may produce carbon monoxide and carbon dioxide.

Hazardous Polymerization: Will not occur.

--- HEALTH HAZARD DATA ---

Effects of Overexposure: This material will not be sold for use in products for which prolonged contact with mucous membranes or abraded skin or implantation within the human body is specifically intended.

Primary routes of exposure:
Dermal, ingestion, and inhalation
Threshold Limit Value: none established.

Emergency Response and First Aid:
Skin: Wash with lots of clean water for at least 15 min.
Eyes: Flush eyes with lots of water. If irritation persists, consult a physician.

Inhalation: Expose to fresh air. Keep warm and quiet. Give artificial respiration if necessary.
Ingestion: None recommended. Consult physician immediately.

Cancer Information: None reported.

--- TOXICITY DATA ---

Animal Effects:
Primary Routes of Exposure: dermal, ingestion, and inhalation.
Oral LD50: (Rats) >10,000 mg/kg
Intraperitoneal LD50: (Rats) >1000 mg/kg
Dermal LD50: (Rabbits) >2000 mg/kg
Skin Irritation: (Rabbits) Minimal (Draize score 0.95/8)
Phototoxicity: negative.
Eye Irritation: (Rabbits) Minimal (Draize score 4/110)
Sensitization:
(Guinea pigs) Considered to possess no skin sensitizing potential.
RIPT (humans) in 4 separate studies, a total of 3 of 183 subjects exhibited reactions indicative of sensitization; concentrations ranged from 25% in petrolatum (25 subjects), 0.5% in dimethyl phthalate (58 subjects), to neat material (100 subjects).
Inhalation LC50:
(Rats) >1.8 mg/l air for a 4-hr dust exposure with approximately 90% of particles <7 mm diameter.
Teratogenicity: (Rats and mice) No teratogenic effect seen.
Mutagenicity:
Ames Test: negative.
Dominant lethal test (mouse): no evidence of a dominant lethal effect.
Signs and Symptoms of Exposure (Acute): *Caution!!* May cause irritation to skin, especially after prolonged or repeated exposure, and especially in body areas of heavy perspiration, based upon use experience.
Signs and Symptoms of Exposure (Chronic): no identified health effects.

--- REACTIVITY DATA ---

Stability and Conditions to Avoid: No specific incompatibility.

Materials to Avoid (Incompatibility): Strong oxidizing agents.

Hazard Decomposition Products: Thermal decomposition may produce carbon monoxide and/or carbon dioxide.

Hazardous Polymerization: Will not occur.

--- SPILL AND EMERGENCY RESPONSE ---

For Spills: Prewet material with water to avoid dust formation. Sweep or vacuum and place in closable container for disposal. Wear protective equipment specified below. Flush residue with water.

Waste Disposal Method: Dispose in accordance with federal, state, and local regulations. *Effluent Data-BOD:* (Sturn test) Partially biodegradable. *Fish toxicity:* Bluegill LC50 (96 hr): >100 ppm. *Invertebrate toxicity:* Daphnia Magna EC50 (24 hr): >100 ppm.

Empty Containers: Burn containers in an approved incinerator or dispose of in an approved chemical landfill in accordance with all applicable local, state, and federal laws and regulations.

--- SPECIAL PROTECTION AND PRECAUTIONS ---

Respiratory Protection: Avoid breathing dust. Dust respirator recommended. Although OSHA and ACGIH have not established specific exposure limits for this material, they have established the following limits for nuisance dusts:

OSHA PEL/8-hr TWA: Total 15 mg/m^3, respirable 5 mg/m^3.
ACGIH TLV/8-hr TWA: Total 10 mg/m^3, respirable 5 mg/m^3.
These limits are stated only to indicate the least stringent airborne dust exposure levels applicable to nuisance dusts.

Type of Ventilation Required: General mechanical. Avoid breathing dusts and vapors. Adequate ventilation should be provided to keep mist and dust concentrations below acceptable exposure limits. Discharge from the ventilation system should comply with applicable air-pollution regulations. Local exhaust ventilation is recommended with a capture velocity of 150 to 200 fpm.

Eye Protection Requirements: Chemical goggles. Eyewash fountains and safety showers should be easily accessible.

Protective Gloves and Other: Wear rubber gloves and goggles.

Storage and Handling Precautions: Containers should be stored in a cool, dry, well-ventilated area. Store away from flammable materials, sources of heat, flame, sparks, and foodstuffs. Exercise due caution to prevent damage to or leakage from the container.

Prevent inhalation of dust. Prevent contact with eyes and skin. Under dusty conditions, static electricity may cause an explosion. Keep containers closed. Store in a cool, dry place.

Use with adequate ventilation. Wash thoroughly after handling before eating and smoking. Work clothing should be frequently laundered and not worn away from work premises.

Other Precautions: Use all standard practices for good personal hygiene. Completely isolate and thoroughly clean all equipment, piping, or vessels before beginning maintenance or repairs. Keep area clean. As with any powder or dustlike material, improper handling can lead to spontaneous combustion (dust explosions).

--- REGULATORY INFORMATION ---

DOT Proper Shipping Name: Not regulated as a hazardous material by the U.S. Department of Transportation (DOT) 49 CFR 172.101 Hazardous Materials Table.
DOT Class: None.
DOT Number: None.
RCRA Status: Not a hazardous waste under RCRA (40 CFR 261).
CERCLA Status: Not listed.
TSCA Inventory Status: Chemical components listed on TSCA inventory.

▼ ▼ ▼ ▼ ▼ ▼ ▼ ▼ ▼ ▼ ▼ ▼ ▼ ▼

--- IDENTIFIERS ---

Chemical Name and Synonyms: TRIS(3,5-DI-TERT-BUTYL-4-HYDROXYBENZYL)ISOCYANURATE

Trade Name and Synonyms: Irganox 3114

Chemical Family: Rubber Antioxidant

Manufacturer: Ciba-Geigy Corp., Hawthorne, New York

UN ID Number: NA

--- INGREDIENTS AND COMPOSITION DATA ---

Material	Cas No.	%	TLV
Product	27676-62-6		Not established

--- PHYSICAL DATA ---

Boiling Point (°F): NA

Specific Gravity (H₂O = 1): 1.03

Vapor Pressure (mm Hg): Not determined, but very low

Percent Volatile by Volume (%): <0.01

Vapor Density (air = 1): NA

Freezing Point: NA

Evaporation Rate: NA

pH: NA

Solubility: Insoluble

Melting Point (°C): 220

Appearance/Color/Odor: Fine white crystalline powder with no discernable odor.

--- FIRE AND EXPLOSION HAZARDS ---

Flash Point: 552°F (Cleveland closed cup)

Flammability Limits: NA

Autoignition Temperature: Above normal processing temperatures

Decomposition Temperature: NA

Extinguishing Media: Carbon dioxide, dry chemical, or water spray.

Special Fire Fighting Procedures: Use self-contained breathing apparatus.

Unusual Fire and Explosion Hazards: Decomposition and combustion products may be toxic. When this product was heated to various temperatures and rats exposed for 4-hr to the emitted fumes and vapors, the following results were observed:

1. Exposure Temperature 316°C: Adverse body weight gains

2. Exposure Temperature 316°C: 100% Mortality after 1 day

3. Exposure Temperature 288°C: No reactions

The manufacturer notes that the differing results at 316°C cannot be explained.

Hazardous Products of Combustion and Decomposition: Thermal decomposition and burning may produce carbon monoxide and carbon dioxide, nitrogen oxides, and small amounts of aromatic and aliphatic hydrocarbons.

Hazardous Polymerization: Will not occur.

--- HEALTH HAZARD DATA ---

Effects of Overexposure: This material will not be sold for use in products for which prolonged contact with mucous membranes or abraded skin or implantation within the human body is specifically intended. There are no reported medical conditions aggravated by exposure.

Primary routes of exposure:
  Dermal, ingestion, and inhalation
  Threshold Limit Value: none established.

Emergency Response and First Aid:
Skin: Wash with lots of clean water for at least 15 min.
Eyes: Flush eyes with lots of water. If irritation persists, consult a physician.
Inhalation: Expose to fresh air. Keep warm and quiet. Give artificial respiration if necessary.
Ingestion: None recommended. Consult physician immediately.

Cancer Information: None reported.

--- **TOXICITY DATA** ---

Animal Effects:
Oral LD50: (Rats) >6800 mg/kg; >5000 mg/kg
Dermal LD50: (Rabbits) >10,000 mg/kg
Skin Irritation: (Rabbits) Slight erythema, which cleared after 24 hr (OECD 404)
Eye Irritation: (Rabbits) (1) Nonirritant; (Rabbits) (2) Slight irritation, which cleared after 7 days (OECD 405)
Sensitization:
(Guinea pigs) No animal sensitized in maximization test.
RIPT (humans) not a primary irritant, fatiguing agent, or sensitizer in any of 200 individuals. Tested as a slurry (concentration not specified).
Inhalation LC50:
(Rats) >66.6 mg/l air for a 1-hr dust exposure
Mutagenicity:
Ames Test: negative.

--- **REACTIVITY DATA** ---

Stability and Conditions to Avoid: No specific incompatibility.

Materials to Avoid (Incompatibility): Strong oxidizing agents.

Hazard Decomposition Products: Thermal decomposition may produce carbon monoxide and/or carbon dioxide.

Hazardous Polymerization: Will not occur.

--- **SPILL AND EMERGENCY RESPONSE** ---

For Spills: Prewet material with water to avoid dust formation. Sweep or vacuum and place in closable container for disposal. Wear protective equipment specified below. Flush residue with water.

Waste Disposal Method: Dispose in accordance with federal, state, and local regulations. *Effluent Data-BOD:* (Sturm test) Partially biodegradable. *Fish toxicity:* Bluegill LC50 (96 hr): >100 ppm. *Invertebrate toxicity:* Daphnia Magna EC50 (24 hr): >100 ppm.

Empty Containers: Burn containers in an approved incinerator or dispose of in an approved chemical landfill in accordance with all applicable local, state, and federal laws and regulations.

--- SPECIAL PROTECTION AND PRECAUTIONS ---

Respiratory Protection: Avoid breathing dust. Dust respirator recommended. Although OSHA and ACGIH have not established specific exposure limits for this material, they have established the following limits for nuisance dusts:

OSHA PEL/8-hr TWA: Total 15 mg/m^3, respirable 5 mg/m^3.
ACGIH TLV/8-hr TWA: Total 10 mg/m^3, respirable 5 mg/m^3.
These limits are stated only to indicate the least stringent airborne dust exposure levels applicable to nuisance dusts.

Type of Ventilation Required: General mechanical. Avoid breathing dusts and vapors. Adequate ventilation should be provided to keep mist and dust concentrations below acceptable exposure limits. Discharge from the ventilation system should comply with applicable air-pollution regulations. Local exhaust ventilation is recommended with a capture velocity of 150 to 200 fpm.

Eye Protection Requirements: Chemical goggles. Eyewash fountains and safety showers should be easily accessible.

Protective Gloves and Other: Wear rubber gloves and goggles.

Storage and Handling Precautions: Containers should be stored in a cool, dry, well-ventilated area. Store away from flammable materials, sources of heat, flame, sparks, and foodstuffs. Exercise due caution to prevent damage to or leakage from the container.

Prevent inhalation of dust. Prevent contact with eyes and skin. Under dusty conditions, static electricity may cause an explosion. Keep containers closed. Store in a cool, dry place.

Use with adequate ventilation. Wash thoroughly after handling before eating and smoking. Work clothing should be frequently laundered and not worn away from work premises.

Other Precautions: Use all standard practices for good personal hygiene. Completely isolate and thoroughly clean all equipment, piping, or vessels

before beginning maintenance or repairs. Keep area clean. As with any powder or dustlike material, improper handling can lead to spontaneous combustion (dust explosions).

--- REGULATORY INFORMATION ---

DOT Proper Shipping Name: Not regulated as a hazardous material by the U.S. Department of Transportation (DOT) 49 CFR 172.101 Hazardous Materials Table.
DOT Class: None.
DOT Number: None.
RCRA Status: Not a hazardous waste under RCRA (40 CFR 261).
CERCLA Status: Not listed.
TSCA Inventory Status: Chemical components listed on TSCA inventory.

▼   ▼   ▼   ▼   ▼   ▼   ▼   ▼   ▼   ▼   ▼   ▼   ▼   ▼

--- IDENTIFIERS ---

Name: *N,N'*-DIPHENYL-*p*-PHENYLENEDIAMINE
Synonyms: Agerite; Agerite DPPD; 1,4-Bis(Phenylamino)Benzene; *N,N'*-Difenyl-*p*-Fenylendiamin (Czech); Diphenyl-*p*-Phenylenediamine; *N,N'*-Diphenyl-*p*-Phenylenediamine; DPPD; Flexamine G; JZF; Nonox DPPD; *p*-Phenylaminodiphenylamine; 4-Phenylaminodiphenylamine; USAF GY-2
CAS: 74-31-7; **RTECS:** ST2275000
Formula: $C_{18}H_{16}N_2$; **Mol Wt:** 260.36
WLN: RMR DMR
Chemical Class: Aromatic amine

See other identifiers listed below under Regulations.

--- PROPERTIES ---

Boiling Point: -498.16 K at 0.5 mm; 0 to 225°C at 0.5 mm; 8 to 437°F at 0.5 mm
Melting Point: 419.16 to 421.16 K; 146 to 148°C; 294.8 to 298.4°F
Flash Point: NA
Autoignition: NA
UEL: NA
LEL: NA
Vapor Density: No data
Specific Gravity: No data

Reactivity with Water: No data on water reactivity
Reactivity with Common Materials: No data
Stability during Transport: No data
Neutralizing Agents: No data
Polymerization Possibilities: No data

Toxic Fire Gases: None reported other than possible unburned vapors
Odor Detected at (ppm): Unknown
Odor Description: No data
100% Odor Detection: No data

--- REGULATIONS ---

DOT hazard class: Not given
Packaging exceptions: 173.
Nonbulk packaging: 173.
Bulk packaging: 173.

STCC Number: Not listed

Clean Water Act Sect. 307: No
Clean Water Act Sect. 311: No
Clean Air Act: Not listed
EPA Waste Number: None
CERCLA Ref: Not listed
RQ Designation: Not listed
SARA TPQ Value: Not listed
SARA Sect. 312 Categories:
  Acute toxicity: irritant.
  Chronic toxicity: mutagen.
  Chronic toxicity: reproductive toxin.

U.S. Postal Service Mailability: Not given

--- TOXICITY DATA ---

Short-term Toxicity: Unknown

Long-term Toxicity: Unknown

Conc IDLH: Unknown

NIOSH REL: Not given

ACGIH TLV: Not listed
ACGIH STEL: Not listed

OSHA PEL: Not in Table Z-1-A

MAK Information: Not listed

Carcinogen: N; **Status:** See below

Carcinogen Lists: *IARC*: Not listed; *MAK*: Not listed; *NIOSH*: Not listed; *NTP*: Not listed; *ACGIH*: Not listed; *OSHA*: Not listed.

LD50 Value: orl-rat LD50:2370 mg/kg

Other Species Toxicity Data: (Source: NIOSH RTECS 1992)
orl-rat LD50:2370 mg/kg
orl-mus LD50:18 g/kg
ipr-mus LD50:300 mg/kg

Reproductive Toxicity (1992 RTECS): This chemical is a mammalian reproductive toxin.

Reproductive Toxicity Data (1992 RTECS):
orl-rat TDLo:450 mg/kg (14D pre/1-22D preg) JAFCAU 4, 796, 56
Maternal Effects
Parturition
Effects on Newborn
Stillbirth

orl-rat TDLo:2500 mg/kg (1-22D preg) AJANA2, 110, 29, 62
Effects on Fertility
Postimplantation mortality

--- INITIAL INCIDENT RESPONSE ---

No DOT Guide information for this product.

▼   ▼   ▼   ▼   ▼   ▼   ▼   ▼   ▼   ▼   ▼   ▼   ▼   ▼

--- IDENTIFIERS ---

Chemical Name and Synonyms: OCTYLATED DIPHENYLAMINE AND N,N'-DIPHENYL-p-PHENYLENEDIAMINE

Trade Name and Synonyms: Agerite HP-S

Chemical Family: Rubber Antioxidant

Manufacturer: R. T. Vanderbilt Co., Inc., Norwalk, Connecticut

UN ID Number: NA

--- INGREDIENTS AND COMPOSITION DATA ---

Material	Cas No.	%	TLV
Octylated diphenlamine	68411-46-1		Not established
N,N'-Diphenyl-p-phenylenediamine	74-31-7		Not established

--- PHYSICAL DATA ---

Boiling Point (°F): NA

Specific Gravity (H₂O = 1): 1.11

Vapor Pressure (mm Hg): Not determined, but very low

Percent Volatile by Weight (%): 1.0 max.

Vapor Density (air = 1): NA

Freezing Point: NA

Evaporation Rate: NA

pH: NA

Solubility: Negligible in water; soluble in acetone, toluene, chloroform, carbon disulfide; slightly soluble in gasoline.

Melting Point (°C): 80 to 100

Appearance/Color/Odor: Tan to brown powder with slight phenolic odor.

--- FIRE AND EXPLOSION HAZARDS ---

Flash Point: NA

Flammability Limits: NA

Autoignition Temperature: Above normal processing temperatures

Decomposition Temperature: NA

Extinguishing Media: Carbon dioxide, dry chemical, or water spray. Carbon dioxide may be ineffective on larger fires due to lack of cooling capacity, which may result in reignition.

Special Fire Fighting Procedures: Use positive-pressure SCBA (self-contained breathing apparatus). In enclosed or poorly ventilated areas, use SCBA during cleanup immediately after a fire and during fire fighting.

Unusual Fire and Explosion Hazards: Decomposition and combustion products may be toxic. Dust suspended in the air in critical proportions and in the presence of an ignition source may present an explosion hazard.

Hazardous Products of Combustion and Decomposition: Thermal decomposition and burning may produce carbon monoxide and carbon dioxide, nitrogen oxides, and small amounts of aromatic and aliphatic hydrocarbons.

Hazardous Polymerization: Will not occur.

--- HEALTH HAZARD DATA ---

Effects of Overexposure: None expected for product. Heating may generate vapors, which can irritate the eyes and respiratory passages.

Primary routes of exposure:
Dermal, ingestion, and inhalation.
Threshold Limit Value: none established.

Emergency Response and First Aid:
Skin: Wash with lots of clean water for at least 15 min.
Eyes: Flush eyes with lots of water. If irritation persists, consult a physician.
Inhalation: Expose to fresh air. Keep warm and quiet. Give artificial respiration if necessary.
Ingestion: None recommended. Consult physician immediately.

Cancer Information: None reported.

--- TOXICITY DATA ---

Animal Effects:
For Octylated Diphenylamine: Acute oral LD50 >7000 mg/kg in rats.
For *N,N'*-Diphenyl-*p*-phenylenediamine: Acute oral LD50 ca. 10,000 mg/kg in rats.

--- REACTIVITY DATA ---

Stability and Conditions to Avoid: This product is described as being thermally stable.

Materials to Avoid (Incompatibility): Strong oxidizing agents such as hydrogen peroxide, permanganates, and perchlorates.

Hazard Decomposition Products: Thermal decomposition may produce carbon monoxide and/or carbon dioxide, small amounts of aromatic and aliphatic hydrocarbons, as well as oxides of nitrogen.

Hazardous Polymerization: Will not occur.

--- SPILL AND EMERGENCY RESPONSE ---

For Spills: Prewet material with water to avoid dust formation. Sweep or vacuum and place in closable container for disposal. Wear protective equipment specified below. Flush residue with water.

Waste Disposal Method: Dispose in accordance with federal, state, and local regulations. *Effluent Data-BOD:* (Sturm test) Partially biodegradable. *Fish toxicity:* Bluegill LC50 (96 hr): >100 ppm. *Invertebrate toxicity:* Daphnia Magna EC50 (24 hr): >100 ppm.

Empty Containers: Burn containers in an approved incinerator or dispose of in an approved chemical landfill in accordance with all applicable local, state, and federal laws and regulations.

--- SPECIAL PROTECTION AND PRECAUTIONS ---

Respiratory Protection: Avoid breathing dust. Dust respirator recommended. Although OSHA and ACGIH have not established specific exposure limits for this material, they have established the following limits for nuisance dusts:

OSHA PEL/8-hr TWA: Total 15 mg/m^3, respirable 5 mg/m^3.
ACGIH TLV/8-hr TWA: Total 10 mg/m^3, respirable 5 mg/m^3.
These limits are stated only to indicate the least stringent airborne dust exposure levels applicable to nuisance dusts.

Type of Ventilation Required: General mechanical. Avoid breathing dusts and vapors. Adequate ventilation should be provided to keep mist and dust concentrations below acceptable exposure limits. Discharge from the ventilation system should comply with applicable air-pollution regulations. Local exhaust ventilation is recommended with a capture velocity of 150 to 200 fpm.

Eye Protection Requirements: Chemical goggles. Eyewash fountains and safety showers should be easily accessible.

Protective Gloves and Other: Wear rubber gloves and goggles.

Storage and Handling Precautions: Containers should be stored in a cool, dry, well-ventilated area. Store away from flammable materials, sources of heat, flame, sparks, and foodstuffs. Exercise due caution to prevent damage to or leakage from the container.

Prevent inhalation of dust. Prevent contact with eyes and skin. Under dusty conditions, static electricity may cause an explosion. Keep containers closed. Store in a cool, dry place.

Use with adequate ventilation. Wash thoroughly after handling before eating and smoking. Work clothing should be frequently laundered and not worn away from work premises.

Other Precautions: Use all standard practices for good personal hygiene. Completely isolate and thoroughly clean all equipment, piping, or vessels

before beginning maintenance or repairs. Keep area clean. As with any powder or dustlike material, improper handling can lead to spontaneous combustion (dust explosions).

--- REGULATORY INFORMATION ---

DOT Proper Shipping Name: Not regulated as a hazardous material by the U.S. Department of Transportation (DOT) 49 CFR 172.101 Hazardous Materials Table.
DOT Class: None.
DOT Number: None.
RCRA Status: Not a hazardous waste under RCRA (40 CFR 261).
CERCLA Status: Not listed.
TSCA Inventory Status: Chemical components listed on TSCA inventory.

▼   ▼   ▼   ▼   ▼   ▼   ▼   ▼   ▼   ▼   ▼   ▼   ▼   ▼

--- IDENTIFIERS ---

Chemical Name and Synonyms: MIXTURE OF OCTYLATED DIPHENYLAMINES

Trade Name and Synonyms: Agerite* Stalite* S

Chemical Family: NA

Manufacturer: R. T. Vanderbilt Co., Inc., Norwalk, Connecticut

UN ID Number: NA

--- INGREDIENTS AND COMPOSITION DATA ---

Material	Cas No.	%	TLV
Agerite	68411-46-1		Not established

--- PHYSICAL DATA ---

Boiling Point (°F): NA

Specific Gravity (H$_2$O = 1): 1.02

Vapor Pressure (mm Hg): NA

Percent Volatile by Weight (%): 1.5 max.

Vapor Density (air = 1): NA

Freezing Point: NA

Evaporation Rate: NA

pH: NA

Solubility: Insoluble in water; soluble in toluene, alcohol, and gasoline.

Melting Point (°C): NA

Appearance/Color/Odor: Light tan to brown friable powder, slight amine odor.

--- FIRE AND EXPLOSION HAZARDS ---

Flash Point: 213°C (COC)

Flammability Limits: NA

Autoignition Temperature: Above normal processing temperatures

Decomposition Temperature: NA

Extinguishing Media: Foam, dry chemical, water. Carbon dioxide may be ineffective on large fires due to lack of cooling capacity, which may result in reignition.

Special Fire Fighting Procedures: Use self-contained breathing apparatus (SCBA). In enclosed or poorly ventilated areas, wear SCBA during cleanup immediately after a fire and during fire fighting.

Unusual Fire and Explosion Hazards: None known.

Hazardous Products of Combustion and Decomposition: Oxides of nitrogen and carbon, small amounts of aromatic and aliphatic hydrocarbons.

Hazardous Polymerization: Will not occur.

--- HEALTH HAZARD DATA ---

Effects of Overexposure: None expected for product. Heating may generate vapors, which can irritate the eyes and respiratory passages.
For Diphenylamines: In general, may be irritating to the mucous membranes. Heating may generate vapors, which can irritate eyes and the respiratory passages.
Primary Routes of Exposure: Dermal, ingestion, and inhalation.
Threshold Limit Value: None established.

Emergency Response and First Aid:
Skin: Wash with lots of clean water for at least 15 min.
Eyes: Flush eyes with lots of water. If irritation persists, consult a physician.
Inhalation: Expose to fresh air. Keep warm and quiet. Give artificial respiration if necessary.
Ingestion: None recommended. Consult physician immediately.

Cancer Information: None reported.

--- TOXICITY DATA ---

Animal Effects:
For Residual Diphenylamines (*1.0% max.):
   OSHA TWA: 10 mg/m^3
   ACGIH TLV-TWA: 10 mg/m^3
Acute Oral LD50 (rats): 7580 mg/kg

Ames Test: Negative

Thirteen-Week Feeding Studies in Rats and Dogs:
   Daily administration in the diet of rats produced some toxic effects at levels as low as 1500 ppm. In dogs, no discernible toxic effects at 1500 ppm were reported.

   High levels of this material in the diet caused degenerative changes in the liver, kidney, adrenal medulla, and thyroid of rats and a decrease in blood prothrombin content in the rats and dogs.

Medical Conditions Generally Aggravated By Exposure: Unknown.

--- REACTIVITY DATA ---

Stability and Conditions to Avoid: This product is described as being thermally stable.

Materials to Avoid (Incompatibility): Strong oxidizing agents such as hydrogen peroxide, permanganates, and perchlorates.

Hazard Decomposition Products: Thermal decomposition may produce carbon monoxide and/or carbon dioxide, small amounts of aromatic and aliphatic hydrocarbons, as well as oxides of nitrogen.

Hazardous Polymerization: Will not occur.

--- SPILL AND EMERGENCY RESPONSE ---

For Spills: Vacuum or sweep into a closed container for disposal. Avoid dust generation. Do not flush chemical into public sewer or water system.

Waste Disposal Method: Not classified as a RCRA hazardous waste. Dispose of according to applicable environmental regulations.

Empty Containers: Burn containers in an approved incinerator or dispose of in an approved chemical landfill in accordance with all applicable local, state, and federal laws and regulations.

--- SPECIAL PROTECTION AND PRECAUTIONS ---

Respiratory Protection: Avoid breathing dust. Dust respirator recommended. Although OSHA and ACGIH have not established specific exposure limits for this material, they have established the following limits for nuisance dusts:

OSHA PEL/8-hr TWA: Total 15 mg/m^3, respirable 5 mg/m^3.
ACGIH TLV/8-hr TWA: Total 10 mg/m^3, respirable 5 mg/m^3.
These limits are stated only to indicate the least stringent airborne dust exposure levels applicable to nuisance dusts.

Type of Ventilation Required: General mechanical. Avoid breathing dusts and vapors. Adequate ventilation should be provided to keep mist and dust concentrations below acceptable exposure limits. Discharge from the ventilation system should comply with applicable air-pollution regulations.

Local exhaust ventilation is recommended with a capture velocity of 150 to 200 fpm.

Eye Protection Requirements: Chemical goggles. Eyewash fountains and safety showers should be easily accessible.

Protective Gloves and Other: Wear rubber gloves and goggles.

Storage and Handling Precautions: Containers should be stored in a cool, dry, well-ventilated area. Store away from flammable materials, sources of heat, flame, sparks, and foodstuffs. Exercise due caution to prevent damage to or leakage from the container.

Prevent inhalation of dust. Prevent contact with eyes and skin. Under dusty conditions, static electricity may cause an explosion. Keep containers closed. Store in a cool, dry place.

Use with adequate ventilation. Wash thoroughly after handling before eating and smoking. Work clothing should be frequently laundered and not worn away from work premises.

Other Precautions: Use all standard practices for good personal hygiene. Completely isolate and thoroughly clean all equipment, piping, or vessels before beginning maintenance or repairs. Keep area clean. As with any powder or dustlike material, improper handling can lead to spontaneous combustion (dust explosions).

--- REGULATORY INFORMATION ---

None reported.

▼ ▼ ▼ ▼ ▼ ▼ ▼ ▼ ▼ ▼ ▼ ▼ ▼ ▼

--- IDENTIFIERS ---

Chemical Name and Synonyms: POLYMERIZED 1,2-DIHYDRO-2,2,4-TRIMETHYLQUINOLINE, POLYMERIZED TRIMETHYL DIHYDROQUINOLINE

Trade Name and Synonyms: Agerite* MA (pellets, pastilles, and powder)

Chemical Family: NA

Manufacturer: R. T. Vanderbilt Co., Inc., Norwalk, Connecticut

UN ID Number: NA

--- INGREDIENTS AND COMPOSITION DATA ---

Material	Cas No.	%	TLV
Agerite	NA		Not established

--- PHYSICAL DATA ---

Boiling Point (°F): NA

Specific Gravity (H₂O = 1): 1.06

Vapor Pressure (mm Hg): NA

Percent Volatile by Weight (%): 1.0 max.

Vapor Density (air = 1): NA

Freezing Point: NA

Evaporation Rate: NA

pH: NA

Solubility: Very soluble in acetone, toluene, chloroform, and carbon disulfide. Slightly soluble in petroleum hydrocarbons. Practically insoluble in water.

Melting Point (°C): 105 C min.

Appearance/Color/Odor: Amber or cream to yellow pellets, pastilles, or powder; no odor. *Powder:* slight amine odor.

--- FIRE AND EXPLOSION HAZARDS ---

Flash Point: 204°C (COC)

Flammability Limits: NA

Autoignition Temperature: Above normal processing temperatures

Decomposition Temperature: NA

Extinguishing Media: Foam, dry chemical, water. Carbon dioxide may be ineffective on large fires due to lack of cooling capacity, which may result in reignition.

Special Fire Fighting Procedures: Use self-contained breathing apparatus (SCBA). In enclosed or poorly ventilated areas, wear SCBA during cleanup immediately after a fire and during fire fighting.

Unusual Fire and Explosion Hazards: None known.

Hazardous Products of Combustion and Decomposition: Oxides of nitrogen and carbon, small amounts of aromatic and aliphatic hydrocarbons.

Hazardous Polymerization: Will not occur.

--- HEALTH HAZARD DATA ---

Effects of Overexposure: None expected for product. Heating may generate vapors, which can irritate the eyes and respiratory passages.
For Diphenylamines: In general, may be irritating to the mucous membranes. Heating may generate vapors, which can irritate eyes and the respiratory passages.
Primary Routes of Exposure: Dermal, ingestion, and inhalation.
Threshold Limit Value: None established.

Emergency Response and First Aid:
Skin: Wash with lots of clean water for at least 15 min.
Eyes: Flush eyes with lots of water. If irritation persists, consult a physician.
Inhalation: Expose to fresh air. Keep warm and quiet. Give artificial respiration if necessary.
Ingestion: None recommended. Consult physician immediately.

Cancer Information: None reported.

--- TOXICITY DATA ---

Animal Effects:
For Assessing Exposure, Nuisance Dust Standard Recommended:
OSHA TWA: 15 mg/m^3, total dust; 5 mg/m^3, respirable dust
ACGIH TWA: 10 mg/m^3, total dust; 5 mg/m^3, respirable dust
Acute Oral LD50 (rats): 4900 mg/kg
Acute Dermal LD50 (rabbits): >20,000 mg/kg
No skin or eye irritation observed in rabbits.
The results of a 2-yr chronic toxicity/carcinogenicity rat feeding study, a lifetime hamster intratracheal, and mouse skin painting study were negative.

Ames Test: Negative

--- REACTIVITY DATA ---

Stability and Conditions to Avoid: Stable.

Materials to Avoid (Incompatibility): Strong oxidizing agents such as hydrogen peroxide, permanganates, and perchlorates.

Hazard Decomposition Products: Thermal decomposition may produce carbon monoxide and/or carbon dioxide, small amounts of aromatic and aliphatic hydrocarbons, as well as oxides of nitrogen.

Hazardous Polymerization: Will not occur.

--- SPILL AND EMERGENCY RESPONSE ---

For Spills: Vacuum or sweep into a closed container for disposal. Avoid dust generation. Do not flush chemical into public sewer or water system.

Waste Disposal Method: Not classified as a RCRA hazardous waste. Dispose of according to applicable environmental regulations.

Empty Containers: Burn containers in an approved incinerator or dispose of in an approved chemical landfill in accordance with all applicable local, state, and federal laws and regulations.

--- SPECIAL PROTECTION AND PRECAUTIONS ---

Respiratory Protection: Not typically required. If dusty conditions, wear NIOSH-approved respirator according to OSHA 29 CFR 1910.134.

Type of Ventilation Required: General mechanical. Avoid breathing dusts and vapors. Adequate ventilation should be provided to keep mist and dust concentrations below acceptable exposure limits. Discharge from the ventilation system should comply with applicable air-pollution regulations. Local exhaust ventilation is recommended with a capture velocity of 150 to 200 fpm.

Eye Protection Requirements: Chemical goggles. Eyewash fountains and safety showers should be easily accessible.

Protective Gloves and Other: Wear rubber gloves and goggles.

Storage and Handling Precautions: Containers should be stored in a cool, dry, well-ventilated area. Store away from flammable materials, sources of heat, flame, sparks, and foodstuffs. Exercise due caution to prevent damage to or leakage from the container.

Prevent inhalation of dust. Prevent contact with eyes and skin. Under dusty conditions, static electricity may cause an explosion. Keep containers closed. Store in a cool, dry place.

Use with adequate ventilation. Wash thoroughly after handling before eating and smoking. Work clothing should be frequently laundered and not worn away from work premises.

Other Precautions: Use all standard practices for good personal hygiene. Completely isolate and thoroughly clean all equipment, piping, or vessels before beginning maintenance or repairs. Keep area clean. As with any powder or dustlike material, improper handling can lead to spontaneous combustion (dust explosions).

--- REGULATORY INFORMATION ---

None reported.

▼ ▼ ▼ ▼ ▼ ▼ ▼ ▼ ▼ ▼ ▼ ▼ ▼ ▼

--- IDENTIFIERS ---

Chemical Name and Synonyms: POLYMERIZED 1,2-DIHYDRO-2,2,4-TRIMETHYLQUINOLINE, QUINOLINE

Trade Name and Synonyms: Agerite* Resin* D (pellets, pastilles, flakes, and powder)

Chemical Family: NA

Manufacturer: R. T. Vanderbilt Co., Inc., Norwalk, Connecticut

UN ID Number: NA

--- INGREDIENTS AND COMPOSITION DATA ---

Material	Cas No.	%	TLV
Quinoline	26780-96-12		Not established

--- PHYSICAL DATA ---

Boiling Point (°F): NA

Specific Gravity (H$_2$O = 1): 1.03 to 1.09

Vapor Pressure (mm Hg): NA

Percent Volatile by Weight (%): 1.0 max.

Vapor Density (air = 1): NA

Freezing Point: NA

Evaporation Rate: NA

pH: NA

Solubility in Water: Insoluble

Melting Point (°C): NA

Appearance/Color/Odor: Amber to cream to yellow to tan pastilles, powder, flakes, or pellets. No odor.

--- FIRE AND EXPLOSION HAZARDS ---

Flash Point: 204°C (COC)

Flammability Limits: NA

Autoignition Temperature: Above normal processing temperatures

Decomposition Temperature: NA

Extinguishing Media: Water, foam, dry chemical. Carbon dioxide may be ineffective.

Special Fire Fighting Procedures: Use self-contained breathing apparatus (SCBA). In enclosed or poorly ventilated areas, wear SCBA during cleanup immediately after a fire and during fire fighting.

Unusual Fire and Explosion Hazards: Dust suspended in air in critical proportions and in the presence of an ignition source presents an explosion hazard. The following characteristics apply to powder and, also, are expected to apply to dust from pastilles, flakes, or pellets or if these forms are reduced to a powder.

- Minimum explosive concentration: 0.03 oz/ft^3
- Minimum ignition energy: 0.15 J (dust cloud)
- Maximum rate of pressure rise: 16,800 psi/sec at 0.1 oz/ft^3
- Maximum pressure of explosion: 72 psig at 0.5 oz/ft^3
- Explosion severity: 4.0 (severe)
- Ignition sensitivity: Not determined
- Explosibility index: Not determined
- Volume resistivity: 3.92 x 10 + 17 ohm-cm
- National Fire Protection Association Standard 497 M rating (1986 edition): Class II, Group G

Dust is very sensitive to electrostatic discharge. Dust may be ignited by electrostatic discharge, electrical arcs, sparks, welding torches, open flame, or other significant heat sources. Bond, ground, and properly vent containers, conveyors, dust control devices, and other transfer equipment. Prohibit flow of powder or dust through nonconductive ducts or pipes, etc.; only use grounded, electrically conductive transfer lines when pneumatically conveying product.

Prevent accumulation of dust (e.g., use well-ventilated conditions, promptly vacuum spills, clean overhead horizontal surfaces, etc.). Eliminate ignition sources such as sparks or static buildup (e.g., use humidification). A properly engineered explosion suppression system must be considered where large amounts of product are handled. See standards such as the National Fire Protection Association NFPA 654, "Standard for the Prevention of Dust Explosions in the Plastics Industry"; NFPA 69, "Explosion Prevention Systems"; NFPA 68, "Explosion Venting Protection"; NFPA 77, "Static Electricity"; and other standards as the need exists. Implement other measures as the need exists.

Hazardous Products of Combustion and Decomposition: Carbon monoxide, carbon dioxide, oxides of nitrogen, and aromatic and aliphatic hydrocarbons.

Hazardous Polymerization: Will not occur.

--- HEALTH HAZARD DATA ---

Effects of Overexposure: None expected for product. Heating may generate vapors, which can irritate the eyes and respiratory passages.
Primary Routes of Exposure: Dermal, ingestion, and inhalation.
Threshold Limit Value: None established.

Emergency Response and First Aid:
Skin: Wash with lots of clean water for at least 15 min.
Eyes: Flush eyes with lots of water. If irritation persists, consult a physician.
Inhalation: Expose to fresh air. Keep warm and quiet. Give artificial respiration if necessary.
Ingestion: None recommended. Consult physician immediately.

Cancer Information: None reported.

--- TOXICITY DATA ---

Animal Effects:
For Assessing Exposure, Nuisance Dust Standard Recommended:
OSHA TWA: 15 mg/m^3, total dust; 5 mg/m^3, respirable dust
ACGIH TWA: 10 mg/m^3, total dust; 5 mg/m^3, respirable dust
Acute Oral LD50 (rats): 4900 mg/kg

Acute Dermal LD50 (rabbits): >20,000 mg/kg
No skin or eye irritation observed in rabbits.
A 2-yr chronic feeding study in rats showed no evidence of carcinogenicity.
Mouse skin painting tests were negative.

Ames Test: Negative

Medical Conditions Generally Aggravated By Exposure: Unknown.

--- REACTIVITY DATA ---

Stability and Conditions to Avoid: Stable.

Materials to Avoid (Incompatibility): Strong oxidizing agents such as hydrogen peroxide, permanganates, and perchlorates.

Hazard Decomposition Products: Thermal decomposition may produce carbon monoxide and/or carbon dioxide, small amounts of aromatic and aliphatic hydrocarbons, as well as oxides of nitrogen.

Hazardous Polymerization: Will not occur.

--- SPILL AND EMERGENCY RESPONSE ---

For Spills: Sweep or wet mop into a closed container for disposal. Avoid generation of dust. Do not flush chemical into public sewer or water system.

Waste Disposal Method: Not classified as a RCRA hazardous waste. Dispose of according to applicable environmental regulations.

Empty Containers: Burn containers in an approved incinerator or dispose of in an approved chemical landfill in accordance with all applicable local, state, and federal laws and regulations.

--- SPECIAL PROTECTION AND PRECAUTIONS ---

Respiratory Protection: Not typically required. If dusty conditions, wear NIOSH-approved respirator according to OSHA 29 CFR 1910.134.

Type of Ventilation Required: General mechanical. Avoid breathing dusts and vapors. Adequate ventilation should be provided to keep mist and dust concentrations below acceptable exposure limits. Discharge from the ventilation system should comply with applicable air-pollution regulations. Local exhaust ventilation is recommended with a capture velocity of 150 to 200 fpm.

Eye Protection Requirements: Chemical goggles. Eyewash fountains and safety showers should be easily accessible.

Protective Gloves and Other: Wear rubber gloves and goggles.

Storage and Handling Precautions: Containers should be stored in a cool, dry, well-ventilated area. Store away from flammable materials, sources of heat, flame, sparks, and foodstuffs. Exercise due caution to prevent damage to or leakage from the container.

Prevent inhalation of dust. Prevent contact with eyes and skin. Under dusty conditions, static electricity may cause an explosion. Keep containers closed. Store in a cool, dry place.

Use with adequate ventilation. Wash thoroughly after handling before eating and smoking. Work clothing should be frequently laundered and not worn away from work premises.

Other Precautions: Use all standard practices for good personal hygiene. Completely isolate and thoroughly clean all equipment, piping, or vessels before beginning maintenance or repairs. Keep area clean. As with any powder or dustlike material, improper handling can lead to spontaneous combustion (dust explosions).

--- REGULATORY INFORMATION ---

None reported.

▼ ▼ ▼ ▼ ▼ ▼ ▼ ▼ ▼ ▼ ▼ ▼ ▼ ▼

--- IDENTIFIERS ---

Chemical Name and Synonyms: POWDERED PHENOLIC RESIN

Trade Name and Synonyms: Akrochem Antioxidant 33

Chemical Family: Phenolic Antioxidant

Manufacturer: Akron Chemical Co., Akron, Ohio

UN ID Number: NA

--- INGREDIENTS AND COMPOSITION DATA ---

Material	Cas No.	%	TLV
Powdered phenolic resin	NA		Not established

--- PHYSICAL DATA ---

Boiling Point (°F): NA

Specific Gravity (H$_2$O = 1): 1.04

Vapor Pressure (mm Hg): NA

Percent Volatile by Weight (%): NA

Vapor Density (air = 1): NA

Freezing Point: NA

Evaporation Rate: NA

pH: NA

Solubility in Water: Negligible

Melting Point (°C): 105

Appearance/Color/Odor: White powder; little odor at room temperature.

--- FIRE AND EXPLOSION HAZARDS ---

Flash Point: >200°F

Flammability Limits: NA

Autoignition Temperature: Above normal processing temperatures

Decomposition Temperature: NA

Extinguishing Media: Water, foam, dry chemical. Carbon dioxide may be ineffective.

Special Fire Fighting Procedures: Use self-contained breathing apparatus (SCBA). In enclosed or poorly ventilated areas, wear SCBA during cleanup immediately after a fire and during fire fighting.

Unusual Fire and Explosion Hazards: A high concentration of dust in the air could result in explosive mixture.

Hazardous Products of Combustion and Decomposition: Carbon monoxide, carbon dioxide, and oxides of nitrogen.

Hazardous Polymerization: Will not occur.

--- HEALTH HAZARD DATA ---

Effects of Overexposure: Occasional reaction on persons with extreme skin sensitivity or allergic tendencies.
Primary Routes of Exposure: Dermal, ingestion, and inhalation.
Threshold Limit Value: None established.

Emergency Response and First Aid:
Skin: Wash with lots of clean water for at least 15 min.
Eyes: Flush eyes with lots of water. If irritation persists, consult a physician.
Inhalation: Expose to fresh air. Keep warm and quiet. Give artificial respiration if necessary.
Ingestion: None recommended. Consult physician immediately.

Cancer Information: None reported.

--- TOXICITY DATA ---

Animal Effects: No information reported by manufacturer.

--- REACTIVITY DATA ---

Stability and Conditions to Avoid: Stable.

Materials to Avoid (Incompatibility):   Strong oxidizing agents such as hydrogen peroxide, permanganates, and perchlorates.

Hazard Decomposition Products: Thermal decomposition may produce carbon monoxide and/or carbon dioxide, small amounts of aromatic and aliphatic hydrocarbons, as well as oxides of nitrogen.

Hazardous Polymerization: Will not occur.

--- SPILL AND EMERGENCY RESPONSE ---

For Spills: Sweep or wet mop into a closed container for disposal. Avoid generation of dust. Do not flush chemical into public sewer or water system.

Waste Disposal Method: Not classified as a RCRA hazardous waste. Dispose of according to applicable environmental regulations.

Empty Containers: Burn containers in an approved incinerator or dispose of in an approved chemical landfill in accordance with all applicable local, state, and federal laws and regulations.

--- SPECIAL PROTECTION AND PRECAUTIONS ---

Respiratory Protection: Not typically required. If dusty conditions, wear NIOSH-approved respirator according to OSHA 29 CFR 1910.134.

Type of Ventilation Required: General mechanical. Avoid breathing dusts and vapors. Adequate ventilation should be provided to keep mist and dust concentrations below acceptable exposure limits. Discharge from the ventilation system should comply with applicable air-pollution regulations. Local exhaust ventilation is recommended with a capture velocity of 150 to 200 fpm.

Eye Protection Requirements: Chemical goggles. Eyewash fountains and safety showers should be easily accessible.

Protective Gloves and Other: Wear rubber gloves and goggles.

Storage and Handling Precautions: Containers should be stored in a cool, dry, well-ventilated area. Store away from flammable materials, sources of heat, flame, sparks, and foodstuffs. Exercise due caution to prevent damage to or leakage from the container.

Prevent inhalation of dust. Prevent contact with eyes and skin. Under dusty conditions, static electricity may cause an explosion. Keep containers closed. Store in a cool, dry place.

Use with adequate ventilation. Wash thoroughly after handling before eating and smoking. Work clothing should be frequently laundered and not worn away from work premises.

Other Precautions: Use all standard practices for good personal hygiene. Completely isolate and thoroughly clean all equipment, piping, or vessels before beginning maintenance or repairs. Keep area clean. As with any powder or dustlike material, improper handling can lead to spontaneous combustion (dust explosions).

--- REGULATORY INFORMATION ---

Contains no SARA Title III chemicals present at or above the de minimus concentration.

▼   ▼   ▼   ▼   ▼   ▼   ▼   ▼   ▼   ▼   ▼   ▼   ▼   ▼

--- IDENTIFIERS ---

Chemical Name and Synonyms: 2,6-DI-TERT-BUTYL-p-CRESOL, C_6H_2 $(C_4H_9)_2(CH_3)OH$

Trade Name and Synonyms: BHT (butylated hydroxy toluene)

Chemical Family: Alkylated Cresol

Manufacturer: Koppers Co., Inc., Pittsburgh, Pennsylvania

UN ID Number: NA

--- INGREDIENTS AND COMPOSITION DATA ---

Material	Cas No.	%	TLV
BHT	NA		10 mg/m^3 (TWA)

--- PHYSICAL DATA ---

Boiling Point (°F): 510

Specific Gravity (H_2O = 1): 1.01

Vapor Pressure (mm Hg): < 1

Percent Volatile by Weight (%): NA

Vapor Density (air = 1): 7.6

Freezing Point: NA

Evaporation Rate: NA

pH: NA

Solubility in Water: Negligible

Melting Point (°C): NA

Appearance/Color/Odor: White crystalline solid with typical alkylated cresol odor.

--- FIRE AND EXPLOSION HAZARDS ---

Flash Point: 245°F (closed cup)

Flammability Limits: NA

Autoignition Temperature: NA

Decomposition Temperature: NA

Extinguishing Media: Water, foam, dry chemical. Carbon dioxide may be ineffective.

Special Fire Fighting Procedures: Use self-contained breathing apparatus (SCBA). In enclosed or poorly ventilated areas, wear SCBA during cleanup immediately after a fire and during fire fighting.

Unusual Fire and Explosion Hazards: Sealed (closed) containers can build up pressure if exposed to heat (fire). Water can be used to cool containers.

Hazardous Products of Combustion and Decomposition: Carbon monoxide, carbon dioxide, and oxides of nitrogen.

Hazardous Polymerization: Will not occur.

--- HEALTH HAZARD DATA ---

Effects of Overexposure: Dust or vapors may cause irritation to eyes, nose, and throat. Contact of dust or vapors with skin may cause irritation. *Primary Routes of Exposure:* Dermal, ingestion, and inhalation. *Threshold Limit Value:* None established.

Emergency Response and First Aid:
Skin: Wash with lots of clean water for at least 15 min. For burns from molten BHT, treat as a thermal burn. Call a physician.
Eyes: Flush eyes with lots of water. If irritation persists, consult a physician.
Inhalation: Expose to fresh air. Keep warm and quiet. Give artificial respiration if necessary.
Ingestion: None recommended. Consult physician immediately.

Cancer Information: None reported.

--- TOXICITY DATA ---

Animal Effects: No information reported by manufacturer.

--- REACTIVITY DATA ---

Stability and Conditions to Avoid: Stable.

Materials to Avoid (Incompatibility): Strong oxidizing agents such as hydrogen peroxide, permanganates, and perchlorates.

Hazard Decomposition Products: No information reported by manufacturer.

Hazardous Polymerization: Will not occur.

--- SPILL AND EMERGENCY RESPONSE ---

For Spills: Remove sources of ignition. Avoid breathing dust or vapors. Avoid contact of eyes or skin with dust or vapors. Use protective measures outlined below. Solidified BHT may be picked up with a shovel and placed in container for disposal.

Waste Disposal Method: Use either approved sanitary landfill or incineration. (Do not incinerate closed containers.) Disposal must be carried out in accordance with local, state, and federal regulations.

Empty Containers: Burn containers in an approved incinerator or dispose of in an approved chemical landfill in accordance with all applicable local, state, and federal laws and regulations.

--- SPECIAL PROTECTION AND PRECAUTIONS ---

Respiratory Protection: Use full face unit per OSHA regulations and manufacturer's "Instructions" and "Warnings."

Type of Ventilation Required: Use adequate ventilation to keep inhalation and flammable concentrations to a minimum.

Eye Protection Requirements: Chemical goggles, face shields. Eyewash fountains and safety showers should be easily accessible.

Protective Gloves and Other: Wear rubber gloves and goggles. Coveralls and/or rubber apron. Rubber shoes or boots.

Storage and Handling Precautions: Store in tightly closed, properly labeled containers in cool, well-ventilated area away from all ignition sources. Avoid prolonged and/or repeated contact with skin.

Prevent inhalation of dust. Prevent contact with eyes and skin. Under dusty conditions, static electricity may cause an explosion. Keep containers closed. Store in a cool, dry place.

Use with adequate ventilation. Wash thoroughly after handling before eating and smoking. Work clothing should be frequently laundered and not worn away from work premises.

Other Precautions: Do not take internally. NIOSH-approved respiratory units can be used if oxygen level is above 19.5%. Otherwise, use self-contained units.

--- REGULATORY INFORMATION ---

No information reported by manufacturer.

▼ ▼ ▼ ▼ ▼ ▼ ▼ ▼ ▼ ▼ ▼ ▼ ▼ ▼

--- IDENTIFIERS ---

Chemical Name and Synonyms: NICKEL BIS(O-ETHYL-(3,5-DI-TERT-BUTYL-4-HYDROXYBENZYL)PHOSPHONATE, NI $P_2O_8C_{34}H_{56}$

Trade Name and Synonyms: Irgastab 2002

Chemical Family: Light Stabilizer

Manufacturer: Ciba-Geigy Corp., Ardsley, New York

UN ID Number: NA

--- INGREDIENTS AND COMPOSITION DATA ---

Material	Cas No.	%	TLV
Ni $P_2O_8C_{34}H_{56}$	NA		Not established

--- PHYSICAL DATA ---

Boiling Point (°F): NA

Specific Gravity (H_2O = 1): NA

Vapor Pressure (mm Hg): NA

Percent Volatile by Weight (%): NA

Vapor Density (air = 1): NA

Freezing Point: NA

Evaporation Rate: NA

pH: NA

Solubility in Water: Negligible

Melting Point (°C): NA

Appearance/Color/Odor: Light tan, free-flowing powder.

--- FIRE AND EXPLOSION HAZARDS ---

Flash Point: NA

Flammability Limits: NA

Autoignition Temperature: NA

Decomposition Temperature: NA

Extinguishing Media: Water, foam, dry chemical. Carbon dioxide may be ineffective.

Special Fire Fighting Procedures: None reported by manufacturer.

Unusual Fire and Explosion Hazards: None reported by manufacturer.

Hazardous Products of Combustion and Decomposition: Carbon monoxide, carbon dioxide, and oxides of nitrogen.

Hazardous Polymerization: Will not occur.

--- HEALTH HAZARD DATA ---

Effects of Overexposure: None reported by manufacturer. Refer to toxicity section.
Primary Routes of Exposure: Dermal, ingestion, and inhalation.
Threshold Limit Value: None established.

Emergency Response and First Aid:
Skin: Wash with lots of clean water for at least 15 min. For burns from molten BHT, treat as a thermal burn. Call a physician.

Eyes: Flush eyes with lots of water. If irritation persists, consult a physician.
Inhalation: Expose to fresh air. Keep warm and quiet. Give artificial respiration if necessary.
Ingestion: None recommended. Consult physician immediately.

Cancer Information: None reported.

--- TOXICITY DATA ---

Animal Effects:
Acute Oral LD50 (rats): 3750 mg/kg
Acute Oral LD50 (mice): 965 mg/kg

--- REACTIVITY DATA ---

Stability and Conditions to Avoid: Stable.

Materials to Avoid (Incompatibility): No information reported.

Hazard Decomposition Products: No information reported by manufacturer.

Hazardous Polymerization: Will not occur.

--- SPILL AND EMERGENCY RESPONSE ---

For Spills: Sweep up promptly. Wear an approved dust respirator.

Waste Disposal Method: Normal plant waste, bury or incinerate.

Empty Containers: Burn containers in an approved incinerator or dispose of in an approved chemical landfill in accordance with all applicable local, state, and federal laws and regulations.

--- SPECIAL PROTECTION AND PRECAUTIONS ---

Respiratory Protection: Approved dust respirator.

Type of Ventilation Required: Use adequate ventilation to keep inhalation concentrations to a minimum.

Eye Protection Requirements: Chemical goggles, face shields. Eyewash fountains and safety showers should be easily accessible.

Protective Gloves and Other: Wear rubber gloves and goggles. Coveralls and/or rubber apron. Rubber shoes or boots.

Storage and Handling Precautions: Store in tightly closed, properly labeled containers in cool, well-ventilated area away from all ignition sources. Avoid prolonged and/or repeated contact with skin.

Prevent inhalation of dust. Prevent contact with eyes and skin. Under dusty conditions, static electricity may cause an explosion. Keep containers closed. Store in a cool, dry place.

Use with adequate ventilation. Wash thoroughly after handling before eating and smoking. Work clothing should be frequently laundered and not worn away from work premises.

Other Precautions: In accord with good industrial practice, avoid unnecessary personal contact. Employ bonding, ground, venting, and explosion-relief provisions in accord with accepted engineering practices in process operations capable of generating dust and/or static electricity.

--- REGULATORY INFORMATION ---

No information reported by manufacturer.

▼ ▼ ▼ ▼ ▼ ▼ ▼ ▼ ▼ ▼ ▼ ▼ ▼ ▼

--- IDENTIFIERS ---

Chemical Name and Synonyms: BROMINATED ORGANIC COMPOUND

Trade Name and Synonyms: Citex BN-451

Chemical Family: Flame Retardant Additive

Manufacturer: Cities Service Co., Cranbury, New Jersey

UN ID Number: NA

--- INGREDIENTS AND COMPOSITION DATA ---

Material	Cas No.	%	TLV
Brominated organic	NA		Not established

--- PHYSICAL DATA ---

Boiling Point (°F): NA

Specific Gravity (H₂O = 1): 2.07

Vapor Pressure (mm Hg): NA

Percent Volatile by Weight (%): NA

Vapor Density (air = 1): NA

Freezing Point: NA

Evaporation Rate: NA

pH: NA

Solubility in Water: Negligible

Melting Point (°C): NA

Appearance/Color/Odor: White to off-white solid; odorless.

--- FIRE AND EXPLOSION HAZARDS ---

Flash Point: NA

Flammability Limits: NA

Autoignition Temperature: NA

Decomposition Temperature: NA

Extinguishing Media: No information reported. In absence of data, use carbon dioxide.

Special Fire Fighting Procedures: Hydrogen bromide or possibly other toxic gases may be given off at elevated temperatures; self-contained breathing equipment may be necessary.

Unusual Fire and Explosion Hazards: None reported by manufacturer.

Hazardous Products of Combustion and Decomposition: Carbon monoxide, carbon dioxide, and oxides of nitrogen.

Hazardous Polymerization: Will not occur.

--- HEALTH HAZARD DATA ---

Effects of Overexposure: None reported by manufacturer. Refer to toxicity section.
Primary Routes of Exposure: Dermal, ingestion, and inhalation.
Threshold Limit Value: None established.

Emergency Response and First Aid:
Skin: Wash with lots of clean water for at least 15 min. For burns from molten BHT, treat as a thermal burn. Call a physician.
Eyes: Flush eyes with lots of water. If irritation persists, consult a physician.
Inhalation: Expose to fresh air. Keep warm and quiet. Give artificial respiration if necessary.
Ingestion: None recommended. Consult physician immediately.

Cancer Information: None reported.

--- TOXICITY DATA ---

Animal Effects: LD50 > 10,000 mg/kg body weight.

--- REACTIVITY DATA ---

Stability and Conditions to Avoid: Stable.

Materials to Avoid (Incompatibility): No information reported.

Hazard Decomposition Products: Hydrogen bromide gas.

Hazardous Polymerization: Will not occur.

--- SPILL AND EMERGENCY RESPONSE ---

For Spills: Sweep up promptly. Wear an approved dust respirator.

Waste Disposal Method: Normal plant waste, bury, or incinerate.

Empty Containers: Burn containers in an approved incinerator or dispose of in an approved chemical landfill in accordance with all applicable local, state, and federal laws and regulations.

--- SPECIAL PROTECTION AND PRECAUTIONS ---

Respiratory Protection: Approved dust respirator.

Type of Ventilation Required: Use the guidelines recommended by the American Conference of Governmental Industrial Hygienists in the current edition of "Industrial Ventilation," considering the TLV, lower explosive (Flammable) limit and conditions under which this product is used.

Eye Protection Requirements: Chemical goggles, face shields. Eyewash fountains and safety showers should be easily accessible.

Protective Gloves and Other: Wear rubber gloves and goggles. Coveralls and/or rubber apron. Rubber shoes or boots.

Storage and Handling Precautions: Store in tightly closed, properly labeled containers in cool, well-ventilated area away from all ignition sources. Avoid prolonged and/or repeated contact with skin.

Prevent inhalation of dust. Prevent contact with eyes and skin. Under dusty conditions, static electricity may cause an explosion. Keep containers closed. Store in a cool, dry place.

Use with adequate ventilation. Wash thoroughly after handling before eating and smoking. Work clothing should be frequently laundered and not worn away from work premises.

Other Precautions: In accordance with good industrial practice, avoid unnecessary personal contact. Employ bonding, ground, venting, and explosion-relief provisions in accordance with accepted engineering practices in process operations capable of generating dust and/or static electricity.

--- REGULATORY INFORMATION ---

No information reported by manufacturer.

▼ ▼ ▼ ▼ ▼ ▼ ▼ ▼ ▼ ▼ ▼ ▼ ▼ ▼

--- IDENTIFIERS ---

Name: MIREX
Synonyms: Bichlorendo; CG-1283; Cyclopentadiene, Hexachloro-, Dimer; Decane,Perchloropentacyclo-; Dechlorane; Dechlorane 515; Dechlorane 4070; Dechlorane Plus; Dechlorane Plus 515; Dodecachlorooctahydro-1,3,4-Metheno-2H-Cyclobuta(C,D)Pentalene; Dodecachloropentacyclodecane; Dodecachloropentacyclo(3.2.2.0(Sup 2,6),0(Sup 3,9),0(Sup 5,10))Decane; Ent 25, 719; Ferriamicide; GC 1283; Hexachlorocyclopentadiene Dimer; 1,2,3,4,5,5-Hexachloro-1,3-Cyclopentadiene Dimer; HRS L276; 1,3,4-Metheno-1H-Cyclobuta(CD)Pentalene, Dodecachlorooctahydro-; Mirex; NCI-C06428; Perchlorodihomocubane; Perchloropentacyclodecane; Perchloropentacyclo(5.2.1.0(Sup 2,6).0(Sup 3,9).0(Sup 5,8))Decane
CAS: 2385-85-5; **RTECS:** PC8225000
Formula: $C_{10}Cl_{12}$; **Mol Wt:** 545.50
WLN: L545 B4 C5 D 4ABCE JTJ-/G 1 2
Chemical Class: Halogenated h-carbon

See other identifiers listed below under Regulations.

--- PROPERTIES ---

Physical Description: White crystals. May be dissolved in petroleum-based solvent. (NYDH)
Boiling Point: 758.15 K dec; 485°C dec; 905°F dec
Melting Point: 758.15 K dec; 485°C dec; 905°F dec
Flash Point: NA
Autoignition: NA
UEL: NA
LEL: NA
Vapor Density: No data
Specific Gravity: No data

Reactivity with Water: No data on water reactivity
Reactivity with Common Materials: No data
Stability during Transport: No data

Neutralizing Agents: No data
Polymerization Possibilities: No data

Toxic Fire Gases: None reported other than possible unburned vapors
Odor Detected at (ppm): Unknown
Odor Description: None (Source: NYDH)
100% Odor Detection: No data

--- REGULATIONS ---

DOT hazard class: Not given
Packaging exceptions: 173.
Nonbulk packaging: 173.
Bulk packaging: 173.

STCC Number: Not listed

Clean Water Act Sect. 307: No
Clean Water Act Sect. 311: No
Clean Air Act: Not listed
EPA Waste Number: None
CERCLA Ref: Not listed
RQ Designation: Not listed
SARA TPQ Value: Not listed
SARA Sect. 312 Categories:
  Chronic toxicity: carcinogen.
  Chronic toxicity: mutagen.
  Chronic toxicity: reproductive toxin.

U.S. Postal Service Mailability: Not given

--- TOXICITY DATA ---

Short-term Toxicity: *Inhalation:* No information found. *Skin:* Can cause irritation. *Eyes:* No information found. *Ingestion:* No cases of human toxicity reported. Possible symptoms include nausea, vomiting, restlessness, tremor, weight loss, nervous system and liver abnormalities, skin rash, and reproductive system disorders.

Long-term Toxicity: Mirex has caused cataracts, cancer, and birth defects in both rats and mice. Whether it does so in humans is not known. (NYDH)

Conc IDLH: Unknown

NIOSH REL: Not given.

ACGIH TLV: Not listed
ACGIH STEL: Not listed

OSHA PEL: Not in Table Z-1-A

MAK Information: Not listed

Carcinogen: Y; **Status:** See below
References: Animal positive IARC**20, 283, 79; Animal positive IARC** 5, 203, 74

Carcinogen Lists: *IARC*: Carcinogen defined by IARC to be possibly carcinogenic to humans, but having (usually) no human evidence; *MAK*: Not listed; *NIOSH*: Not listed; *NTP*: Carcinogen defined by NTP as reasonably anticipated to be carcinogenic, with limited evidence in humans or sufficient evidence in experimental animals; *ACGIH*: Not listed; *OSHA*: Not listed.

LD50 Value: orl-rat LD50:235 mg/kg

Other Species Toxicity Data: (Source: NIOSH RTECS 1992)
orl-rat LD50:235 mg/kg
skn-rat LD50: >2 g/kg
skn-rbt LD50:800 mg/kg
orl-ham LD50:125 mg/kg
orl-dck LD50:2400 mg/kg
ihl-brd LC50:1400 ppm

Reproductive Toxicity (1992 RTECS): This chemical is a mammalian reproductive toxin.

Reproductive Toxicity Data (1992 RTECS):
orl-rat TDLo:50 mg/kg (5-9D preg) TJADAB 27, 401, 83
  Maternal Effects
    Ovaries, fallopian tubes
    Uterus, cervix, vagina
  Effects on Embryo or Fetus
    Fetotoxicity (except death, e.g., stunted fetus)

orl-rat TDLo:23750 mg/kg (4-22D preg) TOLED5 4, 263, 79
Effects on Newborn
Stillbirth
Weaning or lactation index (number alive at weaning per number alive at day 4)

orl-rat TDLo:60 mg/kg (6-15D preg) FCTXAV 14, 25, 76
Effects on Fertility
Litter size (number of fetuses per litter; measured before birth)
Specific Developmental Abnormalities
Homeostasis
Musculoskeletal system

orl-rat TDLo:56 mg/kg (8-15D preg) TJADAB 22, 167, 80
Effects on Fertility
Postimplantation mortality
Specific Developmental Abnormalities
Cardiovascular (circulatory) system
Other developmental abnormalities

orl-rat TDLo:48 mg/kg (8-15D preg) TJADAB 21, 40A, 80
Specific Developmental Abnormalities
Blood and lymphatic systems (including spleen and marrow)
Homeostasis

orl-rat TDLo:60 mg/kg (10D male) FCTXAV 14, 25, 76
Effects on Fertility
Male fertility index

orl-rat TDLo:18750 mg/kg (1-15D post) TOLED5 4, 263, 79
Specific Developmental Abnormalities
Eye, ear

orl-rat TDLo:10 mg/kg (1-8D post) TOLED5 4, 263, 79
Effects on Newborn
Growth statistics (e.g., reduced weight gain)

scu-rat TDLo:10 mg/kg (2D pre) ENVRAL 16, 131, 78
Maternal Effects
Uterus, cervix, vagina

scu-rat TDLo:2 mg/kg (1D pre) PSEBAA 148, 414, 75
Effects on Fertility
Other measures of fertility

unr-rat TDLo:8 mg/kg (8-15D preg) TJADAB 26(3), 17A, 82
Specific Developmental Abnormalities
Cardiovascular (circulatory) system
Respiratory system
Effects on Newborn
Weaning or lactation index (number alive at weaning per number alive at
day 4)

unr-rat TDLo:48 mg/kg (8-15D preg) TJADAB 26(3), 17A, 82
Effects on Newborn
Stillbirth
Viability index (number alive at day 4 per number born alive)

orl-mus TDLo:22 mg/kg (1-21D preg) FEPRA7, 37, 938, 78
Effects on Newborn
Growth statistics (e.g., reduced weight gain)
Behavioral

orl-mus TDLo:58 mg/kg (4W male/4W pre-2W post) TXAPA9 10, 54, 67
Effects on Fertility
Other measures of fertility

orl-mus TDLo:37500 mg/kg (8-12D preg) JTEHD6 10, 541, 82
Effects on Newborn
Live birth index (number of fetuses per litter)
Viability index (number alive at day 4 per number born alive)
Growth statistics (e.g., reduced weight gain)

orl-mus TDLo:1775 mg/kg (8-12D preg) TCMUD8 7, 7, 87
Effects on Newborn
Live birth index (number of fetuses per liter)
Viability index (number alive at day 4 per number born alive)

unr-mam TDLo:7 mg/kg (15-21D preg) TJADAB 25(2), 44A, 82
Specific Developmental Abnormalities
Eye, ear
Cardiovascular (circulatory) system
Respiratory system

▼ ▼ ▼ ▼ ▼ ▼ ▼ ▼ ▼ ▼ ▼ ▼ ▼ ▼

--- IDENTIFIERS ---

Name: ANTIMONY OXIDE
Synonyms: A 1530; A 1582; A 1588LP; Antimonious Oxide; Antimony(3+)
Oxide; Antimony Peroxide; Antimony Sesquioxide; Antimony Trioxide;
Antimony Trioxide (ACGIH, DOT); Antimony White; Antox; AP 50;
Chemetron Fire Shield; C.I. 77052; C.I. Pigment White 11; Dechlorane A-O;
Diantimony Trioxide; Exitelite; Extrema; Flowers of Antimony; NA 9201
(DOT); NCI-C55152; Nyacol A 1530; Senarmontite; Thermoguard B;
Thermoguard S; Timonox; Valentinite; Weisspiessglanz (German)
CAS: 1309-64-4; RTECS: CC5650000
Formula: O_3Sb_2; Mol Wt: 291.50
WLN: .SB2.O3
Chemical Class: Metal oxide

See other identifiers listed below under Regulations.

--- PROPERTIES ---

Boiling Point: NA
Melting Point: 928.16 K; 655°C; 1211°F
Flash Point: NA
Autoignition: NA
UEL: NA
LEL: NA
Vapor Density: No data
Specific Gravity: No data
Density: 5.200

Reactivity with Water: No data on water reactivity
Reactivity with Common Materials: No data
Stability during Transport: No data
Neutralizing Agents: No data
Polymerization Possibilities: No data

Toxic Fire Gases: None reported other than possible unburned vapors
Odor Detected at (ppm): Odorless
Odor Description: None (Source: CHRIS)
100% Odor Detection: No data

--- REGULATIONS ---

DOT hazard class: 9 CLASS 9
DOT guide: 31
Identification number: UN3077
DOT shipping name: ENVIRONMENTALLY HAZARDOUS SUBSTANCES, SOLID, N.O.S.
Packing group: III
Label(s) required: CLASS 9
Special Provisions: 8, B54
Packaging exceptions: 173.155
Nonbulk packaging: 173.213
Bulk packaging: 173.240
Quantity limitations:
Passenger air/rail: None
Cargo aircraft only: None
Vessel stowage: A

STCC Number: Not listed

Clean Water Act Sect. 307: Yes
Clean Water Act Sect. 311: Yes
Clean Air Act: CAA '90 by category
EPA Waste Number: None
CERCLA Ref: 100 > >500 > >510
RQ Designation: C 1000 lb (454 kg) CERCLA
SARA TPQ Value: Not listed
SARA Sect. 312 Categories:
Chronic toxicity: carcinogen.
Chronic toxicity: mutagen.
Chronic toxicity: reproductive toxin.

Listed in SARA Sect. 313: Yes
De Minimus Concentration: 1.0%

U.S. Postal Service Mailability: Not given

--- TOXICITY DATA ---

Short-term Toxicity: Unknown

Long-term Toxicity: Unknown

Symptoms: Inhalation causes inflammation of upper and lower respiratory tract, including pneumonitis. Ingestion causes irritation of the mouth, nose, stomach and intestines; vomiting, purging with bloody stools; slow pulse and low blood pressure; slow, shallow breathing; coma and convulsions, sometimes followed by death. Contact with eyes causes conjunctivitis. Contact with skin causes dermatitis and rhinitis. (Source: CHRIS)

Conc IDLH: 80 mg/m^3

NIOSH REL: Not given.

ACGIH TLV: TLV = 0.5 mg/m^3 handling and use as antimony. Production of: Suspect human carcinogen.
ACGIH STEL: Handling and use as antimony. Production of: Suspect human carcinogen.

OSHA PEL: Transitional Limits: PEL = 0.5 mg/m^3; Final Rule Limits: TWA = 0.5 mg/m^3.

MAK Information: Carcinogenic working material without MAK; in the Commission's view, an animal carcinogen.

Carcinogen: Y; **Status:** See below
References:
3V01J60L60 ihl-rat TCLo:4200 mg/m^3/52W-I AIHAM* 20, 1, 80
3V03J60L60 ihl-rat TC:4 mg/m^3/1Y-I PESTC* 8, 16, 80
3V02L60R60 ihl-rat TC:1600 mg/m^3/52W-I AIHAM* 20, 1, 80
2V01J60 ihl-rat TC:50 mg/m^3/7H/52W-I JTEHD6 18, 607, 86

Carcinogen Lists: *IARC*: Not listed; *MAK*: An animal carcinogen; *NIOSH*: Not listed; *NTP*: Not listed; *ACGIH*: Carcinogen defined by ACGIH TLV Committee as a suspected carcinogen, based on either limited epidemiological evidence or demonstration of carcinogenicity in experimental animals; *OSHA*: Not listed.

LD50 Value: orl-rat LD50: >20 g/kg

Other Species Toxicity Data: (Source: NIOSH RTECS 1992)
orl-rat LD50: >20 g/kg
ipr-rat LD50:3250 mg/kg
ipr-mus LD50:172 mg/kg
ivn-dog LDLo:3 mg/kg

scu-rbt LDLo:2500 mg/kg
scu-mam LD :>120 mg/kg

Reproductive Toxicity (1992 RTECS):   This chemical is a mammalian reproductive toxin.

Reproductive Toxicity Data (1992 RTECS):
ihl-rat TCLo:270 mg/m^3 (1-21D preg) GISAAA 52(10), 85, 87
  Effects on Fertility
    Postimplantation mortality
  Effects on Embryo or Fetus
    Fetal death

ihl-rat TCLo:82 mg/m^3 (1-21D preg) GISAAA 52(10), 85, 87
  Effects on Fertility
    Preimplantation mortality
  Effects on Embryo or Fetus
    Fetotoxicity (except death, e.g., stunted fetus)

--- PROTECTION AND FIRST AID ---

First Aid Source: CHRIS Manual 1991.
If any of the symptoms of poisoning, even slight, are noticed, the affected individual should be removed from contact with the chemical and placed under care of a physician.
Ingestion: Induce vomiting.
Skin: Wash well with soap and water.
Eyes: Flush with water for at least 15 min.

First Aid Source: DOT Emergency Response Guide 1993.
In case of contact with material, immediately flush eyes with running water for at least 15 min. Wash skin with soap and water. Remove and isolate contaminated clothing and shoes at the site.

--- INITIAL INCIDENT RESPONSE ---

U.S. Department of Transportation Guide to Hazardous Materials Transport Information - Publication DOT 5800.5 (1990).
DOT Shipping Name: ENVIRONMENTALLY HAZARDOUS SUBSTANCES, SOLID, N.O.S.
DOT ID Number: UN3077

* POTENTIAL HAZARDS *

*Health Hazards
Contact may cause burns to skin and eyes.
Fire may produce irritating or poisonous gases.
Runoff from fire control or dilution water may cause pollution.

*Fire or Explosion
Some of these materials may burn, but none of them ignites readily.

* EMERGENCY ACTION *

Keep unnecessary people away; isolate hazard area and deny entry. Positive-pressure self-contained breathing apparatus (SCBA) and structural firefighters' protective clothing will provide limited protection. **CALL CHEMTREC AT 1-800-424-9300 FOR EMERGENCY ASSISTANCE.** If water pollution occurs, notify the appropriate authorities.

*Fire
Small fires: Dry chemical, CO_2, water spray, or regular foam.
Large fires: Water spray, fog, or regular foam.
Move container from fire area if you can do it without risk.
Do not scatter spilled material with high-pressure water streams.
Dike fire-control water for later disposal.

*Spill or Leak
Stop leak if you can do it without risk.
Small dry spills: With clean shovel, place material into clean, dry container and cover loosely; move containers from spill area.
Small spills: Take up with sand or other noncombustible absorbent material and place into containers for later disposal.
Large spills: Dike far ahead of liquid spill for later disposal. Cover powder spill with plastic sheet or tarp to minimize spreading.

*First Aid
In case of contact with material, immediately flush eyes with running water for at least 15 min. Wash skin with soap and water. Remove and isolate contaminated clothing and shoes at the site.

7 RETARDERS

--- IDENTIFIERS ---

Name: *N*-NITROSODIPHENYLAMINE
Synonyms: Benzenamine, *N*-Nitroso-*N*-Phenyl- (9CI); Curetard A; Delac J; Diphenylnitrosamin (German); Diphenylnitrosamine; Diphenyl *N*-Nitrosoamine; *N,N*-Diphenylnitrosamine; Naugard TJB; NCI-C02880; NDPA; NDPhA; *N*-Nitrosodifenylamin (Czech); Nitrosodiphenylamine; *N*-Nitrosodiphenylamine; *N*-Nitroso-*N*-Phenylaniline; Nitrous Diphenylamide; Redax; Retarder J; TJB; Vulcalent A; Vulcatard; Vulcatard A; Vulkalent A (Czech); Vultrol
CAS: 86-30-6; **RTECS:** JJ9800000
Formula: $C_{12}H_{10}N_2O$; **Mol Wt:** 198.24
WLN: ONNR
Chemical Class: Nitro compound; aromatic amine

See other identifiers listed below under Regulations.

--- PROPERTIES ---

Boiling Point: NA
Melting Point: 418 K; 144.8°C; 292.7°F
Flash Point: NA
Autoignition: NA
UEL: NA
LEL: NA
Vapor Density: No data
Specific Gravity: No data

Reactivity with Water: No data on water reactivity
Reactivity with Common Materials: No data
Stability during Transport: No data

Neutralizing Agents: No data
Polymerization Possibilities: No data

Toxic Fire Gases: None reported other than possible unburned vapors
Odor Detected at (ppm): Unknown
Odor Description: No data
100% Odor Detection: No data

--- REGULATIONS ---

DOT hazard class: Not given
Packaging exceptions: 173
Nonbulk packaging: 173
Bulk packaging: 173

STCC Number: Not listed

Clean Water Act Sect. 307: Yes
Clean Water Act Sect. 311: No
Clean Air Act: Not listed
EPA Waste Number: None
CERCLA Ref: Y
RQ Designation: B, 100 lb (45.4 kg) CERCLA
SARA TPQ Value: Not listed
SARA Sect. 312 Categories:
  Acute toxicity: adverse effect to target organs.
Listed in SARA Sect. 313: Yes
De Minimus Concentration: 1.0%

U.S. Postal Service Mailability: Not given

NFPA Codes:
  Health Hazard (blue): Unspecified
  Flammability (red): Unspecified
  Reactivity (yellow): Unspecified
  Special: Unspecified

--- TOXICITY DATA ---

Short-term Toxicity: Unknown

Long-term Toxicity: Unknown

Target Organs: Eyes

Symptoms: An eye irritant (Source: SAX)

Conc IDLH: Unknown

NIOSH REL: Not given

ACGIH TLV: Not listed
ACGIH STEL: Not listed

OSHA PEL: Not in Table Z-1-A

MAK Information: Not listed

Carcinogen: N; **Status:** See below
References: Animal suspected IARC** 27, 213, 82; Animal Positive IARC** 28, 151, 82; Human Indefinite IARC** 27, 213, 82

Carcinogen Lists: *IARC*: Not classified as to human carcinogenicity or probably not carcinogenic to humans; *MAK*: Not listed; *NIOSH*: Not listed; *NTP*: Not listed; *ACGIH*: Not listed; *OSHA*: Not listed.

LD50 Value: orl-rat LD50:2500 mg/kg

Other Species Toxicity Data: (Source: NIOSH RTECS 1992)
orl-rat LD50:2500 mg/kg
unr-rat LD50:3000 mg/kg
orl-mus LD50:3850 mg/kg
ipr-mus LD50:1000 mg/kg

Irritation Data: (Source: NIOSH RTECS 1992)
eye-rbt 500 mg/24H SEV

Reproductive Toxicity (1992 RTECS): This chemical has no known mammalian reproductive toxicity.

No Significant Risk Level (Ca P65): N80 mg/day

▼ ▼ ▼ ▼ ▼ ▼ ▼ ▼ ▼ ▼ ▼ ▼ ▼ ▼

--- IDENTIFIERS ---

Name: SALICYCLIC ACID
Synonyms: O-Hydroxybenzoic Acid; Retarder W
CAS: 69-72-7; RTECS: VO0525000
Formula: $C_7H_6O_3$; Mol Wt: 138.13
WLN: QVR BQ
Chemical Class: Organic acid

See other identifiers listed below under Regulations.

--- PROPERTIES ---

Physical Description: White needlelike crystals or powder. (NYDH)
Boiling Point: 484.16 K at 20 mm; 211°C at 20 mm; 411.8°F at 20 mm
Melting Point: 430 K; 156.8°C; 314.3°F
Flash Point: NA
Autoignition: NA
UEL: NA
LEL: NA
Vapor Density: No data
Specific Gravity: 1.44 at 20°C
Density: 1.44 g/cc or 13.392 lb/gal

Reactivity with Water: No data on water reactivity
Reactivity with Common Materials: No data
Stability during Transport: Stable
Neutralizing Agents: No data
Polymerization Possibilities: No data

Toxic Fire Gases: Irritating vapors of unburned material and phenol may form
    in fire.
Odor Detected at (ppm): Data not available.
Odor Description: None (Source: CHRIS)
100% Odor Detection: No data

--- REGULATIONS ---

DOT hazard class: Not given
Packaging exceptions: 173
Nonbulk packaging: 173
Bulk packaging: 173

Retarders

421

STCC Number: Not listed

Clean Water Act Sect. 307: No
Clean Water Act Sect. 311: No
Clean Air Act: Not listed
EPA Waste Number: None
CERCLA Ref: Not listed
RQ Designation: Not listed
SARA TPQ Value: Not listed
SARA Sect. 312 Categories:
Acute toxicity: irritant.
Chronic toxicity: mutagen.
Chronic toxicity: reproductive toxin.

U.S. Postal Service Mailability: Not given

NFPA Codes:
Health Hazard (blue): (0) No unusual health hazard.
Flammability (red): (1) This material must be preheated before ignition can occur.
Reactivity (yellow): (0) Stable even under fire conditions.
Special: Unspecified.

--- TOXICITY DATA ---

Short-term Toxicity: *Inhalation:* May cause ringing in the ears, confusion, rapid pulse and breathing, headache, dizziness, nausea and vomiting. *Eyes:* May cause severe irritation. *Skin:* May be very irritating and cause skin sores. *Ingestion:* Ten grams may cause headache, dizziness, nausea, and vomiting. Death may occur from ingestion of about 1 ounce. (NYDH)

Long-term Toxicity: Repeated large doses may cause, in addition to the symptoms listed above, abdominal pain, loss of appetite, heartburn, poor digestion, stomach ulcers, bleeding of the stomach, iron-deficiency anemia, acnelike skin sores, restlessness, incoherent speech, tremor, kidney damage, coma, convulsions, and death. (NYDH)

Target Organs: Skin, eyes.

Symptoms: Inhalation of dust irritates nose and throat. Vomiting may occur spontaneously if large amounts are swallowed. Contact with eyes causes irritation, marked pain, and corneal injury, which should heal. Prolonged or repeated skin contact may cause marked irritation or even a mild burn. (Source: CHRIS)

Conc IDLH: Unknown

NIOSH REL: Not given

ACGIH TLV: Not listed
ACGIH STEL: Not listed

OSHA PEL: Not in Table Z-1-A

MAK Information: Not listed

Carcinogen: N; **Status:** See below

Carcinogen Lists: *IARC*: Not listed; *MAK*: Not listed; *NIOSH*: Not listed; *NTP*: Not listed; *ACGIH*: Not listed; *OSHA*: Not listed.

Human Toxicity Data: (Source: NIOSH RTECS)
skn-man TDLo:57 mg/kg JAMAAP 244, 660, 80
  Sense Organs
    Ear
      Tinnitus

LD50 Value: orl-rat LD50:891 mg/kg

Other Species Toxicity Data: (Source: NIOSH RTECS 1992)
orl-rat LD50:891 mg/kg
scu-rat LD50:1250 mg/kg
orl-mus LD50:480 mg/kg
ipr-mus LD50:300 mg/kg
scu-mus LD60:520 mg/kg
ivn-mus LD50:184 mg/kg
orl-cat LD50:400 mg/kg
orl-rbt LD50:1300 mg/kg
scu-rbt LDLo:6 g/kg

Reproductive Toxicity (1992 RTECS): This chemical is a mammalian reproductive toxin.

Reproductive Toxicity Data (1992 RTECS):
orl-rat TDLo:40 mg/kg (20-21D preg) PRGLBA 4, 93, 73
Maternal Effects
Parturition

orl-rat TDLo:1050 mg/kg (8-14D preg) SEIJBO 13, 73, 73
Maternal Effects
Uterus, cervix, vagina
Effects on Fertility
Postimplantation mortality
Litter size (number of fetuses per litter; measured before birth)

orl-rat TDLo:1050 mg/kg (8-14D preg) SEIJBO 13, 73, 73
Specific Developmental Abnormalities
Central nervous system
Craniofacial (including nose and tongue)
Musculoskeletal system

orl-rat TDLo:700 mg/kg (8-14D preg) SKEZAP 14, 549, 73
Effects on Embryo or Fetus
Fetotoxicity (except death, e.g., stunted fetus)
Specific Developmental Abnormalities
Musculoskeletal system

orl-rat TDLo:350 mg/kg (8-14D preg) SKEZAP 14, 549, 73
Effects on Embryo or Fetus
Extra embryonic features (e.g., placenta, umbilical cord)

scu-rat TDLo:380 mg/kg (9D preg) BCPCA6 22, 407, 73
Effects on Fertility
Postimplantation mortality
Effects on Embryo or Fetus
Fetotoxicity (except death, e.g., stunted fetus)
Specific Developmental Abnormalities
Other developmental abnormalities

orl-mus TDLo:1 g/kg (17D preg) APTOA6 35, 107, 74
Effects on Fertility
Postimplantation mortality
Effects on Embryo or Fetus
Fetotoxicity (except death, e.g., stunted fetus)

--- PROTECTION AND FIRST AID ---

Protection Suggested from the CHRIS Manual: Gloves; goggles; respirator for dust; clean body-covering clothing.

First Aid Source: CHRIS Manual, 1991
Inhalation: Move to fresh air.
Ingestion: Induce vomiting and get medical attention promptly.
Skin: Wash with soap and water.
Eyes: Promptly flush with water for 15 min and get medical attention.

--- INITIAL INCIDENT RESPONSE ---

Fire Extinguishment: Water, foam, dry chemical, carbon dioxide. *Note:* Water or foam may cause frothing. (CHRIS 91)

No DOT Guide information for this product.

▼ ▼ ▼ ▼ ▼ ▼ ▼ ▼ ▼ ▼ ▼ ▼ ▼ ▼

--- IDENTIFIERS ---

Name: PHTHALIC ANHYDRIDE
Synonyms: Anhydride Phtalique (French); Anidride Ftalica (Italian); 1,2-Benzenedicarboxylic Acid Anhydride; 1,3-Dioxophthalan; Esen; Ftalowy Bezwodnik (Polish); Ftaalzuuranhydride (Dutch); Isobenzofuran, 1,3-Dihydro-1,3-Dioxo-; 1,3-Isobenzofurandione; NCI-C03601; Phthalandione; 1,3-Phthalandione; Phthalic Acid Anhydride; Phthalsaeureanhydrid (German); Retarder AK; Retarder Esen; Retarder PD; PAN
CAS: 85-44-9; **RTECS:** TI3150000
Formula: $C_8H_4O_3$; **Mol Wt:** 148.12
WLN: T56 BVOVJ
Chemical Class: Organic anhydride

See other identifiers listed below under Regulations.

--- PROPERTIES ---

Physical Description: Colorless or pale yellow solid flakes or liquid (heated) with a choking odor
Boiling Point: 545.93 K; 272.7°C; 523°F
Melting Point: 404.26 K; 131.1°C; 267.9°F

Flash Point: 424 K; 150.8°C; 303.5°F
Autoignition: 843 K; 569.8°C; 1057.7°F
Vapor Pressure: <0.05 mm
UEL: 10.4%
LEL: 1.7%
Vapor Density: No data
Specific Gravity: 1.20; 1.53 at 135°C; 20°C
Water Solubility: 0.62%
Incompatibilities: Strong oxidizers

Reactivity with Water: Solid has very slow reaction; no hazard. Liquid spatters when in contact with water.
Reactivity with Common Materials: No data
Stability during Transport: No data
Neutralizing Agents: Water and sodium bicarbonate
Polymerization Possibilities: No data

Toxic Fire Gases: None reported other than possible unburned vapors
Odor Detected at (ppm): 0.32 to 0.72
Odor Description: Characteristic choking odor; choking, acrid. (Source: CHRIS)
100% Odor Detection: No data

--- REGULATIONS ---

DOT hazard class: 8 CORROSIVE
DOT guide: 60
Identification number: UN2214
DOT shipping name: Phthalic anhydride (with more than 0.05% maleic anhydride)
Packing group: III
Label(s) required: CORROSIVE
Special Provisions: T7, T38
Packaging exceptions: 173.154
Nonbulk packaging: 173.213
Bulk packaging: 173.240
Quantity limitations:
  Passenger air/rail: 25 kg
  Cargo aircraft only: 100 kg
  Vessel stowage: A

STCC Number: Not listed

Clean Water Act Sect. 307: No
Clean Water Act Sect. 311: No
Clean Air Act: CAA '90 listed
EPA Waste Number: U190
CERCLA Ref: Y
RQ Designation: D 5000 lb (2270 kg) CERCLA
SARA TPQ Value: Not listed
SARA Sect. 312 Categories:
Acute toxicity: corrosive.
Acute toxicity: irritant.
Chronic toxicity: adverse effect to target organ after long period of exposure.
Chronic toxicity: reproductive toxin.
Listed in SARA Sect 313: Yes
De Minimus Concentration: 1.0%

U.S. Postal Service Mailability: Not given

NFPA Codes:
Health Hazard (blue): (2) Hazardous to health. Area may be entered with self-contained breathing apparatus.
Flammability (red): (1) This material must be preheated before ignition can occur.
Reactivity (yellow): (0) Stable even under fire conditions.
Special: Unspecified.

--- TOXICITY DATA ---

Short-term Toxicity: *Inhalation:* May cause irritation of nose, throat, and mouth with sneezing, and excessive discharge and bleeding from nose. Studies suggest that this will occur at about 4 ppm. *Eyes:* May cause severe irritation and chemical burns on contact or at dust levels above 5 ppm. *Skin:* Rapid chemical burns may occur on contact with wet skin. Molten material may cause severe burns. *Ingestion:* May cause severe irritation to mouth and throat. Animal studies suggest that death may occur from ingestion of 4 to 8 oz. (NYDH)

Long-term Toxicity: May cause irritation of nose, mouth, throat, and lungs. Allergy may develop in sensitive individuals, which can lead to bronchial asthma. (NYDH)

Target Organs: Skin, eye, upper respiratory tract, mucous membrane, liver, kidneys.

Symptoms: Solid irritates skin and eyes, causing coughing and sneezing. Liquid causes severe thermal burns. (Source: CHRIS)

Conc IDLH: 10,000 ppm

ACGIH TLV: TLV = 1 ppm (6 mg/m^3)
ACGIH STEL: Not listed

OSHA PEL: Transitional Limits: PEL = 2 ppm (12 mg/m^3); Final Rule Limits: TWA = 1 ppm (6 mg/m^3)

MAK Information: 5 calculated as total dust mg/m^3
Local irritant: Peak = 2xMAK for 5 min, 8 times per shift. Causes allergic reactions

Carcinogen: N; **Status:** See below

Carcinogen Lists: *IARC*: Not listed; *MAK*: Not listed; *NIOSH*: Not listed; *NTP*: Not listed; *ACGIH*: Not listed; *OSHA*: Not listed.

LD50 Value: orl-rat LD50:4020 mg/kg

Other Species Toxicity Data: (Source: NIOSH RTECS 1992)
orl-rat LD50:4020 mg/kg
unr-rat LD50:1100 mg/kg
orl-mus LD50:1500 mg/kg
orl-cat LD50:800 mg/kg
ipr-gpg LD50:100 mg/kg

Irritation Data: (Source: NIOSH RTECS 1992)
skn-rbt 500 mg/24H
eye-rbt 100 mg

Reproductive Toxicity (1992 RTECS): This chemical is a mammalian reproductive toxin.

Reproductive Toxicity Data (1992 RTECS):
ihl-rat TCLo:1 mg/m^3 (45D male) GISAAA 35(1), 105, 70
Paternal Effects
Spermatogenesis
Testes, epididymis, sperm duct

--- **PROTECTION AND FIRST AID** ---

NIOSH Pocket Guide to Chemical Hazards:
**Wear Appropriate Equipment to Prevent: Repeated or prolonged skin contact.

**Wear Eye Protection to Prevent: Reasonable probability of eye contact.

**Exposed Personnel Should Wash: Promptly when skin becomes contaminated.

**Work Clothing Should Be Changed Daily: If there is any reasonable possibility that the clothing may be contaminated.

**Remove Clothing: Promptly remove nonimpervious clothing that becomes contaminated.

**Reference: NIOSH

Recommended Respiration Protection Source: NIOSH Pocket Guide (85-114)
ACGIH (Phthalic Anhydride)
5 ppm: Any dust and mist respirator except single-use respirators. Substance reported to cause eye irritation or damage; may require eye protection.
10 ppm: Any dust and mist respirator except single-use and quarter-mask respirators. Any supplied-air respirator. Any self-contained breathing apparatus. Substance reported to cause eye irritation or damage may require eye protection.
25 ppm: Any supplied-air respirator operated in a continuous-flow mode. Any powered air-purifying respirator with a dust and mist filter. Substance reported to cause eye irritation or damage may require eye protection.
50 ppm: Any supplied-air respirator with a full facepiece. Any self-contained breathing apparatus with a full facepiece. Any air-purifying full facepiece respirator with a high-efficiency particulate filter.
2000 ppm: Any supplied-air respirator with a full facepiece and operated in a pressure-demand or other positive-pressure mode.

Emergency or Planned Entry in Unknown Concentrations or IDLH Conditions:
Any self-contained breathing apparatus with a full facepiece and operated in
a pressure-demand or other positive-pressure mode. Any supplied-air
respirator with a full facepiece and operated in a pressure-demand or other
positive-pressure mode in combination with an auxiliary self-contained
breathing apparatus operated in a pressure-demand or other positive-pressure
mode.
Escape: Any air-purifying full facepiece respirator with a high-efficiency
particulate filter. Any appropriate escape-type self-contained breathing
apparatus.

First Aid Source: NIOSH
None given.

First Aid Source: CHRIS Manual 1991.
Inhalation: Gargle with water and use a sedative cough mixture.
Ingestion: Induce vomiting and give water, milk, or vegetable oil.
Skin or Eye Contact: Flush with water for at least 15 min; if burned by
molten material, remove as much solid as possible, soak off the remainder in
cold water, and then treat the burn.

First Aid Source: DOT Emergency Response Guide 1993.
Move victim to fresh air; call emergency medical care. In case of contact
with material, immediately flush skin or eyes with running water for at least
15 min. Remove and isolate contaminated clothing and shoes at the site.
Keep victim quiet and maintain normal body temperature.

--- INITIAL INCIDENT RESPONSE ---

Fire Extinguishment: Water, fog, dry chemical, carbon dioxide, or foam.
Note: Water may cause frothing. (CHRIS 91)

U.S. Department of Transportation Guide to Hazardous Materials Transport
Information - Publication DOT 5800.5 (1990).
DOT Shipping Name: Phthalic anhydride (with more than 0.05% maleic
anhydride)
DOT ID Number: UN2214

* POTENTIAL HAZARDS *

*Health Hazards
Contact causes burns to skin and eyes.
If inhaled, may be harmful.
Fire may produce irritating or poisonous gases.
Runoff from fire control or dilution water may cause pollution.

*Fire or Explosion
Some of these materials may burn, but none of them ignites readily.
Flammable/poisonous gases may accumulate in tanks and hopper cars.
Some of these materials may ignite combustibles (wood, paper, oil, etc.).

* EMERGENCY ACTION *

Keep unnecessary people away; isolate hazard area and deny entry.
Stay upwind; keep out of low areas.
Positive-pressure self-contained breathing apparatus (SCBA) and structural firefighters' protective clothing will provide limited protection.
CALL CHEMTREC AT 1-800-424-9300 FOR EMERGENCY ASSISTANCE. If water pollution occurs, notify the appropriate authorities.

*Fire
Some of these materials may react violently with water.
Small fires: Dry chemical, CO_2, water spray, or regular foam.
Large fires: Water spray, fog, or regular foam.
Move container from fire area if you can do it without risk.
Apply cooling water to sides of containers that are exposed to flames until well after fire is out. Stay away from ends of tanks.

*Spill or Leak
Do not touch spilled material; stop leak if you can do it without risk.
Small spills: Take up with sand, or other noncombustible absorbent material and place into containers for later disposal.
Small dry spills: With clean shovel, place material into clean, dry container and cover loosely; move containers from spill area.
Large spills: Dike far ahead of liquid spill for later disposal.

*First Aid

Move victim to fresh air; call emergency medical care.

In case of contact with material, immediately flush skin or eyes with running water for at least 15 min.

Remove and isolate contaminated clothing and shoes at the site.

Keep victim quiet and maintain normal body temperature.

8   BLOWING AGENTS

Name: FORMAMIDE, 1,1'-AZOBIS-
Synonyms: 1,1'-Azobiscarbamide; Azobiscarbonamide; Azobiscarboxamide; 1,1'-Azobis(Formamide); Azodicarbamide; Azodicarboamide; Azodicarbonamide; Azodicarboxamide; Azodicarboxylic Acid Diamide; delta(1,1'-Biurea; Celosen AZ; ChKhZ 21; ChKhZ 21R; Diazenedicarboxamide; Genitron AC; Genitron AC 2; Genitron AC 4; Kempore; Kempore 125; Kempore R 125; Lucel ADA; NCI-C55981; Nitropore; Pinhole AK 2; Porofor 505; Porofor ADC/R; Porofor ChKhZ 21; Porofor ChKhZ 21R; Unifoam AZ; Uniform AZ; Yunihomu AZ
CAS: 123-77-3; **RTECS:** LQ1040000
Formula: $C_2H_4N_4O_2$; **Mol Wt:** 116.10
WLN: ZVNUNVZ
Chemical Class: Amide

See other identifiers listed below under Regulations.

--- PROPERTIES ---

Boiling Point: NA
Melting Point: NA
Flash Point: NA
Autoignition: NA
UEL: NA
LEL: NA
Vapor Density: No data
Specific Gravity: No data

Reactivity with Water: No data on water reactivity
Reactivity with Common Materials: No data

Stability during Transport: No data
Neutralizing Agents: No data
Polymerization Possibilities: No data

Toxic Fire Gases: None reported other than possible unburned vapors
Odor Detected at (ppm): Unknown
Odor Description: No data
100% Odor Detection: No data

--- REGULATIONS ---

DOT hazard class: Not given
Packaging exceptions: 173
Nonbulk packaging: 173
Bulk packaging: 173

STCC Number: Not listed

Clean Water Act Sect. 307: No
Clean Water Act Sect. 311: No
Clean Air Act: Not listed
EPA Waste Number: None
CERCLA Ref: Not listed
RQ Designation: Not listed
SARA TPQ Value: Not listed
SARA Sect. 312 Categories:
    Chronic toxicity: mutagen.

U.S. Postal Service Mailability: Not given

--- TOXICITY DATA ---

Short-term Toxicity: Unknown

Long-term Toxicity: Unknown

Conc IDLH: Unknown

NIOSH REL: Not given

ACGIH TLV: Not listed
ACGIH STEL: Not listed

OSHA PEL: Not in Table Z-1-A

MAK Information: Not listed

Carcinogen: N; **Status:** See below

Carcinogen Lists: *IARC*: Not listed; *MAK*: Not listed; *NIOSH*: Not listed; *NTP*: Not listed; *ACGIH*: Not listed; *OSHA*: Not listed.

LD50 Value: No LD50 in RTECS 1992

Reproductive Toxicity (1992 RTECS): This chemical has no known mammalian reproductive toxicity.

--- INITIAL INCIDENT RESPONSE ---

No DOT Guide information for this product.

▼ ▼ ▼ ▼ ▼ ▼ ▼ ▼ ▼ ▼ ▼ ▼ ▼ ▼

--- IDENTIFIERS ---

Name: DINITROSOPENTAMETHYLENETETRAMINE
Synonyms: Aceto DNPT 40; Aceto DNPT 80; Aceto DNPT 100; CHKHZ 18; Dinitrosopentamethylenetetramine; *N,N*-Dinitrosopentamethylenetetramine; *N*(sup 1),*N*(sup 3)-Dinitrosopentamethylenetetramine; *N,N'*-Dinitrosopentamethylenetetramine; 3,4-Di-*N*-Nitrosopentamethylenetetramine; 3,7-Di-*N*-Nitrosopentamethylenetetramine; 3,7-Dinitroso-1,3,5,7-Tetraazabicyclo-(3,3,1)-Nonane; DNPMT; DNPT; 1,5-Endomethylene-3,7-Dinitroso-1,3,5,7-Tetraazacyclooctane; 1,5-Methylene-3,7-Dinitroso-1,3,5,7-Tetraazacyclooctane; Micropor; Mikrofor N; NSC 73599; Opex; Pentamethylenetetramine, Dinitroso-; Porofor CHKHC-18; Porophor B; Unicel-ND; Unicel NDX; Vulcacel B-40; Vulcacel BN
CAS: 101-25-7; **RTECS:** XA5250000
Formula: $C_5H_{10}N_6O_2$; **Mol Wt:** 186.21
WLN: T66 A BN DN FN HNTJ DNO HNO
Chemical Class: Nitro compound; Aliphatic amine

See other identifiers listed below under Regulations.

--- PROPERTIES ---

Boiling Point: NA
Melting Point: NA
Flash Point: NA
Autoignition: NA
UEL: NA
LEL: NA
Vapor Density: No data
Specific Gravity: No data

Reactivity with Water: No data on water reactivity
Reactivity with Common Materials: No data
Stability during Transport: No data
Neutralizing Agents: No data
Polymerization Possibilities: No data

Toxic Fire Gases: None reported other than possible unburned vapors
Odor Detected at (ppm): Unknown
Odor Description: No data
100% Odor Detection: No data

--- REGULATIONS ---

DOT hazard class: 4.1 FLAMMABLE SOLID
DOT guide: 71
Identification number: UN2972
DOT shipping name: N,N'-Dinitrosopentamethylenetetramine (not more than 82% with phlegmatizer)
Packing group: II
Label(s) required: FLAMMABLE SOLID, EXPLOSIVE
Special Provisions: 41, 53
Packaging exceptions: 173.None
Nonbulk packaging: 173.224
Bulk packaging: 173.None
Quantity limitations:
   Passenger air/rail: Forbidden
   Cargo aircraft only: Forbidden
   Vessel stowage: D
   Other stowage provisions: 12, 61

STCC Number: Not listed

Clean Water Act Sect. 307: No
Clean Water Act Sect. 311: No
Clean Air Act: Not listed
EPA Waste Number: None
CERCLA Ref: Not listed
RQ Designation: Not listed
SARA TPQ Value: Not listed
SARA Sect. 312 Categories:
    Chronic toxicity: mutagen.

U.S. Postal Service Mailability: Not given

--- TOXICITY DATA ---

Short-term Toxicity: Unknown

Long-term Toxicity: Unknown

Conc IDLH: Unknown

NIOSH REL: Not given

ACGIH TLV: Not listed
ACGIH STEL: Not listed

OSHA PEL: Not in Table Z-1-A

MAK Information: Not listed

Carcinogen: N; Status: See below

Carcinogen Lists: *IARC*: Not classified as to human carcinogenicity or probably not carcinogenic to humans; *MAK*: Not listed; *NIOSH*: Not listed; *NTP*: Not listed; *ACGIH*: Not listed; *OSHA*: Not listed.

LD50 Value: orl-rat LD50:940 mg/kg

Other Species Toxicity Data: (Source: NIOSH RTECS 1992)
    orl-rat LD50:940 mg/kg
    ipr-rat LD50:220 mg/kg
    scu-rat LD50:220 mg/kg
    ipr-mus LD50:130 mg/kg

scu-mus LD50:140 mg/kg
ivn-mus LD50:120 mg/kg
ivn-rbt LD50:130 mg/kg

Reproductive Toxicity (1992 RTECS):   This chemical has no known
mammalian reproductive toxicity.

--- PROTECTION AND FIRST AID ---

First Aid Source: DOT Emergency Response Guide 1993.
Move victim to fresh air. In case of contact with material, immediately flush
eyes with running water for at least 15 min. Wash skin with soap and water.
Remove and isolate contaminated clothing and shoes at the site. Keep victim
quiet and maintain normal body temperature.

--- INITIAL INCIDENT RESPONSE ---

U.S. Department of Transportation Guide to Hazardous Materials Transport
Information - Publication DOT 5800.5 (1990).
DOT Shipping Name: N,N'-Dinitrosopentamethylenetetramine (not more than
82% with phlegmatizer)
DOT ID Number: UN2972

ERG90   GUIDE 71
* POTENTIAL HAZARDS *

*Health Hazards
Contact may cause burns to skin and eyes.
Fire may produce irritating or poisonous gases.
Runoff from fire control or dilution water may cause pollution.

*Fire or Explosion
May be ignited by heat, sparks, or flames.
May burn rapidly.
Container may explode violently in heat of fire.
May explode from friction, heat, or contamination.

* EMERGENCY ACTION *

Keep unnecessary people away; isolate hazard area and deny entry.
Stay upwind; keep out of low areas.

Positive-pressure self-contained breathing apparatus (SCBA) and structural firefighters' protective clothing will provide limited protection. **CALL CHEMTREC AT 1-800-424-9300 FOR EMERGENCY ASSISTANCE.** If water pollution occurs, notify the appropriate authorities.

*Fire
Small fires: Dry chemical, CO_2, water spray, or regular foam.
Large fires: Flood fire area with water.
Fight fire from maximum distance. Stay away from ends of tanks.
If fire can be controlled, cool container with water from unmanned hose holder or monitor nozzles until well after fire is out.
If this is impossible, withdraw from fire area and let fire burn.

*Spill or Leak
Shut off ignition sources; no flares, smoking, or flames in hazard area.
Do not touch or walk through spilled material.
Spills: Moisten material with water and place it into loosely covered plastic or fiberboard containers for later disposal.

*First Aid
Move victim to fresh air.
In case of contact with material, immediately flush eyes with running water for at least 15 min. Wash skin with soap and water.
Remove and isolate contaminated clothing and shoes at the site.
Keep victim quiet and maintain normal body temperature.

▼ ▼ ▼ ▼ ▼ ▼ ▼ ▼ ▼ ▼ ▼ ▼ ▼ ▼

--- IDENTIFIERS ---

Name: BENZENESULFONIC ACID, 4,4'-OXYBIS-, DIHYDRAZIDE
Synonyms: Benzenesulfonic Acid, Oxybis-, Dihydrazide (9CI); Cellmic S; Celogen OT; Cenitron OB; Nitropore OBSH; OBSH; *p,p'*-Oxybisbenzene Disulfonylhydrazide; Oxybis(Benzenesulfonylhydrazide); *p,p'*-Oxybis(Benzenesulfonyl Hydrazide)
CAS: 80-51-3; **RTECS:** DB7321000
Formula: $C_{12}H_{14}N_4O_5S_2$; **Mol Wt:** 358.42
WLN: ZMSWR DOR DSWMZ
Chemical Class: Organic acid salt

See other identifiers listed below under Regulations.

--- PROPERTIES ---

Boiling Point: NA
Melting Point: 3.16 to 434.16 K dec; 160 to 161°C dec; 320 to 321.8°F dec
Flash Point: NA
Autoignition: NA
UEL: NA
LEL: NA
Vapor Density: No data
Specific Gravity: No data

Reactivity with Water: No data on water reactivity
Reactivity with Common Materials: No data
Stability during Transport: No data
Neutralizing Agents: No data
Polymerization Possibilities: No data

Toxic Fire Gases: None reported other than possible unburned vapors
Odor Detected at (ppm): Unknown
Odor Description: No data
100% Odor Detection: No data

--- REGULATIONS ---

DOT hazard class: Not given
Packaging exceptions: 173
Nonbulk packaging: 173
Bulk packaging: 173

STCC Number: Not listed

Clean Water Act Sect. 307: No
Clean Water Act Sect. 311: No
Clean Air Act: Not listed
EPA Waste Number: None
CERCLA Ref: Not listed
RQ Designation: Not listed
SARA TPQ Value: Not listed
SARA Sect. 312 Categories:
   Chronic toxicity: mutagen.

U.S. Postal Service Mailability: Not given

--- TOXICITY DATA ---

Short-term Toxicity: Unknown

Long-term Toxicity: Unknown

Conc IDLH: Unknown

NIOSH REL: Not given

ACGIH TLV: Not listed
ACGIH STEL: Not listed

OSHA PEL: Not in Table Z-1-A

MAK Information: Not listed

Carcinogen: N; Status: See below

Carcinogen Lists: *IARC*: Not listed; *MAK*: Not listed; *NIOSH*: Not listed; *NTP*: Not listed; *ACGIH*: Not listed; *OSHA*: Not listed.

LD50 Value: No LD50 in RTECS 1992

Reproductive Toxicity (1992 RTECS): This chemical has no known mammalian reproductive toxicity.

--- INITIAL INCIDENT RESPONSE ---

No DOT Guide information for this product.

▼ ▼ ▼ ▼ ▼ ▼ ▼ ▼ ▼ ▼ ▼ ▼ ▼ ▼

--- IDENTIFIERS ---

Chemical Name and Synonyms: *p*-TOLUENE SULFONYL
SEMICARBAZIDE

Trade Name and Synonyms: Celogen RA

Chemical Family: Blowing Agent for Rubbers and Other Polymers

Manufacturer: Uniroyal Chemical Co., Naugatuck, Connecticut

UN ID Number: NA

--- INGREDIENTS AND COMPOSITION DATA ---

Material	Cas No.	%	TLV
p-Toluene sulfonyl	NA		Not established

--- PHYSICAL DATA ---

Boiling Point (°F): NA

Specific Gravity (H$_2$O = 1): 1.44 at 25°C

Vapor Pressure (mm Hg): NA

Percent Volatile by Weight (%): Relatively nonvolatile below decomposition temperature

Vapor Density (air = 1): NA

Freezing Point: NA

Evaporation Rate: NA

pH: Noncorrosive

Solubility: Relatively insoluble in organic solvents and water. Soluble in aqueous bases.

Melting Point (°C): Decomposes at about 225.

Appearance/Color/Odor: White powder

--- FIRE AND EXPLOSION HAZARDS ---

Flash Point: NA

Flammability Limits: NA

Autoignition Temperature: Decomposes at about 430°F without ignition when diluted or in small amounts. Large amounts may ignite at this temperature due to heat buildup from decomposition.

Decomposition Temperature: 430°F

Extinguishing Media: Water in copious amounts.

Special Fire Fighting Procedures: Use self-contained breathing apparatus.

Unusual Fire and Explosion Hazards: Fires involving large amounts of material should not be approached because individual containers may rupture abruptly causing a fireball effect. In case of fire in nearby materials, soak containers with water. Fires involving Celogen RA should be fought only with fixed automatic protection equipment. *DO NOT ATTEMPT MANUAL FIRE FIGHTING.*

Hazardous Products of Combustion and Decomposition: Carbon monoxide, carbon dioxide, and oxides of nitrogen.

Hazardous Polymerization: Will not occur.

--- HEALTH HAZARD DATA ---

Effects of Overexposure: None reported by manufacturer.
Primary Routes of Exposure: Dermal, ingestion, and inhalation.
Threshold Limit Value: None established.

Emergency Response and First Aid:
Skin: Wash with lots of clean water for at least 15 min.
Eyes: Flush eyes with lots of water. If irritation persists, consult a physician.
Inhalation: Expose to fresh air. Keep warm and quiet. Give artificial respiration if necessary.
Ingestion: None recommended. Consult physician immediately.

Cancer Information: None reported.

--- TOXICITY DATA ---

Animal Effects:
Oral, LD50 (rats): Over 10 g/kg.
Skin, LD50 (rabbits): No evidence of systemic toxicity or irritation effects.

Manufacturer notes that there are no health problems associated with normal handling and use.

--- REACTIVITY DATA ---

Stability and Conditions to Avoid: Stable.

Materials to Avoid (Incompatibility): Oxidizing agents; e.g., peroxides will reduce decomposition temperature.

Hazard Decomposition Products: Gaseous decomposition products are nitrogen, carbon dioxide, and carbon monoxide.

Hazardous Polymerization: Will not occur.

--- SPILL AND EMERGENCY RESPONSE ---

For Spills: Sweep up promptly. Wear an approved dust respirator.

Waste Disposal Method: Burn cautiously in a remote, open area.

Empty Containers: Store in cool, dry place in closed containers. Keep away from heat, sparks, and open flames. Separate from combustible storage by 8 ft aisles or isolate in a detached building. Do not store over 4 ft high unless sprinklers are hydraulically designed to produce a density of 0.5 GPM/ft^2 over the entire area of the building.

--- SPECIAL PROTECTION AND PRECAUTIONS ---

Respiratory Protection: Approved dust respirator.

Type of Ventilation Required: Use the guidelines recommended by the American Conference of Governmental Industrial Hygienists in the current edition of "Industrial Ventilation," considering the TLV, lower explosive (Flammable) limit, and conditions under which this product is used.

Eye Protection Requirements: Chemical goggles, face shields. Eyewash fountains and safety showers should be easily accessible.

Protective Gloves and Other: Wear rubber gloves and goggles. Coveralls and/or rubber apron. Rubber shoes or boots.

Storage and Handling Precautions: Store in tightly closed, properly labeled containers in cool, well-ventilated area away from all ignition sources. Avoid prolonged and/or repeated contact with skin.

Prevent inhalation of dust. Prevent contact with eyes and skin. Under dusty conditions, static electricity may cause an explosion. Keep containers closed. Store in a cool, dry place.

Use with adequate ventilation. Wash thoroughly after handling before eating and smoking. Work clothing should be frequently laundered and not worn away from work premises.

Other Precautions: In accord with good industrial practice, avoid unnecessary personal contact. Employ bonding, ground, venting, and explosion-relief provisions in accord with accepted engineering practices in process operations capable of generating dust and/or static electricity.

Avoid excessive heat and pressure buildup during any mixing or processing operations.

--- REGULATORY INFORMATION ---

No information reported by manufacturer.

9 OILS, PLASTICIZERS, LUBRICANTS, SOLVENTS, AND MOLD RELEASE AGENTS

--- IDENTIFIERS ---

Chemical Name and Synonyms: THERMOPLASTIC PHENOLIC RESINS

Trade Name and Synonyms: Durez 12686

Chemical Family: Phenolic Resins

Manufacturer: Hooker Chemicals & Plastics Corp., North Tonawanda, New York

UN ID Number: NA

--- INGREDIENTS AND COMPOSITION DATA ---

Material	Cas No.	%	TLV
Phenolic resins	NA		Not established

--- PHYSICAL DATA ---

Boiling Point (°F): NA

Specific Gravity (H$_2$O = 1): 1.05

Vapor Pressure (mm Hg): NA

Percent Volatile by Weight (%): NA

Vapor Density (air = 1): NA

Freezing Point: NA

Evaporation Rate: NA

pH: NA

Solubility in Water: Negligible

Melting Point (°C): NA

Appearance/Color/Odor: Available in flake or crushed form; phenolic in odor.

--- FIRE AND EXPLOSION HAZARDS ---

Flash Point: NA

Flammability Limits: NA

Autoignition Temperature: NA

Decomposition Temperature: NA

Extinguishing Media: Water fog, carbon dioxide, dry chemical.

Special Fire Fighting Procedures: Use self-contained breathing apparatus.

Unusual Fire and Explosion Hazards: Ignition temperature of dust cloud >400°C.

Hazardous Products of Combustion and Decomposition: Carbon monoxide, carbon dioxide, and oxides of nitrogen.

Hazardous Polymerization: Will not occur.

--- HEALTH HAZARD DATA ---

Effects of Overexposure: None reported by manufacturer.
Primary routes of exposure: Dermal, ingestion, and inhalation.
Threshold Limit Value: None established.

Emergency Response and First Aid:
Skin: Wash with lots of clean water for at least 15 min. Call a physician.
Eyes: Flush eyes with lots of water. If irritation persists, consult a physician.
Inhalation: Expose to fresh air. Keep warm and quiet. Give artificial respiration if necessary.
Ingestion: None recommended. Consult physician immediately.

Cancer Information: None reported.

--- TOXICITY DATA ---

Animal Effects: No data reported.

--- REACTIVITY DATA ---

Stability and Conditions to Avoid: Stable.

Materials to Avoid (Incompatibility): No information reported.

Hazard Decomposition Products: No information reported.

Hazardous Polymerization: Will not occur.

--- SPILL AND EMERGENCY RESPONSE ---

For Spills: Sweep up promptly. Wear an approved dust respirator.

Waste Disposal Method: Normal plant waste, bury, or incinerate.

Empty Containers: Burn containers in an approved incinerator or dispose of in an approved chemical landfill in accordance with all applicable local, state, and federal laws and regulations.

--- SPECIAL PROTECTION AND PRECAUTIONS ---

Respiratory Protection: Approved dust respirator.

Type of Ventilation Required: Use the guidelines recommended by the American Conference of Governmental Industrial Hygienists in the current edition of "Industrial Ventilation," considering the TLV, lower explosive (flammable) limit, and conditions under which this product is used.

Eye Protection Requirements: Chemical goggles, face shields.   Eyewash fountains and safety showers should be easily accessible.

Protective Gloves and Other: Wear rubber gloves and goggles.   Coveralls and/or rubber apron.   Rubber shoes or boots.

Storage and Handling Precautions: Store in tightly closed, properly labeled containers in cool, well-ventilated area away from all ignition sources.   Avoid prolonged and/or repeated contact with skin.

Prevent inhalation of dust.   Prevent contact with eyes and skin.   Under dusty conditions, static electricity may cause an explosion.   Keep containers closed.   Store in a cool, dry place.

Use with adequate ventilation.   Wash thoroughly after handling before eating and smoking.   Work clothing should be frequently laundered and not worn away from work premises.

Other Precautions: In accordance with good industrial practice, avoid unnecessary personal contact.   Employ bonding, ground, venting, and explosion-relief provisions in accordance with accepted engineering practices in process operations capable of generating dust and/or static electricity.

--- REGULATORY INFORMATION ---

No information reported by manufacturer.

▼   ▼   ▼   ▼   ▼   ▼   ▼   ▼   ▼   ▼   ▼   ▼   ▼   ▼

--- IDENTIFIERS ---

Chemical Name and Synonyms: SYNTHETIC WAX ADDITIVE

Trade Name and Synonyms: Castorwax

Chemical Family: NA

Manufacturer: NL Industries, Hightstown, New Jersey

UN ID Number: NA

--- INGREDIENTS AND COMPOSITION DATA ---

Material	Cas No.	%	TLV
Synthetic wax additive	NA		30 mpp cf or 10 mg/m^3

--- PHYSICAL DATA ---

Boiling Point (°F): NA

Specific Gravity (H$_2$O = 1): 1.01

Vapor Pressure (mm Hg): NA

Percent Volatile by Weight (%): NA

Vapor Density (air = 1): NA

Freezing Point: NA

Evaporation Rate: NA

pH: NA

Solubility in Water: NA

Melting Point: NA

Appearance/Color/Odor: Dust or powdery material.

--- FIRE AND EXPLOSION HAZARDS ---

Flash Point: NA

Flammability Limits: NA

Autoignition Temperature: NA

Extinguishing Media: Foam or water spray, dry chemical, carbon dioxide.

Special Fire Fighting Procedures: Use water to keep fire-exposed containers cool. Use self-contained breathing apparatus. If fire is large, evacuate area and fight fire from a safe distance. Cool surrounding area with water.

If confined during exposure to a fire, can decompose with force. Material is subject to decomposition under friction or heavy shock.

Unusual Fire and Explosion Hazards: Exposure of containers to fire results in rapid product decomposition, container pressure buildup, and failure, followed by vigorous burning with flare effect. Cleanup should not be attempted until all the product has cooled completely. May form flammable dust-air mixtures.

Hazardous Products of Combustion and Decomposition: No information provided by manufacturer.

Hazardous Polymerization: Will not occur.

--- HEALTH HAZARD DATA ---

Effects of Overexposure: Unknown; this is not considered a hazardous material.

Emergency Response and First Aid:
Skin: Wash thoroughly with soap and running water. Remove contaminated clothing and wash before reuse.
Eyes: In case of contact, immediately flush with plenty of low-pressure water for at least 15 min. Remove any contact lenses to ensure thorough flushing. Call a physician.
Inhalation: Remove to fresh air and, if indicated, give artificial respiration. If breathing is difficult, give oxygen. Call a physician. Treat immediately.
Ingestion: If conscious, the person should immediately drink large quantities of liquid to dilute this product. *Never* give liquids to an unconscious person. Call a physician.

Cancer Information: Not listed as a carcinogen by NTP (National Toxicology Program); not regulated as a carcinogen by OSHA; not evaluated by IARC (International Agency for Research on Cancer).

--- TOXICITY DATA ---

Animal Effects: No information reported.

--- REACTIVITY DATA ---

Stability and Conditions to Avoid: No information reported.

Materials to Avoid (Incompatibility): No information reported.

Hazard Decomposition Products: No information reported.

Hazardous Polymerization: Will not occur.

--- SPILL AND EMERGENCY RESPONSE ---

For Spills: No special requirements reported.

Waste Disposal Method: Dispose of in accordance with federal, state, and local regulations.

Empty Containers: The container for this product can present explosion or fire hazards, even when emptied. To avoid the risk of injury, do not cut, puncture, or weld on or near this container.

--- SPECIAL PROTECTION AND PRECAUTIONS ---

Respiratory Protection: Where TLV is exceeded, use of an approved dust respirator is recommended (U.S. Bureau of Mines).

Type of Ventilation Required: Provide ventilation as required to keep TLV at or below 30 mpp cf or 10 mg/m^3 (nuisance dust).

Eye Protection Requirements: Chemical goggles. Eyewash fountains and safety showers should be easily accessible.

Protective Gloves and Other: Impervious gloves and clothing should be worn.

Storage and Handling Precautions: Store in cool, dry location away from all sources of heat, spark, and open flames. Avoid ignition sources such as sparks and flames. Ground all equipment. In addition, when emptying bags where flammable vapors may be present, blanket vessel with inert gas, ground operator, and pour material slowly into conductive, grounded chute.

Other Precautions: Use all standard practices for good personal hygiene. Completely isolate and thoroughly clean all equipment, piping, or vessels before beginning maintenance or repairs. Keep area clean.

--- REGULATORY INFORMATION ---

No information reported.

▼ ▼ ▼ ▼ ▼ ▼ ▼ ▼ ▼ ▼ ▼ ▼ ▼ ▼

--- IDENTIFIERS ---

Chemical Name and Synonyms: SYNTHETIC WAX ADDITIVE, $C_{24}H_{28}Cl_{22}$

Trade Name and Synonyms: Chlorowax 70

Chemical Family: NA

Manufacturer: Diamond Shamrock Chemical Co., Cleveland, Ohio

UN ID Number: NA

--- INGREDIENTS AND COMPOSITION DATA ---

Material	Cas No.	%	TLV
Wax	NA		Not established

--- PHYSICAL DATA ---

Boiling Point (°F): None known, decomposes

Specific Gravity (H_2O = 1): Approximately 1.35 at 25/25°C

Vapor Pressure (mm Hg): Essentially nonvolatile

Percent Volatile by Weight (%): 0% at 70°F

Vapor Density (air = 1): Essentially zero

Freezing Point: Essentially nonvolatile

Evaporation Rate: Essentially zero (relative to butyl acetate)

pH: NA

Solubility in Water: Negligible

Melting Point: NA

Appearance/Color/Odor: White to tan powder with characteristic odor.

--- FIRE AND EXPLOSION HAZARDS ---

Flash Point: None under 400°F

Flammability Limits: NA

Autoignition Temperature: NA

Extinguishing Media: Foam or water spray, dry chemical, carbon dioxide. Although not considered flammable, liquid chlorowax products flow freely when hot and should, therefore, be treated as an oil when exposed in a fire area.

Special Fire Fighting Procedures: Treat as a hot oil.

Unusual Fire and Explosion Hazards: Gaseous decomposition products generated with excess heat, which could result in rupture of nonvented containing vessel.

Hazardous Products of Combustion and Decomposition: Hydrogen chloride and traces of fragmented short-chain hydrocarbons.

Hazardous Polymerization: Will not occur.

--- HEALTH HAZARD DATA ---

Effects of Overexposure: None known.

Emergency Response and First Aid:
Skin: Wash thoroughly with soap and running water. Remove contaminated clothing and wash before reuse.

Eyes: In case of contact, immediately flush with plenty of low-pressure water for at least 15 min. Remove any contact lenses to ensure thorough flushing. Call a physician.

Inhalation: Remove to fresh air and, if indicated, give artificial respiration. If breathing is difficult, give oxygen. Call a physician. Treat immediately.

Ingestion: If conscious, the person should immediately drink large quantities of liquid to dilute this product. *Never* give liquids to an unconscious person. Call a physician.

Cancer Information: Not listed as a carcinogen by NTP (National Toxicology Program); not regulated as a carcinogen by OSHA; not evaluated by IARC (International Agency for Research on Cancer).

--- TOXICITY DATA ---

Animal Effects: Clinical laboratory tests have indicated that Chlorowax 70 is not acutely toxic as defined in the Regulations of the Federal Hazardous Substances Labeling Act.

Test Results

Oral: Based on laboratory toxicological investigations, the LD50 for white rats is assumed to be approximately 50 g/kg and 25 g/kg for guinea pigs.

Dermal: Patch tests on 200 unselected human subjects (100 male and 100 female) indicated that Chlorowax 70 is neither a primary irritant nor a sensitizer and may be used safely in contact with human skin.

1. Whether the test material is classed as toxic as this term is defined in the Regulations could not be determined since a sufficiently high dosage concentration for LD50 and LD100 could not be attained before death would be caused due to mechanical injury and not toxicity.
2. Smyth Laboratories, Philadelphia, Pennsylvania

Other Studies

Clinical Toxicology of Commercial Products by Gleason, Gosselin, and Hodge, published by Williams & Wilkins Co. (1963). Chlorowax type products are rated Class I, "no injury in test animals short of doses which produce intestinal obstruction" and "do not appear to be irritants or sensitizers." Class I is practically nontoxic with probable lethal dose for humans being greater than 1 qt (above 15 g/kg), page 90.

--- REACTIVITY DATA ---

Stability and Conditions to Avoid: Temperatures exceeding 400°F; strong bases.

Materials to Avoid (Incompatibility): Metals of the third and fourth periods of the periodic chart, when hot, especially in finely divided free state.

Hazard Decomposition Products: Unknown

Hazardous Polymerization: Will not occur.

--- SPILL AND EMERGENCY RESPONSE ---

For Spills: Cover liquid with oil-absorbent-type materials; sweep up and dispose.

Waste Disposal Method: Follow federal, state, and local regulations for health and pollution.

Empty Containers: Storage temperatures should not exceed 66°C for more than several hours in vented containers. Storage temperatures should not exceed 40°C in contained vessels.

Aqueous systems of finely divided metals of the third and fourth periods of the periodic chart, strongly basic chemicals in undiluted, aqueous form and especially in concentrated organic solvent solutions.

--- SPECIAL PROTECTION AND PRECAUTIONS ---

Respiratory Protection: Utilization of respiratory protection equipment when working in aerosol mist of liquid chlorowax.

Type of Ventilation Required: Not necessary unless overheated.

Eye Protection Requirements: Chemical goggles. Eyewash fountains and safety showers should be easily accessible.

Protective Gloves and Other: Impervious gloves and clothing should be worn. Manufacturer recommends rubber gloves be used.

Storage and Handling Precautions: Store in cool, dry location away from all sources of heat, spark, and open flames. Avoid ignition sources such as sparks and flames.

This material may react with acids and acidlike materials and strong oxidizing and reducing agents and should not be stored near such materials.

Other Precautions: Use all standard practices for good personal hygiene. Completely isolate and thoroughly clean all equipment, piping, or vessels before beginning maintenance or repairs. Keep area clean. *Note* that this product will burn.

--- REGULATORY INFORMATION ---

No information found.

▼   ▼   ▼   ▼   ▼   ▼   ▼   ▼   ▼   ▼   ▼   ▼   ▼   ▼

--- IDENTIFIERS ---

Chemical Name and Synonyms: DI-(2-ETHYL HEXYL) ADIPATE

Trade Name and Synonyms: Dioctyl Adipate C-497

Chemical Family: Ester

Manufacturer: C. P. Hall Co., Chicago, Illinois

UN ID Number: NA

--- INGREDIENTS AND COMPOSITION DATA ---

Material	Cas No.	%	TLV
Di-(2ethyl hexyl) adipate	NA		Not established

--- PHYSICAL DATA ---

Boiling Point (°F): 210 to 221 at 4 mm Hg

Specific Gravity (H$_2$O = 1): 0.923

Vapor Pressure (mm Hg): <0.15 at 150°C

Percent Volatile by Weight (%): Nonvolatile

Vapor Density (air = 1): NA

Freezing Point: NA

Evaporation Rate: NA

pH: NA

Solubility in Water: <0.01

Melting Point: NA

Appearance/Color/Odor: Mild, oxatyl type

--- FIRE AND EXPLOSION HAZARDS ---

Flash Point: 400°F (COC)

Flammability Limits: NA

Autoignition Temperature: NA

Extinguishing Media: Foam or water spray, dry chemical, carbon dioxide.

Special Fire Fighting Procedures: None reported.

Unusual Fire and Explosion Hazards: None reported.

Hazardous Products of Combustion and Decomposition: None reported.

Hazardous Polymerization: Will not occur.

--- HEALTH HAZARD DATA ---

Effects of Overexposure: None known.

Emergency Response and First Aid:
Skin: Wash thoroughly with soap and running water. Remove contaminated clothing and wash before reuse.
Eyes: In case of contact, immediately flush with plenty of low-pressure water for at least 15 min. Remove any contact lenses to ensure thorough flushing. Call a physician.
Inhalation: Remove to fresh air and, if indicated, give artificial respiration. If breathing is difficult, give oxygen. Call a physician. Treat immediately.
Ingestion: If conscious, the person should immediately drink large quantities of liquid to dilute this product. *Never* give liquids to an unconscious person. Call a physician.

Cancer Information: Not listed as a carcinogen by NTP (National Toxicology Program); not regulated as a carcinogen by OSHA; not evaluated by IARC (International Agency for Research on Cancer).

--- TOXICITY DATA ---

Animal Effects: None reported.

--- REACTIVITY DATA ---

Stability and Conditions to Avoid: Temperatures exceeding 400°F; strong bases.

Materials to Avoid (Incompatibility): None reported.

Hazard Decomposition Products: Unknown

Hazardous Polymerization: Will not occur.

--- SPILL AND EMERGENCY RESPONSE ---

For Spills: Cover liquid with oil-absorbent-type materials; sweep up and dispose.

Waste Disposal Method: Follow federal, state, and local regulations for health and pollution.

Empty Containers: Storage temperatures should not exceed 66°C for more than several hours in vented containers. Storage temperatures should not exceed 40°C in contained vessels.

Aqueous systems of finely divided metals of the third and fourth periods of the periodic chart, strongly basic chemicals in undiluted, aqueous form and especially in concentrated organic solvent solutions.

--- SPECIAL PROTECTION AND PRECAUTIONS ---

Respiratory Protection: Utilization of respiratory protection equipment when working in aerosol mist of liquid chlorowax.

Type of Ventilation Required: Not necessary unless overheated.

Eye Protection Requirements: Chemical goggles. Eyewash fountains and safety showers should be easily accessible.

Protective Gloves and Other: Impervious gloves and clothing should be worn. Manufacturer recommends rubber gloves be used.

Storage and Handling Precautions: Store in cool, dry location away from all sources of heat, spark, and open flames. Avoid ignition sources such as sparks and flames.

This material may react with acids and acidlike materials and strong oxidizing and reducing agents and should not be stored near such materials.

Other Precautions: Use all standard practices for good personal hygiene. Completely isolate and thoroughly clean all equipment, piping, or vessels before beginning maintenance or repairs. Keep area clean. *Note* that this product will burn.

--- REGULATORY INFORMATION ---

No information found.

▼   ▼   ▼   ▼   ▼   ▼   ▼   ▼   ▼   ▼   ▼   ▼   ▼   ▼

--- IDENTIFIERS ---

Name: DI-*N*-OCTYL PHTHALATE
Synonyms: 1,2-Benzenedicarboxylic acid, Di-*N*-Octyl Ester; Dioctyl Phthalate
CAS: 117-84-0; **RTECS:** TI1925000
Formula: $C_{24}H_{38}O_4$; **Mol Wt:** 390.54
WLN: 8OVR BVO8

Chemical Class: Ester

See other identifiers listed below under Regulations.

--- PROPERTIES ---

Boiling Point: 657 K; 383.8°C; 722.9°F
Melting Point: 223 K; -50.2°C; -58.3°F
Flash Point: 480 K; 206.8°C; 404.3°F
Autoignition: NA
UEL: NA
LEL: NA
Vapor Density: No data
Specific Gravity: No data
Density: 0.981

Reactivity with Water: No data on water reactivity
Reactivity with Common Materials: No data
Stability during Transport: No data
Neutralizing Agents: No data
Polymerization Possibilities: No data

Toxic Fire Gases: None reported other than possible unburned vapors
Odor Detected at (ppm): Unknown
Odor Description: No data
100% Odor Detection: No data

--- REGULATIONS ---

DOT hazard class: 9 CLASS 9
DOT guide: 31
Identification number: UN3077
DOT shipping name: ENVIRONMENTALLY HAZARDOUS
  SUBSTANCES, SOLID, N.O.S.
Packing group: III
Label(s) required: CLASS 9
Special Provisions: 8, B54
Packaging exceptions: 173.155
Nonbulk packaging: 173.213
Bulk packaging: 173.240
Quantity limitations:
  Passenger air/rail: None

Cargo aircraft only: None
Vessel stowage: A

STCC Number: Not listed

Clean Water Act Sect. 307: Yes
Clean Water Act Sect. 311: No
Clean Air Act: Not listed
EPA Waste Number: U107
CERCLA Ref: Y
RQ Designation: D, 5000 lb (2270 kg) CERCLA
SARA TPQ Value: Not listed
SARA Sect. 312 Categories:
  Acute toxicity: irritant.
  Chronic toxicity: carcinogen.
  Chronic toxicity: reproductive toxin.
Listed in SARA Sect. 313: Yes
De Minimus Concentration: 1.0%

U.S. Postal Service Mailability: Not given

--- TOXICITY DATA ---

Short-term Toxicity: Unknown

Long-term Toxicity: Unknown

Conc IDLH: Unknown

NIOSH REL: Not given

ACGIH TLV: Not listed
ACGIH STEL: Not listed

OSHA PEL: Not in Table Z-1-A

MAK Information: Not listed

Carcinogen: Y; **Status:** See below

Carcinogen Lists: *IARC*: Not classified as to human carcinogenicity or probably not carcinogenic to humans; *MAK*: Not listed; *NIOSH*: Carcinogen defined by NIOSH with no further categorization; *NTP*: Carcinogen defined by NTP as reasonably anticipated to be carcinogenic, with limited evidence in humans or sufficient evidence in experimental animals; *ACGIH*: Not listed; *OSHA*: Not listed.

LD50 Value: orl-rat LD50: >15 g/kg

Other Species Toxicity Data: (Source: NIOSH RTECS 1992)
orl-rat LD50: >15 g/kg
orl-mus LD50:6513 mg/kg
ipr-mus LD50:65 g/kg
skn-gpg LD50: >5 g/kg

Reproductive Toxicity (1992 RTECS): This chemical is a mammalian reproductive toxin.

Reproductive Toxicity Data (1992 RTECS):
ipr-rat TDLo:5 g/kg (5-15D preg) JPMSAE 61, 51, 72
    Effects on Embryo or Fetus
        Fetotoxicity (except death, e.g., stunted fetus)
    Specific Developmental Abnormalities
        Eye, ear
        Other developmental abnormalities

orl-mus TDLo:78 g/kg (7-14D preg) NTIS** PB85-220143
    Effects on Newborn
        Live birth index (number of fetuses per litter)
        Growth statistics (e.g., reduced weight gain)

--- PROTECTION AND FIRST AID ---

Recommended Respiration Protection Source: NIOSH Pocket Guide (85-114)
NIOSH (Di-*N*-Octyl Phthalate)
Greater at any detectable concentration: Any self-contained breathing apparatus with a full facepiece and operated in a pressure-demand or other positive-pressure mode. Any supplied-air respirator with a full facepiece and operated in a pressure-demand or other positive-pressure mode in combination with an auxiliary self-contained breathing apparatus operated in a pressure-demand or other positive-pressure mode.

Escape: Any air-purifying full facepiece respirator with a high-efficiency particulate filter. Any appropriate escape-type self-contained breathing apparatus.

First Aid Source: DOT Emergency Response Guide 1993.
In case of contact with material, immediately flush eyes with running water for at least 15 min. Wash skin with soap and water. Remove and isolate contaminated clothing and shoes at the site.

--- INITIAL INCIDENT RESPONSE ---

U.S. Department of Transportation Guide to Hazardous Materials Transport Information - Publication DOT 5800.5 (1990).
DOT Shipping Name: ENVIRONMENTALLY HAZARDOUS SUBSTANCES, SOLID, N.O.S.
DOT ID Number: UN3077

ERG90   GUIDE 31
* POTENTIAL HAZARDS *

*Health Hazards
Contact may cause burns to skin and eyes.
Fire may produce irritating or poisonous gases.
Runoff from fire control or dilution water may cause pollution.

*Fire or Explosion
Some of these materials may burn, but none of them ignites readily.

* EMERGENCY ACTION *

Keep unnecessary people away; isolate hazard area and deny entry.
Positive-pressure self-contained breathing apparatus (SCBA) and structural firefighters' protective clothing will provide limited protection.
CALL CHEMTREC AT 1-800-424-9300 FOR EMERGENCY ASSIS-TANCE. If water pollution occurs, notify the appropriate authorities.

*Fire
Small fires: Dry chemical, CO_2, water spray, or regular foam.
Large fires: Water spray, fog, or regular foam.
Move container from fire area if you can do it without risk.
Do not scatter spilled material with high-pressure water streams.
Dike fire-control water for later disposal.

*Spill or Leak
Stop leak if you can do it without risk.
Small dry spills: With clean shovel, place material into clean, dry container and cover loosely; move containers from spill area.
Small spills: Take up with sand or other noncombustible absorbent material and place into containers for later disposal.
Large spills: Dike far ahead of liquid spill for later disposal.
Cover powder spill with plastic sheet or tarp to minimize spreading.

*First Aid
In case of contact with material, immediately flush eyes with running water for at least 15 min. Wash skin with soap and water. Remove and isolate contaminated clothing and shoes at the site.

▼   ▼   ▼   ▼   ▼   ▼   ▼   ▼   ▼   ▼   ▼   ▼   ▼   ▼

--- IDENTIFIERS ---

Chemical Name and Synonyms: PETROLEUM HYDROCARBON
LUBRICANT

Trade Name and Synonyms: Penreco A.S.T.M. Reference Oil No. 1

Chemical Family: Petroleum Hydrocarbon Product

Manufacturer: PENRECO, Los Angeles, California

UN ID Number: NA

--- INGREDIENTS AND COMPOSITION DATA ---

Material	Cas No.	%	TLV
Petroleum hydrocarbon lubricant	NA		Not established

--- PHYSICAL DATA ---

Boiling Point (°F): >600

Specific Gravity (H₂O = 1): 0.882

Vapor Pressure (mm Hg): <0.0001

Percent Volatile by Weight (%): NA

Vapor Density (air = 1): >1

Freezing Point: NA

Evaporation Rate: <0.01

pH: NA

Solubility in Water: Soluble in hydrocarbons; insoluble in water.

Melting Point: NA

Appearance/Color/Odor: Petroleum hydrocarbon odor.

--- FIRE AND EXPLOSION HAZARDS ---

Flash Point: >500°F (COC)

Flammability Limits: NA

Autoignition Temperature: >600°F

Extinguishing Media: Use dry chemical, foam, or carbon dioxide.

Special Fire Fighting Procedures: Water may be ineffective but can be used to cool containers exposed to heat or flame. Caution should be exercised when using water or foam as frothing may occur, especially if sprayed into containers of hot, burning liquid.

Unusual Fire and Explosion Hazards: Dense smoke may be generated while burning. Carbon monoxide, carbon dioxide, and other oxides may be generated as products of combustion.

Hazardous Products of Combustion and Decomposition: None reported.

Hazardous Polymerization: Will not occur.

--- HEALTH HAZARD DATA ---

Effects of Overexposure: On rare occasions, prolonged and repeated exposure to oil mist poses a risk of pulmonary disease such as chronic lung inflammation. This condition is usually asymptomatic as a result of repeated small aspirations. Shortness of breath and cough are the most common symptoms. Aspiration may lead to chemical pneumonitis, which is characterized by pulmonary edema and hemorrhage and may be fatal. Signs of lung involvement include increased choking and gagging and are often noted at the time of aspiration. Gastrointestinal discomfort may develop, followed by vomiting, with a further risk of aspiration.

Emergency Response and First Aid:
Skin: Remove contaminated clothing. Wash contaminated area thoroughly with soap and water. If redness or irritation occurs, seek medical attention. If material is hot, submerge injured area in cold water. If victim is severely burned, remove to a hospital immediately.
Eyes: Immediately flush eyes with large amounts of water and continue flushing for 15 min. If material is hot, treat for thermal burns and take victim to hospital immediately.
Inhalation: This material has a low vapor pressure and is not expected to present an inhalation exposure at ambient conditions. If vapor or mist is generated when the material is heated or handled, remove victim from exposure. If breathing has stopped or is irregular, administer artificial respiration and supply oxygen if it is available. If victim is unconscious, remove to fresh air and seek medical attention.
Ingestion: Do not induce vomiting. Seek medical attention.

Cancer Information: Not listed as a carcinogen by NTP (National Toxicology Program); not regulated as a carcinogen by OSHA; not evaluated by IARC (International Agency for Research on Cancer).

--- TOXICITY DATA ---

Animal Effects: None reported.

--- REACTIVITY DATA ---

Stability and Conditions to Avoid: Stable.

Materials to Avoid (Incompatibility): None reported.

Hazard Decomposition Products: Unknown.

Hazardous Polymerization: Will not occur.

--- SPILL AND EMERGENCY RESPONSE ---

For Spills: Consult health effect information section, personal health protection information section, fire protection section, and reactivity data. Notify appropriate authorities of spill. Contain spill immediately. Do not allow spill to enter sewers or watercourses. Remove all sources of ignition. Absorb with appropriate inert material such as sand and clay. Large spills may be picked up using vacuum pumps, shovels, buckets, or other means and placed in drums or other suitable containers.

Waste Disposal Method: All disposals must comply with federal, state, and local regulations. The material, if spilled or discarded, may be a regulated waste. Refer to state and local regulations. *Caution:* If regulated solvents are used to clean up spilled material, the resulting waste mixture may be regulated. Department of Transportation (DOT) regulations may apply for transporting this material when spilled. Waste material may be landfilled or incinerated at an approved facility. Materials should be recycled if possible.

Empty Containers: Storage temperatures should not exceed 66°C for more than several hours in vented containers. Storage temperatures should not exceed 40°C in contained vessels.

Aqueous systems of finely divided metals of the third and fourth periods of the periodic chart, strongly basic chemicals in undiluted, aqueous form and especially in concentrated organic solvent solutions.

--- SPECIAL PROTECTION AND PRECAUTIONS ---

Respiratory Protection: Respiratory protection is not required under conditions of normal use. If vapor or mist is generated when the material is heated or handled, use an organic vapor respirator with a dust and mist filter. All respirators must be NIOSH certified. Do not use compressed oxygen in hydrocarbon atmospheres.

Type of Ventilation Required: If vapor or mist is generated when the material is heated or handled, adequate ventilation in accordance with good engineering practice must be provided to maintain concentrations below the specified exposure or flammable limits.

Eye Protection Requirements: Eye protection not required under conditions of normal use. If material is handled such that it could be splashed into eyes, wear plastic face shield or splash-proof safety goggles.

Protective Gloves and Other: No skin protection is required for single, short-duration exposures. For prolonged or repeated exposures, use impervious synthetic rubber clothing (boots, gloves, aprons, etc.) over parts of the body subject to exposure. If handling hot material, use insulated protective clothing (boots, gloves, aprons, etc.).

Storage and Handling Precautions: Store in cool, dry location away from all sources of heat, spark, and open flames. Avoid ignition sources such as sparks and flames.

This material may react with acids and acidlike materials and strong oxidizing and reducing agents and should not be stored near such materials.

Other Precautions: Use all standard practices for good personal hygiene. Completely isolate and thoroughly clean all equipment, piping, or vessels before beginning maintenance or repairs. Keep area clean. *Note* that this product will burn.

--- REGULATORY INFORMATION ---

Do not transfer to unmarked containers. Store in closed containers away from heat, sparks, open flame, or oxidizing materials. This product is not classified as hazardous under DOT regulations. Fire extinguishers should be kept readily available. See NFPA 30 and OSHA 1910.106 Flammable and Combustible Liquids.

▼   ▼   ▼   ▼   ▼   ▼   ▼   ▼   ▼   ▼   ▼   ▼   ▼

--- IDENTIFIERS ---

Chemical Name and Synonyms: PETROLEUM HYDROCARBON

Trade Name and Synonyms: Califlux 550

Chemical Family: Petroleum Hydrocarbon

Manufacturer: Witco Chemical Corp., Oildale, California

UN ID Number: NA

--- INGREDIENTS AND COMPOSITION DATA ---

Material	Cas No.	%	TLV
Dibenzoyl p-quinonedioxime	NA		Not established

--- PHYSICAL DATA ---

Boiling Point (°F): NA

Specific Gravity (H$_2$O = 1): 1.02

Vapor Pressure (mm Hg): NA

Percent Volatile by Weight (%): Not volatile

Vapor Density (air = 1): NA

Freezing Point: NA

Evaporation Rate: NA

pH: Not corrosive

Solubility in Water: Insoluble

Melting Point: NA

Appearance/Color/Odor: Oily mobile (viscous) liquid; no objectionable odor.

--- FIRE AND EXPLOSION HAZARDS ---

Flash Point: 465°F (COC)

Flammability Limits: NA

Autoignition Temperature: NA

Extinguishing Media: Foam, water fog, CO$_2$, dry chemical, vaporizing-liquid-type extinguishing agent.

Special Fire Fighting Procedures: Standard petroleum procedures.

Unusual Fire and Explosion Hazards:
Do not use welding or cutting torch on or near any container of this material, even empty, because an explosion could occur.
Do not use, pour, spill, or store near heat or open flame.

Hazardous Products of Combustion and Decomposition: Toxic fumes.

Hazardous Polymerization: Will not occur.

--- HEALTH HAZARD DATA ---

Effects of Overexposure: Principal routes of exposure are skin contact and inhalation.   Slight irritation potential reported by the manufacturer.

Ingestion of the product in large quantities may cause mild digestive disorders, including diarrhea.

There are no data available that address medical conditions that are generally recognized as being aggravated by exposure to this product.

Emergency Response and First Aid:
Skin: Flush all affected areas with plenty of water for several minutes. Remove and clean any contaminated clothing and shoes.   Seek medical attention if skin irritation occurs.
Eyes: Flush eyes with plenty of running water for at least 15 min.   Hold the eyelids apart during the flushing to ensure rinsing of the surface of the eye and lids with water.   Seek medical attention if eye irritation occurs.
Inhalation: Remove to fresh air and, if indicated, give artificial respiration. If breathing is difficult, give oxygen.   Call a physician.   Treat immediately.
Ingestion: If swallowed, immediately give several glasses of water and induce vomiting by gagging the victim with a finger placed on the back of the victim's tongue.   Give fluids until vomitus is clear.   If victim is unconscious or convulsing, do not induce vomiting or give anything by mouth.

Cancer Information: Not listed as a carcinogen by NTP (National Toxicology Program); not regulated as a carcinogen by OSHA; not evaluated by IARC (International Agency for Research on Cancer).

--- TOXICITY DATA ---

Animal Effects: Observe current ACGIH TLV of 5 mg/m³ for oil mists. Dermatitis in very sensitive individuals after extended exposure (days).

--- REACTIVITY DATA ---

Stability and Conditions to Avoid: Heat, open flames.

Materials to Avoid (Incompatibility): None reported by the manufacturer.

Hazard Decomposition Products: Unknown

Hazardous Polymerization: Will not occur.

--- SPILL AND EMERGENCY RESPONSE ---

For Spills: Sweep spillage; wash residuals with soap and water and transfer to a closed container.

Waste Disposal Method: Same as for organic chemicals.

Empty Containers: The container for this product can present explosion or fire hazards, even when emptied. To avoid the risk of injury, do not cut, puncture, or weld on or near this container.

--- SPECIAL PROTECTION AND PRECAUTIONS ---

Respiratory Protection: Appropriate respirator selected and used in accordance with OSHA Subpart I (29 CFR 1910.134) and manufacturer's recommendations is required if airborne dust is not adequately controlled or if excessive fumes are generated from product decomposition. Self-contained breathing apparatus should be worn in high vapor and dust concentrations. Under normal use, manufacturer recommends a dust mask.

Type of Ventilation Required: General mechanical. Avoid breathing dusts and vapors. Adequate ventilation should be provided to keep mist and dust concentrations below acceptable exposure limits. Discharge from the ventilation system should comply with applicable air-pollution regulations.

Eye Protection Requirements: Chemical goggles. Eyewash fountains and safety showers should be easily accessible.

Protective Gloves and Other: Impervious gloves and clothing should be worn. Manufacturer recommends rubber gloves be used.

Storage and Handling Precautions: Containers should be stored in a cool, dry, well-ventilated area. Store away from flammable materials, sources of heat, flame, sparks, and foodstuffs. Exercise due caution to prevent damage to or leakage from the container. Avoid any conditions that might tend to create a dust explosion. Maintain good housekeeping practices to minimize dust buildup.

Other Precautions: Use all standard practices for good personal hygiene. Completely isolate and thoroughly clean all equipment, piping, or vessels before beginning maintenance or repairs. Keep area clean. *Note* that this product will burn.

--- REGULATORY INFORMATION ---

None reported.

▼   ▼   ▼   ▼   ▼   ▼   ▼   ▼   ▼   ▼   ▼   ▼   ▼   ▼

--- IDENTIFIERS ---

Chemical Name and Synonyms: PETROLEUM HYDROCARBON

Trade Name and Synonyms: Cyclolube 2310

Chemical Family: Petroleum Hydrocarbon

Manufacturer: Witco Chemical Corp., Oildale, California

UN ID Number: NA

--- INGREDIENTS AND COMPOSITION DATA ---

Material	Cas No.	%	TLV
Petroleum hydrocarbon	NA		Not established

--- PHYSICAL DATA ---

Boiling Point (°F): NA

Specific Gravity (H₂O = 1): 0.92

Vapor Pressure (mm Hg): NA

Percent Volatile by Weight (%): Not volatile

Vapor Density (air = 1): NA

Freezing Point: NA

Evaporation Rate: NA

pH: Not corrosive

Solubility in Water: Insoluble

Melting Point: NA

Appearance/Color/Odor: Oily mobile liquid light odor.

--- FIRE AND EXPLOSION HAZARDS ---

Flash Point: 440°F (COC)

Autoignition Temperature: NA

Extinguishing Media: Foam, water fog, CO_2, dry chemical, vaporizing liquid-type extinguishing agent.

Special Fire Fighting Procedures: Standard petroleum procedures.

Unusual Fire and Explosion Hazards:
Do not use welding or cutting torch on or near any container of this material, even empty, because an explosion could occur.
Do not use, pour, spill, or store near heat or open flame.

Hazardous Products of Combustion and Decomposition: Toxic fumes.

Hazardous Polymerization: Will not occur.

--- HEALTH HAZARD DATA ---

Effects of Overexposure: Principal routes of exposure are skin contact and inhalation. Slight irritation potential reported by the manufacturer.

Ingestion of the product in large quantities may cause mild digestive disorders, including diarrhea.

There are no data available that address medical conditions that are generally recognized as being aggravated by exposure to this product.

Emergency Response and First Aid:
Skin: Flush all affected areas with plenty of water for several minutes. Remove and clean any contaminated clothing and shoes. Seek medical attention if skin irritation occurs.
Eyes: Flush eyes with plenty of running water for at least 15 min. Hold the eyelids apart during the flushing to ensure rinsing of the surface of the eye and lids with water. Seek medical attention if eye irritation occurs.
Inhalation: Remove to fresh air and, if indicated, give artificial respiration. If breathing is difficult, give oxygen. Call a physician. Treat immediately.
Ingestion: If swallowed, immediately give several glasses of water and induce vomiting by gagging the victim with a finger placed on the back of the victim's tongue. Give fluids until vomitus is clear. If victim is unconscious or convulsing, do not induce vomiting or give anything by mouth.

Cancer Information: Not listed as a carcinogen by NTP (National Toxicology Program); not regulated as a carcinogen by OSHA; not evaluated by IARC (International Agency for Research on Cancer).

--- TOXICITY DATA ---

Animal Effects: No data listed.

--- REACTIVITY DATA ---

Stability and Conditions to Avoid: Heat, open flames.

Materials to Avoid (Incompatibility): None reported by the manufacturer.

Hazard Decomposition Products: Unknown

Hazardous Polymerization: Will not occur.

--- SPILL AND EMERGENCY RESPONSE ---

For Spills: Sweep spillage; wash residuals with soap and water and transfer to a closed container.

Waste Disposal Method: Same as for organic chemicals.

Empty Containers: The container for this product can present explosion or fire hazards, even when emptied. To avoid the risk of injury, do not cut, puncture, or weld on or near this container.

--- SPECIAL PROTECTION AND PRECAUTIONS ---

Respiratory Protection: Appropriate respirator selected and used in accordance with OSHA Subpart I (29 CFR 1910.134) and manufacturer's recommendations is required if airborne dust is not adequately controlled or if excessive fumes are generated from product decomposition. Self-contained breathing apparatus should be worn in high vapor and dust concentrations. Under normal use, manufacturer recommends a dust mask.

Type of Ventilation Required: General mechanical. Avoid breathing dusts and vapors. Adequate ventilation should be provided to keep mist and dust concentrations below acceptable exposure limits. Discharge from the ventilation system should comply with applicable air-pollution regulations.

Eye Protection Requirements: Chemical goggles. Eyewash fountains and safety showers should be easily accessible.

Protective Gloves and Other: Impervious gloves and clothing should be worn. Manufacturer recommends rubber gloves be used.

Storage and Handling Precautions: Containers should be stored in a cool, dry, well-ventilated area. Store away from flammable materials, sources of heat, flame, sparks, and foodstuffs. Exercise due caution to prevent damage to or leakage from the container. Avoid any conditions that might tend to create a dust explosion. Maintain good housekeeping practices to minimize dust buildup.

Other Precautions: Use all standard practices for good personal hygiene. Completely isolate and thoroughly clean all equipment, piping, or vessels before beginning maintenance or repairs. Keep area clean. *Note* that this product will burn.

--- REGULATORY INFORMATION ---

None reported.

▼   ▼   ▼   ▼   ▼   ▼   ▼   ▼   ▼   ▼   ▼   ▼   ▼   ▼

--- IDENTIFIERS ---

Chemical Name and Synonyms: PETROLEUM HYDROCARBON OIL

Trade Name and Synonyms: Flexon 641; similar products are Flexxon 766 and Flexxon 580

Chemical Family: Rubber Process Oil

Manufacturer: Exxon Company, Houston, Texas

UN ID Number: NA

--- INGREDIENTS AND COMPOSITION DATA ---

Material	Cas No.	%	TLV
Petroleum hydrocarbon	NA		Not established

--- PHYSICAL DATA ---

Boiling Point (°F): 545 to 884

Specific Gravity (H$_2$O = 1): 0.898

Vapor Pressure (mm Hg): <0.01 at 100°F

Percent Volatile by Weight (%): 2.5 (3 hr at 325°F)

Vapor Density (air = 1): >12

Freezing Point: NA

Evaporation Rate: NA

pH: Not corrosive

Solubility in Water: Negligible

Melting Point: NA

Appearance/Color/Odor: Clear light straw-colored oil; bland odor.

--- FIRE AND EXPLOSION HAZARDS ---

Flash Point: 275°F (COC)

Flammability Limits: LEL, 1%; UEL, 6%

Autoignition Temperature: NA

Extinguishing Media: Foam, water fog, CO_2, dry chemical, vaporizing liquid-type extinguishing agent.

Special Fire Fighting Procedures: Use air-supplied breathing equipment for enclosed areas. Cool exposed containers with water spray. Avoid breathing vapor or fumes.

Unusual Fire and Explosion Hazards:
Do not mix or store with strong oxidants like liquid chlorine or concentrated oxygen.
Do not use, pour, spill, or store near heat or open flame.

Hazardous Products of Combustion and Decomposition: Toxic fumes.

Hazardous Polymerization: Will not occur.

--- HEALTH HAZARD DATA ---

Effects of Overexposure: Prolonged or repeated skin contact may cause mild skin irritation.

Emergency Response and First Aid:
Skin: Flush all affected areas with plenty of water for several minutes. Remove and clean any contaminated clothing and shoes. Seek medical attention if skin irritation occurs.
Eyes: Flush eyes with plenty of running water for at least 15 min. Hold the eyelids apart during the flushing to ensure rinsing of the surface of the eye and lids with water. Seek medical attention if eye irritation occurs.
Inhalation: Remove to fresh air and, if indicated, give artificial respiration. If breathing is difficult, give oxygen. Call a physician. Treat immediately.
Ingestion: If swallowed, immediately give several glasses of water and induce vomiting by gagging the victim with a finger placed on the back of the victim's tongue. Give fluids until vomitus is clear. If victim is unconscious or convulsing, do not induce vomiting or give anything by mouth.

Cancer Information: Not listed as a carcinogen by NTP (National Toxicology Program); not regulated as a carcinogen by OSHA; not evaluated by IARC (International Agency for Research on Cancer).

--- TOXICITY DATA ---

Animal Effects: None reported.

--- REACTIVITY DATA ---

Stability and Conditions to Avoid: Heat, open flames. This product is considered stable.

Materials to Avoid (Incompatibility): Strong oxidizers like liquid chlorine, concentrated oxygen, sodium, or calcium hypochlorite.

Hazard Decomposition Products: Fumes, smoke, carbon monoxide, and sulfur oxides, in the case of incomplete combustion.

Hazardous Polymerization: Will not occur.

--- SPILL AND EMERGENCY RESPONSE ---

For Spills: Recover free liquid. Add absorbent (sand, earth, sawdust, etc.) to spill area. Keep petroleum products out of sewers and watercourses by diking or impounding. Advise authorities if product has entered or may enter sewers, watercourses, or extensive land areas.

Waste Disposal Method: Assure conformity with applicable disposal regulations. Dispose of absorbed material at an approved waste disposal site or facility.

Empty Containers: Keep containers closed when not in use. Do not handle or store near heat, sparks, flame, or strong oxidants.

--- SPECIAL PROTECTION AND PRECAUTIONS ---

Respiratory Protection: Normally not needed. Use supplied-air respiratory protection in confined or enclosed spaces.

Type of Ventilation Required: Use local exhaust to capture fumes and vapors. Provide greater than 60-fpm hood or face velocity for confined spaces.

Eye Protection Requirements: Chemical goggles. Eyewash fountains and safety showers should be easily accessible.

Protective Gloves and Other: Impervious gloves and clothing should be worn. Manufacturer recommends rubber gloves be used.

Storage and Handling Precautions: Containers should be stored in a cool, dry, well-ventilated area. Store away from flammable materials, sources of heat, flame, sparks, and foodstuffs. Exercise due caution to prevent damage to or leakage from the container. Avoid any conditions that might tend to create a dust explosion. Maintain good housekeeping practices to minimize dust buildup.

Other Precautions: Use all standard practices for good personal hygiene. Completely isolate and thoroughly clean all equipment, piping, or vessels before beginning maintenance or repairs. Keep area clean. *Note* that this product will burn.

--- REGULATORY INFORMATION ---

None reported.

▼   ▼   ▼   ▼   ▼   ▼   ▼   ▼   ▼   ▼   ▼   ▼   ▼   ▼

--- IDENTIFIERS ---

Chemical Name and Synonyms: PETROLEUM HYDROCARBON OIL

Trade Name and Synonyms: Flexon 340

Chemical Family: Rubber Process Oil

Manufacturer: Exxon Company, Houston, Texas

UN ID Number: NA

--- INGREDIENTS AND COMPOSITION DATA ---

Material	Cas No.	%	TLV
Extracts (petroleum), heavy naphthenic distillate solvent	64742-11-6	100	Not established

--- PHYSICAL DATA ---

Boiling Point (°F): IBP approximately 251°C (485°F) by ASTM D 2887

Specific Gravity (H$_2$O = 1): (15.6°C/15.6°C) 0.95

Vapor Pressure (mm Hg): <0.01 at 20°F

Percent Volatile by Weight (%): Negligible from open container in 4 hr at 38°C (100°F)

Vapor Density (air = 1): >5

Freezing Point: NA

Evaporation Rate: At 1 atm and 25°C (77°F) (n-butyl acetate = 1) <0.01

pH: Essentially neutral

Solubility in Water: Negligible: less than 0.1%

Melting Point/Pour or Congealing Point: -33°C (-28°F), pour point by ASTM D 97

Viscosity: 12 SSU at 100°F

Appearance/Color/Odor: Clear dark orange liquid; mild, bland petroleum odor.

--- FIRE AND EXPLOSION HAZARDS ---

Flash Point: NA

Flammability Limits: NA

Autoignition Temperature: NA

Extinguishing Media: Foam, water fog, CO_2, dry chemical, vaporizing liquid-type extinguishing agent.

Special Fire Fighting Procedures: Use air-supplied breathing equipment for enclosed areas. Cool exposed containers with water spray. Avoid breathing vapor or fumes.

Unusual Fire and Explosion Hazards:
Do not mix or store with strong oxidants like liquid chlorine or concentrated oxygen.
Do not use, pour, spill, or store near heat or open flame.

Hazardous Products of Combustion and Decomposition: Toxic fumes.

Hazardous Polymerization: Will not occur.

--- HEALTH HAZARD DATA ---

Effects of Overexposure: None recognized.

Emergency Response and First Aid:
Skin: Flush all affected areas with plenty of water for several minutes. Remove and clean any contaminated clothing and shoes. Seek medical attention if skin irritation occurs.

Eyes: Flush eyes with plenty of running water for at least 15 min. Hold the eyelids apart during the flushing to ensure rinsing of the surface of the eye and lids with water. Seek medical attention if eye irritation occurs.
Inhalation: Remove to fresh air and, if indicated, give artificial respiration. If breathing is difficult, give oxygen. Call a physician. Treat immediately.
Ingestion: If swallowed, immediately give several glasses of water and induce vomiting by gagging the victim with a finger placed on the back of the victim's tongue. Give fluids until vomitus is clear. If victim is unconscious or convulsing, do not induce vomiting or give anything by mouth.

Cancer Information: Long-term, repeated exposure may cause skin cancer.

--- TOXICITY DATA ---

Animal Effects: Prolonged or repeated skin contact with this product tends to remove skin oils, possibly leading to irritation and dermatitis; however, based on human experience and available toxicological data, this product is judged to be neither a "corrosive" nor an "irritant" by OSHA criteria.

Product contacting the eyes may cause eye irritation.

Contains greater than 45% aromatic hydrocarbons boiling above 600°F. Lifetime skin painting studies conducted by Exxon and others, and reports by IARC, have shown that this type of product causes skin cancer in test animals. The substance was applied to the shaved backs of mice at regular intervals, without cleanup between applications. In view of these findings, there may be a potential risk of skin cancer in humans from prolonged or repeated skin contact with this product in the absence of good personal hygiene.

Limited studies on oils that are very active carcinogens have shown that washing the animals' skin with soap and water between applications greatly reduces tumor formation. These studies demonstrate the effectiveness of cleansing the skin after contact.

Potential risks to humans can be minimized by observing good work practices and personal hygiene procedures generally recommended for petroleum products. See Section I for recommended protection and precautions.

Product has a low order of acute oral and dermal toxicity, but minute amounts aspirated into the lungs during ingestion or vomiting may cause mild to severe pulmonary injury and possibly death.

This product is judged to have an acute oral LD50 (rat) greater than 5 g/kg of body weight and an acute LD50 (rabbit) greater than 3.16 g/kg of body weight.

--- REACTIVITY DATA ---

Stability and Conditions to Avoid: Heat, open flames. This product is considered stable.

Materials to Avoid (Incompatibility): None reported.

Hazard Decomposition Products: Fumes, smoke, carbon monoxide, and sulfur oxides, in the case of incomplete combustion.

Hazardous Polymerization: Will not occur.

--- SPILL AND EMERGENCY RESPONSE ---

For Spills: Recover free liquid. Add absorbent (sand, earth, sawdust, etc.) to spill area. Keep petroleum products out of sewers and watercourses by diking or impounding. Advise authorities if product has entered or may enter sewers, watercourses, or extensive land areas.

Waste Disposal Method: Assure conformity with applicable disposal regulations. Dispose of absorbed material at an approved waste disposal site or facility.

Empty Containers: Keep containers closed when not in use. Do not handle or store near heat, sparks, flame, or strong oxidants.

--- SPECIAL PROTECTION AND PRECAUTIONS ---

Respiratory Protection: Normally not needed. Use supplied-air respiratory protection in confined or enclosed spaces.

Type of Ventilation Required: Use local exhaust to capture fumes and vapors. Provide greater than 60-fpm hood or face velocity for confined spaces.

Eye Protection Requirements: Chemical goggles. Eyewash fountains and safety showers should be easily accessible.

Protective Gloves and Other: Impervious gloves and clothing should be worn. Manufacturer recommends rubber gloves be used.

Storage and Handling Precautions: Containers should be stored in a cool, dry, well-ventilated area. Store away from flammable materials, sources of heat, flame, sparks, and foodstuffs. Exercise due caution to prevent damage to or leakage from the container. Avoid any conditions that might tend to create a dust explosion. Maintain good housekeeping practices to minimize dust buildup.

Other Precautions: Use all standard practices for good personal hygiene. Completely isolate and thoroughly clean all equipment, piping, or vessels before beginning maintenance or repairs. Keep area clean. *Note* that this product will burn.

--- REGULATORY INFORMATION ---

Transportation Incident Information: For further information relative to spills resulting from transportation incidents, refer to latest Department of Transportation Emergency Response Guidebook for Hazardous Materials Incidents, DOT P 5800.3.

DOT Identification Number: Not applicable.

▼ ▼ ▼ ▼ ▼ ▼ ▼ ▼ ▼ ▼ ▼ ▼ ▼ ▼

--- IDENTIFIERS ---

Chemical Name and Synonyms: RUBBER PROCESS AND EXTENDER OIL

Trade Name and Synonyms: Flexon 391

Chemical Family: Petroleum Hydrocarbon Oil

Manufacturer: Exxon Company, Houston, Texas

UN ID Number: NA

--- INGREDIENTS AND COMPOSITION DATA ---

Material	Cas No.	%	TLV
Contains approximately 66% aromatic hydrocarbons	NA	100	Not established

--- PHYSICAL DATA ---

Boiling Point (°F): 640 to 950 (IBP-FBP)

Specific Gravity (H₂O = 1): 0.975

Vapor Pressure (mm Hg): 1.2E-5 at 100°F

Percent Volatile by Weight (%): 0.21 (3 hr at 325°F)

Vapor Density (air = 1): >12

Freezing Point: NA

Evaporation Rate: <0.01 (*n*-butylacetate = 1)

pH: NA

Solubility in Water: Negligible

Melting Point/Pour or Congealing Point: NA

Viscosity: NA

Appearance/Color/Odor: Clear, orange-red oil; bland odor.

--- FIRE AND EXPLOSION HAZARDS ---

Flash Point: 420°F (Cleveland open cup)

Flammability Limits: LEL, 1%; UEL, 6%

Autoignition Temperature: NA

Extinguishing Media: Foam, dry chemical, carbon dioxide, or water fog or spray.

Special Fire Fighting Procedures: Use air-supplied breathing equipment for enclosed areas. Cool exposed containers with water spray. Avoid breathing vapor or fumes.

Unusual Fire and Explosion Hazards:
Do not mix or store with strong oxidants like liquid chlorine or concentrated oxygen.

Hazardous Products of Combustion and Decomposition: Toxic fumes.

Hazardous Polymerization: Will not occur.

--- HEALTH HAZARD DATA ---

Effects of Overexposure: Prolonged or repeated skin contact may cause skin irritation or more serious skin disorder.

Emergency Response and First Aid:
Skin: Flush all affected areas with plenty of water for several minutes. Remove and clean any contaminated clothing and shoes. Seek medical attention if skin irritation occurs.
Eyes: Flush eyes with plenty of running water for at least 15 min. Hold the eyelids apart during the flushing to ensure rinsing of the surface of the eye and lids with water. Seek medical attention if eye irritation occurs.
Inhalation: Remove to fresh air and, if indicated, give artificial respiration. If breathing is difficult, give oxygen. Call a physician. Treat immediately.
Ingestion: If swallowed, immediately give several glasses of water and induce vomiting by gagging the victim with a finger placed on the back of the victim's tongue. Give fluids until vomitus is clear. If victim is unconscious or convulsing, do not induce vomiting or give anything by mouth.

Cancer Information: Refer to toxicity data.

--- TOXICITY DATA ---

Threshold Limit Value: 5 mg/m^3 for oil mist in air (OSHA Regulation 29 CFR 1910.1000).

Animal Effects: Contains greater than 45% aromatic hydrocarbons boiling above 600°F. Lifetime skin painting studies conducted by Exxon and others, and reports by IARC, have shown that this type of product causes skin cancer in test animals. The substance was applied to the shaved backs of mice at regular intervals, without cleanup between applications. In view of these findings, there may be a potential risk of skin cancer in humans from prolonged or repeated skin contact with this product in the absence of good personal hygiene.

Limited studies on oils that are very active carcinogens have shown that washing the animals' skin with soap and water between applications greatly reduces tumor formation. These studies demonstrate the effectiveness of cleansing the skin after contact.

Potential risks to humans can be minimized by observing good work practices and personal hygiene procedures generally recommended for petroleum products. See Section I for recommended protection and precautions.

Product has a low order of acute oral and dermal toxicity, but minute amounts aspirated into the lungs during ingestion or vomiting may cause mild to severe pulmonary injury and possibly death.

--- REACTIVITY DATA ---

Stability and Conditions to Avoid: Heat, open flames. This product is considered stable.

Materials to Avoid (Incompatibility): Strong oxidants like liquid chlorine, concentrated oxygen, sodium, or calcium hypochlorite.

Hazard Decomposition Products: Fumes, smoke, carbon monoxide, and sulfur oxides, in the case of incomplete combustion.

Hazardous Polymerization: Will not occur.

--- SPILL AND EMERGENCY RESPONSE ---

For Spills: Recover free liquid. Add absorbent (sand, earth, sawdust, etc.) to spill area. Keep petroleum products out of sewers and watercourses by diking or impounding. Advise authorities if product has entered or may enter sewers, watercourses, or extensive land areas.

Waste Disposal Method: Assure conformity with applicable disposal regulations. Dispose of absorbed material at an approved waste disposal site or facility.

Empty Containers: Keep containers closed when not in use. Do not handle or store near heat, sparks, flame, or strong oxidants.

--- SPECIAL PROTECTION AND PRECAUTIONS ---

Respiratory Protection: Normally not needed. Use supplied-air respiratory protection in confined or enclosed spaces.

Type of Ventilation Required: Use local exhaust to capture fumes and vapors. Provide greater than 60-fpm hood or face velocity for confined spaces.

Eye Protection Requirements: Use splash goggles or face shield when eye contact may occur.

Protective Gloves and Other: Use chemical-resistant gloves to avoid skin contact. Use chemical-resistant apron or other clothing if needed to avoid skin contact.

Storage and Handling Precautions: Containers should be stored in a cool, dry, well-ventilated area. Store away from flammable materials, sources of heat, flame, sparks, and foodstuffs. Exercise due caution to prevent damage to or leakage from the container. Avoid any conditions that might tend to create a dust explosion. Maintain good housekeeping practices to minimize dust buildup.

Other Precautions: Avoid breathing oil mist. Remove oil-soiled clothing and launder before reuse. Discard oil-soaked shoes. Wash skin thoroughly with soap and water after handling.

--- REGULATORY INFORMATION ---

Transportation Incident Information: For further information relative to spills resulting from transportation incidents, refer to latest Department of Transportation Emergency Response Guidebook for Hazardous Materials Incidents, DOT P 5800.3.

DOT Identification Number: Not applicable.

▼   ▼   ▼   ▼   ▼   ▼   ▼   ▼   ▼   ▼   ▼   ▼   ▼   ▼

--- IDENTIFIERS ---

Chemical Name and Synonyms: PETROLEUM PROCESS OIL

Trade Name and Synonyms: Flexon 815

Chemical Family: Petroleum Hydrocarbon Oil

Manufacturer: Exxon Company, Houston, Texas

UN ID Number: NA

--- INGREDIENTS AND COMPOSITION DATA ---

Material	Cas No.	%	TLV
Hydrotreated residual oil, petroleum or	64742-57-0 or	100	5 mg/m^3 (OSHA 29 CFR 1910.1000,
Solvent dewaxed residual oil	64742-62-7		ACGIH)

--- PHYSICAL DATA ---

Boiling Point (°F): 700 (371°C) IBP by ASTM D 2887

Specific Gravity (H$_2$O = 1): 0.90 (15.6°C/15.6°C)

Vapor Pressure (mm Hg): <0.01 at 20°C

Percent Volatile by Weight (%): Negligible from open container in 4 hr at 38°C (100°F)

Vapor Density (air = 1): >5

Freezing Point: NA

Evaporation Rate: <0.01 at 1 atm and 25°C (77°F) (*n*-butyl acetate = 1)

pH: Essentially neutral

Solubility in Water: Negligible: <0.1% at 1 atm and 25°C (77°F)

Melting Point/Pour or Congealing Point: -9°C (15°F), pour point by
ASTM D 97

Viscosity: 153 SSU at 210°F

Appearance/Color/Odor: Clear, dark brown liquid; bland petroleum odor.

--- FIRE AND EXPLOSION HAZARDS ---

Flash Point: 420°F (Cleveland open cup)

Flammability Limits: LEL, 1%; UEL, 6%

Autoignition Temperature: NA

Extinguishing Media: Foam, dry chemical, carbon dioxide, or water fog or
spray.

Special Fire Fighting Procedures: Use air-supplied breathing equipment for
enclosed areas. Cool exposed containers with water spray. Avoid breathing
vapor or fumes.

Unusual Fire and Explosion Hazards:
Do not mix or store with strong oxidants like liquid chlorine or concentrated
oxygen.

Hazardous Products of Combustion and Decomposition: Toxic fumes.

Hazardous Polymerization: Will not occur.

--- HEALTH HAZARD DATA ---

Effects of Overexposure: Prolonged or repeated skin contact may cause skin
irritation. Health studies have shown that many petroleum hydrocarbons and
synthetic lubricants pose potential human health risks, which may vary from
person to person. As a precaution, exposure to liquids, vapors, mists or
fumes should be minimized.

Emergency Response and First Aid:
Skin: Flush all affected areas with plenty of water for 15 min.
Remove and clean any contaminated clothing and shoes. Seek medical
attention if skin irritation occurs.

Eyes: Flush eyes with plenty of running water for at least 15 min. Hold the eyelids apart during the flushing to ensure rinsing of the surface of the eye and lids with water. Seek medical attention if eye irritation occurs.

Inhalation: Remove to fresh air and, if indicated, give artificial respiration. If breathing is difficult, give oxygen. Call a physician. Treat immediately.

Ingestion: If swallowed, immediately give several glasses of water and induce vomiting by gagging the victim with a finger placed on the back of the victim's tongue. Give fluids until vomitus is clear. If victim is unconscious or convulsing, do not induce vomiting or give anything by mouth.

Cancer Information: Refer to toxicity data.

--- TOXICITY DATA ---

Threshold Limit Value: 5 mg/m^3 for oil mist in air (OSHA Regulation 29 CFR 1910.1000).

Animal Effects: In accordance with the current OSHA Hazard Communication standard criteria, this product does not require a cancer hazard warning. This is because the product is formulated from base stocks that are severely hydrotreated, severely solvent extracted, and/or processed by mild hydrotreatment and extraction. Alternatively, it may consist of components not otherwise affected by IARC criteria, such as atmosphere distillates or synthetically derived materials, and as such is not characterized by current IARC classification criteria.

Prolonged or repeated skin contact with this product tends to remove skin oils, possibly leading to irritation and dermatitis; however, based on human experience and available toxicological data, this product is judged to be neither a "corrosive" nor an "irritant" by OSHA criteria.

Product contacting the eyes may cause eye irritation.

Product has a low order of acute oral and dermal toxicity, but minute amounts aspirated into the lungs during ingestion or vomiting may cause mild to severe pulmonary injury and possibly death.

This product is judged to have an acute oral LD50 (rat) greater than 5 g/kg of body weight and an acute LD50 (rabbit) greater than 3.16 g/kg of body weight.

--- REACTIVITY DATA ---

Stability and Conditions to Avoid: Heat, open flames. This product is considered stable.

Materials to Avoid (Incompatibility): Strong oxidants like liquid chlorine, concentrated oxygen, sodium, or calcium hypochlorite.

Hazard Decomposition Products: Fumes, smoke, carbon monoxide, and sulfur oxides, in the case of incomplete combustion.

Hazardous Polymerization: Will not occur.

--- SPILL AND EMERGENCY RESPONSE ---

For Spills: Recover free liquid. Add absorbent (sand, earth, sawdust, etc.) to spill area. Keep petroleum products out of sewers and watercourses by diking or impounding. Advise authorities if product has entered or may enter sewers, watercourses, or extensive land areas.

Waste Disposal Method: Assure conformity with applicable disposal regulations. Dispose of absorbed material at an approved waste disposal site or facility.

Empty Containers: Keep containers closed when not in use. Do not handle or store near heat, sparks, flame, or strong oxidants.

--- SPECIAL PROTECTION AND PRECAUTIONS ---

Respiratory Protection: Normally not needed. Use supplied-air respiratory protection in confined or enclosed spaces.

Type of Ventilation Required: Use local exhaust to capture fumes and vapors. Provide greater than 60-fpm hood or face velocity for confined spaces.

Eye Protection Requirements: Use splash goggles or face shield when eye contact may occur.

Protective Gloves and Other: Use chemical-resistant gloves to avoid skin contact. Use chemical-resistant apron or other clothing if needed to avoid skin contact.

Storage and Handling Precautions: Containers should be stored in a cool, dry, well-ventilated area. Store away from flammable materials, sources of heat, flame, sparks, and foodstuffs. Exercise due caution to prevent damage to or leakage from the container. Avoid any conditions that might tend to create a dust explosion. Maintain good housekeeping practices to minimize dust buildup.

Other Precautions: Avoid breathing oil mist. Remove oil-soiled clothing and launder before reuse. Discard oil-soaked shoes. Wash skin thoroughly with soap and water after handling.

--- REGULATORY INFORMATION ---

Transportation Incident Information: For further information relative to spills resulting from transportation incidents, refer to latest Department of Transportation Emergency Response Guidebook for Hazardous Materials Incidents, DOT P 5800.3.

DOT Identification Number: Not applicable.

▼   ▼   ▼   ▼   ▼   ▼   ▼   ▼   ▼   ▼   ▼   ▼   ▼   ▼

--- IDENTIFIERS ---

Chemical Name and Synonyms: BLEND OF HYDROCARBONS, HIGHER ALCOHOLS, ESTERS AND SOAPS WITH SURFACE-ACTIVE PROPERTIES

Trade Name and Synonyms: Aflux 42

Chemical Family: Plasticizer

Manufacturer: Mobay Chemical Corporation, Pittsburgh, Pennsylvania

UN ID Number: NA

--- INGREDIENTS AND COMPOSITION DATA ---

Material	Cas No.	%	TLV
Product	NA		Not established

--- PHYSICAL DATA ---

Boiling Point (°F): NA

Specific Gravity (H$_2$O = 1): 0.91

Vapor Pressure (mm Hg): NA

Percent Volatile by Weight (%): NA

Vapor Density (air = 1): NA

Freezing Point: NA

Evaporation Rate: NA

pH: NA

Solubility in Water: Slightly

Melting Point/Pour or Congealing Point: NA

Viscosity: NA

Appearance/Color/Odor: Light brown lenticular pellets; slight odor.

--- FIRE AND EXPLOSION HAZARDS ---

Flash Point: >200°C

Flammability Limits: NA

Autoignition Temperature: NA

Extinguishing Media: Foam, dry chemical, carbon dioxide, or water fog or spray.

Special Fire Fighting Procedures: Use air-supplied breathing equipment for enclosed areas. Cool exposed containers with water spray. Avoid breathing vapor or fumes.

Unusual Fire and Explosion Hazards: None reported by manufacturer.

Hazardous Products of Combustion and Decomposition: Toxic fumes.

Hazardous Polymerization: Will not occur.

--- HEALTH HAZARD DATA ---

Effects of Overexposure: None reported by manufacturer.

Emergency Response and First Aid:
Skin: Flush all affected areas with plenty of water for 15 min. Remove and clean any contaminated clothing and shoes. Seek medical attention if skin irritation occurs.
Eyes: Flush eyes with plenty of running water for at least 15 min. Hold the eyelids apart during the flushing to ensure rinsing of the surface of the eye and lids with water. Seek medical attention if eye irritation occurs.
Inhalation: Remove to fresh air and, if indicated, give artificial respiration. If breathing is difficult, give oxygen. Call a physician. Treat immediately.
Ingestion: If swallowed, immediately give several glasses of water and induce vomiting by gagging the victim with a finger placed on the back of the victim's tongue. Give fluids until vomitus is clear. If victim is unconscious or convulsing, do not induce vomiting or give anything by mouth.

Cancer Information: Refer to toxicity data.

--- TOXICITY DATA ---

Animal Effects: No information reported.

--- REACTIVITY DATA ---

Stability and Conditions to Avoid: Heat, open flames. This product is considered stable.

Materials to Avoid (Incompatibility): Strong oxidants like liquid chlorine, concentrated oxygen, sodium, or calcium hypochlorite.

Hazard Decomposition Products: Fumes, smoke, carbon monoxide, and sulfur oxides, in the case of incomplete combustion.

Hazardous Polymerization: Will not occur.

--- SPILL AND EMERGENCY RESPONSE ---

For Spills: Recover free liquid. Add absorbent (sand, earth, sawdust, etc.) to spill area. Keep petroleum products out of sewers and watercourses by diking or impounding. Advise authorities if product has entered or may enter sewers, watercourses, or extensive land areas.

Waste Disposal Method: Assure conformity with applicable disposal regulations. Dispose of absorbed material at an approved waste disposal site or facility. Waste material may be incinerated under conditions which meet federal, state, and local environmental control regulations.

Empty Containers: Keep containers closed when not in use. Do not handle or store near heat, sparks, flame, or strong oxidants.

--- SPECIAL PROTECTION AND PRECAUTIONS ---

Respiratory Protection: Normally not needed. Use supplied-air respiratory protection in confined or enclosed spaces.

Type of Ventilation Required: Use local exhaust to capture fumes and vapors. Provide greater than 60-fpm hood or face velocity for confined spaces.

Eye Protection Requirements: Use splash goggles or face shield when eye contact may occur.

Protective Gloves and Other: Use chemical-resistant gloves to avoid skin contact. Use chemical-resistant apron or other clothing if needed to avoid skin contact.

Storage and Handling Precautions: Containers should be stored in a cool, dry, well-ventilated area. Store away from flammable materials, sources of heat, flame, sparks, and foodstuffs. Exercise due caution to prevent damage to or leakage from the container. Avoid any conditions that might tend to create a dust explosion. Maintain good housekeeping practices to minimize dust buildup. *Note:* This product has a 1-yr shelf life.

Other Precautions: Avoid breathing oil mist. Remove oil-soiled clothing and launder before reuse. Discard oil-soaked shoes. Wash skin thoroughly with soap and water after handling.

--- REGULATORY INFORMATION ---

Transportation Incident Information: For further information relative to spills resulting from transportation incidents, refer to latest Department of Transportation Emergency Response Guidebook for Hazardous Materials Incidents, DOT P 5800.3.

DOT Identification Number: Not applicable.

▼   ▼   ▼   ▼   ▼   ▼   ▼   ▼   ▼   ▼   ▼   ▼   ▼

--- IDENTIFIERS ---

Chemical Name and Synonyms: MBTS, BENZOTHIAZOLE DISULFIDE, DITHIOBISBENZOTHIAZOLE

Trade Name and Synonyms: Altax

Chemical Family: Plasticizer

Manufacturer: R. T. Vanderbilt Co., Norwalk, Connecticut

UN ID Number: NA

--- INGREDIENTS AND COMPOSITION DATA ---

Material	Cas No.	%	TLV
Benzothiazole	120-78-5		Not established
Zinc stearate	557-05-1		Not established
Petroleum process oil	64742-55-8 or 64742-56-9		Not established

--- PHYSICAL DATA ---

Boiling Point (°F): NA

Specific Gravity (H$_2$O = 1): 1.50

Vapor Pressure (mm Hg): NA

Percent Volatile by Weight (%): NA

Vapor Density (air = 1): NA

Freezing Point: NA

Evaporation Rate: NA

pH: NA

Solubility in Water: Negligible

Melting Point/Pour or Congealing Point: NA

Viscosity: NA

Appearance/Color/Odor: Gray-white to cream powder or pellets; little odor.

--- FIRE AND EXPLOSION HAZARDS ---

Flash Point: 271°F (COC)

Flammability Limits: NA

Autoignition Temperature: NA

Extinguishing Media: Foam, dry chemical, carbon dioxide, or water fog or spray.

Special Fire Fighting Procedures: Use NIOSH-approved positive-pressure self-contained respirator and protective clothing.

Unusual Fire and Explosion Hazards: When exposed to flame, emits acrid fumes. Dust may form explosive mixture with air.

Hazardous Products of Combustion and Decomposition: Oxides of carbon, sulfur, and nitrogen upon combustion.

Hazardous Polymerization: Will not occur.

--- HEALTH HAZARD DATA ---

Effects of Overexposure: May cause allergic skin reaction. Inhalation may irritate upper respiratory tract. May irritate eyes. In laboratory animals fed relatively high amounts of product, mutagenic, teratogenic, and embryotoxic effects were observed, but no carcinogenic effects were shown.

Emergency Response and First Aid:
Skin: Flush all affected areas with plenty of water for 15 min. Remove and clean any contaminated clothing and shoes. Seek medical attention if skin irritation occurs.
Eyes: Flush the eyes with plenty of running water for at least 15 min. Hold the eyelids apart during the flushing to ensure rinsing of the surface of the eye and lids with water. Seek medical attention if eye irritation occurs.
Inhalaticn: Remove to fresh air and, if indicated, give artificial respiration. If breathing is difficult, give oxygen. Call a physician. Treat immediately.
Ingestion: If swallowed, immediately give several glasses of water and induce vomiting by gagging the victim with a finger placed on the back of the victim's tongue. Give fluids until vomitus is clear. If victim is unconscious or convulsing, do not induce vomiting or give anything by mouth.

Cancer Information: None listed under OSHA/NTP/IARC.

--- TOXICITY DATA ---

Acute oral LD50 (rats): 7000 mg/kg

For Product - Ames Test: negative
E. coli WP 2 UVRA: negative
Cell transformation (BALB/3T3) assay: negative
CHO chromosome aberration assay: negative
Mouse lymphoma L5178Y assay: positive
E. coli pol A+/A- assay: indefinite w/wo activation

Feeding of large amounts to laboratory animals has been reported to cause mutagenic effects.

Carcinogenicity: The National Toxicology Program (NTP), a coordinating body for federal health agencies, reported its results (1987) of MBT toxicology and carcinogenicity studies. In the NTP studies, MBT in corn oil was force fed through a stomach tube to rats and mice for 2 yr. An increased incidence of tumors in a number of tissues was seen in rats. No increase in the incidence of tumors (i.e., no effect) was observed in mice. The strength of the data was evaluated as "some," "equivocal," "no," or "inadequate" evidence of carcinogenicity. Because only a limited response occurred, NTP interpreted these studies as showing "some" evidence of carcinogenicity. The nature of the tumor response (e.g., no effect in mice; some effect in rats) and other concerns about the conduct of these studies (e.g., force feeding an amount of MBT that may have exceeded the maximum tolerable dose) makes it difficult to clearly assess the significance of the results to those who work with MBT. We recommend that worker exposure to MBT should be minimized.

Reproductive Toxicity: Teratology study in rats with mercaptobenzothiazole (MBT) showed negative results.

Medical Conditions Generally Aggravated By Exposure: Unknown.

--- REACTIVITY DATA ---

Stability and Conditions to Avoid: The thermal stability of this material is rated by the manufacturer as stable. However, they note: Contact with strong oxidizing agents will generate heat. Contact with acids will produce toxic sulfur compounds.

Materials to Avoid (Incompatibility): Strong oxidants like liquid chlorine, concentrated oxygen, sodium, or calcium hypochlorite.

Hazard Decomposition Products: Fumes, smoke, carbon monoxide, and sulfur oxides, in the case of incomplete combustion.

Hazardous Polymerization: Will not occur.

--- SPILL AND EMERGENCY RESPONSE ---

For Spills: Recover free liquid. Add absorbent (sand, earth, sawdust, etc.) to spill area. Keep petroleum products out of sewers and watercourses by diking or impounding. Advise authorities if product has entered or may enter sewers, watercourses, or extensive land areas.

Waste Disposal Method: Assure conformity with applicable disposal regulations. Dispose of absorbed material at an approved waste disposal site or facility. Waste material may be incinerated under conditions that meet federal, state, and local environmental control regulations. At present this is not classified as a RCRA hazardous waste.

Empty Containers: Keep containers closed when not in use. Do not handle or store near heat, sparks, flame, or strong oxidants.

--- SPECIAL PROTECTION AND PRECAUTIONS ---

Respiratory Protection: Normally not needed. Use supplied-air respiratory protection in confined or enclosed spaces.

Type of Ventilation Required: Use local exhaust to capture fumes and vapors. Provide greater than 60-fpm hood or face velocity for confined spaces.

Eye Protection Requirements: Use splash goggles or face shield when eye contact may occur.

Protective Gloves and Other: Use chemical-resistant gloves to avoid skin contact. Use chemical-resistant apron or other clothing if needed to avoid skin contact.

Storage and Handling Precautions: Containers should be stored in a cool, dry, well-ventilated area. Store away from flammable materials, sources of heat, flame, sparks, and foodstuffs. Exercise due caution to prevent damage to or leakage from the container. Avoid any conditions that might tend to create a dust explosion. Maintain good housekeeping practices to minimize dust buildup. *Note:* This product has a 1-yr shelf life.

Other Precautions: Avoid breathing oil mist. Remove oil-soiled clothing and launder before reuse. Discard oil-soaked shoes. Wash skin thoroughly with soap and water after handling.

--- REGULATORY INFORMATION ---

Transportation Incident Information: For further information relative to spills resulting from transportation incidents, refer to latest Department of Transportation Emergency Response Guidebook for Hazardous Materials Incidents, DOT P 5800.3.

DOT Identification Number: Not applicable.

▼ ▼ ▼ ▼ ▼ ▼ ▼ ▼ ▼ ▼ ▼ ▼ ▼ ▼

--- IDENTIFIERS ---

Name: CASTOR OIL
Synonyms: Aromatic Caster Oil; Castor Oil Aromatic; Cosmetol; Crystal O;
   Gold Bond; NCI-C55163; Neoloid; Oil of Palma Christi; Phorbyol; Ricinus
   Oil; Ricirus Oil; Tangantangan Oil
CAS: 8001-79-4; **RTECS:** FI4100000
Chemical Class: Ester

See other identifiers listed below under Regulations.

--- PROPERTIES ---

Boiling Point: 586.16 K; 313°C; 595.4°F
Melting Point: NA
Flash Point: >383.16 K; >110°C; >230°F
Autoignition: NA
UEL: NA
LEL: NA
Vapor Density: No data
Specific Gravity: No data
Density: 0.961

Reactivity with Water: No data on water reactivity
Reactivity with Common Materials: No data
Stability during Transport: No data
Neutralizing Agents: No data
Polymerization Possibilities: No data

Toxic Fire Gases: None reported other than possible unburned vapors
Odor Detected at (ppm): Unknown
Odor Description: No data
100% Odor Detection: No data

--- REGULATIONS ---

DOT hazard class: Not given
Packaging exceptions: 173.
Nonbulk packaging: 173.
Bulk packaging: 173.

STCC Number: Not listed

Clean Water Act Sect. 307: No
Clean Water Act Sect. 311: No
Clean Air Act: Not listed
EPA Waste Number: None
CERCLA Ref: Not listed
RQ Designation: Not listed
SARA TPQ Value: Not listed
SARA Sect. 312 Categories:
  Acute toxicity: irritant.

U.S. Postal Service Mailability: Not given

--- TOXICITY DATA ---

Short-term Toxicity: Unknown

Long-term Toxicity: Unknown

Conc IDLH: Unknown

NIOSH REL: Not given

ACGIH TLV: Not listed
ACGIH STEL: Not listed

OSHA PEL: Not in Table Z-1-A

MAK Information: Not listed

Carcinogen: N; **Status:** See below

Carcinogen Lists: *IARC*: Not listed; *MAK*: Not listed; *NIOSH*: Not listed; *NTP*: Not listed; *ACGIH*: Not listed; *OSHA*: Not listed.

LD50 Value: No LD50 in RTECS 1992

Reproductive Toxicity (1992 RTECS): This chemical has no known mammalian reproductive toxicity.

--- INITIAL INCIDENT RESPONSE ---

No DOT Guide information for this product.

▼   ▼   ▼   ▼   ▼   ▼   ▼   ▼   ▼   ▼   ▼   ▼   ▼   ▼

--- IDENTIFIERS ---

Name: GLYCERIN
Synonyms: Glycerol; Glycerin; Glycerin, Anhydrous; Glycerin, Synthetic;
  Glycerine; Glycyl Alcohol; 1,2,3-Propanetriol; Synthetic Glycerin; 90
  Technical Glycerin; Trihydroxypropane; 1,2,3-Trihydroxypropane
CAS: 56-81-5; **RTECS:** MA8050000
Formula: $C_3H_8O_3$; **Mol Wt:** 92.11
WLN: Q1YQ1Q
Chemical Class: Alcohol

See other identifiers listed below under Regulations.

--- PROPERTIES ---

Physical Description: Colorless odorless oily liquid
Boiling Point: 455 K; 181.8°C; 359.3°F
Melting Point: 293 K; 19.8°C; 67.7°F
Flash Point: 433 K; 159.8°C; 319.7°F
Autoignition: 643 K; 369.8°C; 697.7°F
UEL: NA
LEL: NA
Ionization Potential (eV): 10.51
Vapor Density: No data
Specific Gravity: 1.261, 20°C
Density: 1.261

Reactivity with Water: No data on water reactivity
Reactivity with Common Materials: No data
Stability during Transport: No data
Neutralizing Agents: No data
Polymerization Possibilities: No data

Toxic Fire Gases: None reported other than possible unburned vapors
Odor Detected at (ppm): Not pertinent
Odor Description: Odorless (Source: CHRIS)
100% Odor Detection: No data

--- REGULATIONS ---

DOT hazard class: Not given
Packaging exceptions: 173.
Nonbulk packaging: 173.
Bulk packaging: 173.

STCC Number: Not listed

Clean Water Act Sect. 307: No
Clean Water Act Sect. 311: No
Clean Air Act: Not listed
EPA Waste Number: None
CERCLA Ref: Not listed
RQ Designation: Not listed
SARA TPQ Value: Not listed
SARA Sect. 312 Categories:
  Acute toxicity: irritant.
  Acute toxicity: adverse effect to target organs.
  Chronic toxicity: mutagen.
  Chronic toxicity: reproductive toxin.

U.S. Postal Service Mailability: Not given

NFPA Codes:
  Health Hazard (blue): (1) Slightly hazardous to health. As a precaution, wear self-contained breathing apparatus.
  Flammability (red): (1) This material must be preheated before ignition can occur.
  Reactivity (yellow): (0) Stable even under fire conditions.
  Special: Unspecified.

--- TOXICITY DATA ---

Short-term Toxicity: Unknown

Long-term Toxicity: Unknown

Target Organs: Eyes, respiratory tract

Symptoms: No hazard (Source: CHRIS)

Conc IDLH: Unknown

ACGIH TLV: TLV = 10 mg/m^3 as mist
ACGIH STEL: Not listed

OSHA PEL: Transitional Limits: PEL = total dust, 15 mg/m^3; respirable fraction, 5 mg/m^3; Final Rule Limits: TWA = total dust, 10 mg/m^3; respirable fraction, 5 mg/m^3.

MAK Information: Not listed

Carcinogen: N; **Status:** See below

Carcinogen Lists: *IARC*: Not listed; *MAK*: Not listed; *NIOSH*: Not listed; *NTP*: Not listed; *ACGIH*: Not listed; *OSHA*: Not listed.

Human Toxicity Data: (Source: NIOSH RTECS)
orl-hmn TDLo:1428 mg/kg 34ZIAG - 288, 69
   Behavioral
     Headache
   Gastrointestinal
     Nausea or vomiting

LD50 Value: orl-rat LD50:12600 mg/kg

Other Species Toxicity Data: (Source: NIOSH RTECS 1992)
orl-rat LD50:12600 mg/kg
ipr-rat LD50:4420 mg/kg
scu-rat LD50:100 mg/kg
ivn-rat LD50:5566 mg/kg
orl-mus LD50:4090 mg/kg
ipr-mus LD50:8700 mg/kg
scu-mus LD50:91 mg/kg
ivn-mus LD50:4250 mg/kg
orl-rbt LD50:27 g/kg
ivn-rbt LD50:53 g/kg
orl-gpg LD50:7750 mg/kg

Reproductive Toxicity (1992 RTECS): This chemical is a mammalian reproductive toxin.

Reproductive Toxicity Data (1992 RTECS):
  orl-rat TDLo:100 mg/kg (1D male) TGANAK 19, 436, 85
    Effects on Fertility
      Postimplantation mortality

  itt-rat TDLo:280 mg/kg (2D male) CCPTAY 29, 291, 84
    Paternal Effects
      Spermatogenesis
      Testes, epididymis, sperm duct

  itt-rat TDLo:1600 mg/kg (1D male) CCPTAY 29, 291, 84
    Effects on Fertility
      Male fertility index

  itt-rat TDLo:862 mg/kg (1D male) LIFSAK 34, 1747, 84
    Paternal Effects
      Spermatogenesis

--- PROTECTION AND FIRST AID ---

Protection Suggested from the CHRIS Manual: Rubber gloves, goggles.

First Aid Source: CHRIS Manual 1991.
  No hazard

--- INITIAL INCIDENT RESPONSE ---

Fire Extinguishment: Alcohol foam, dry chemical, carbon dioxide, water fog.
  Note: Water or foam may cause frothing. (CHRIS 91)

No DOT Guide information for this product.

▼   ▼   ▼   ▼   ▼   ▼   ▼   ▼   ▼   ▼   ▼   ▼   ▼   ▼

--- IDENTIFIERS ---

Chemical Name and Synonyms: EPOXIDIZED SOYBEAN OIL

Trade Name and Synonyms: Drapex 6.8

Chemical Family: Epoxies

Manufacturer: Argus Chemical Corp., Brooklyn, New York

UN ID Number: NA

--- INGREDIENTS AND COMPOSITION DATA ---

Material	Cas No.	%	TLV
Product (proprietary)	NA		Not established

--- PHYSICAL DATA ---

Boiling Point (°F): Not distillable

Specific Gravity (H$_2$O = 1): 0.992

Vapor Pressure (mm Hg): <1

Percent Volatile by Volume (%): <0.1

Vapor Density (air = 1): NA

Freezing Point: NA

Evaporation Rate: 0

pH: NA

Solubility in Water: Negligible

Melting Point/Pour or Congealing Point: NA

Viscosity: NA

Appearance/Color/Odor: Almost colorless liquid with a faint fatty acid odor.

--- FIRE AND EXPLOSION HAZARDS ---

Flash Point: 575°F (COC)

Flammability Limits: NA

Autoignition Temperature: NA

Extinguishing Media: Foam, dry chemical, carbon dioxide, or water fog or spray.

Special Fire Fighting Procedures: Use NIOSH-approved positive-pressure self-contained respirator and protective clothing.

Unusual Fire and Explosion Hazards: When exposed to flame, emits acrid fumes. Dust may form explosive mixture with air.

Hazardous Products of Combustion and Decomposition: None reported by manufacturer.

Hazardous Polymerization: Will not occur.

--- HEALTH HAZARD DATA ---

Effects of Overexposure: No effects from overexposure reported by the manufacturer. This product has been cleared by FDA under Section 181.22 of the "Food Additives Regulations."

Emergency Response and First Aid:
Skin: Flush all affected areas with plenty of water for 15 min. Remove and clean any contaminated clothing and shoes. Seek medical attention if skin irritation occurs.
Eyes: Flush eyes with plenty of running water for at least 15 min. Hold the eyelids apart during the flushing to ensure rinsing of the surface of the eye and lids with water. Seek medical attention if eye irritation occurs.
Inhalation: There are no effects or recommendations reported.
Ingestion: Consult physician immediately.

Cancer Information: None listed under OSHA/NTP/IARC.

--- TOXICITY DATA ---

No information reported.

--- REACTIVITY DATA ---

Stability and Conditions to Avoid: Manufacturer recommends pumping this product at 40°C. Use low-pressure steam to avoid hot spots that might cause polymerization.

Materials to Avoid (Incompatibility): Strong acids and excessive heat.

Hazard Decomposition Products: Fumes, smoke, carbon monoxide, and carbon dioxide, in the case of incomplete combustion.

Hazardous Polymerization: Will not occur except above 40°C.

--- SPILL AND EMERGENCY RESPONSE ---

For Spills: Use any inert absorbent such as sand, earth, or vermiculite. The product is described as being slippery. Use proper safety equipment during cleanup. Dispose of in a manner that is consistent with local, state, and federal regulations.

Waste Disposal Method: Assure conformity with applicable disposal regulations. Dispose of absorbed material at an approved waste disposal site or facility. Waste material may be incinerated under conditions that meet federal, state, and local environmental control regulations. At present this is not classified as a RCRA hazardous waste.

Empty Containers: Keep containers closed when not in use. Do not handle or store near heat, sparks, flame, or strong oxidants.

--- SPECIAL PROTECTION AND PRECAUTIONS ---

Respiratory Protection: Normally not needed. Use supplied-air respiratory protection in confined or enclosed spaces.

Type of Ventilation Required: Use local exhaust to capture fumes and vapors. Provide greater than 60-fpm hood or face velocity for confined spaces.

Eye Protection Requirements: Use splash goggles or faceshield when eye contact may occur.

Protective Gloves and Other: Use neoprene gloves, neoprene apron, and chemical goggles.

Storage and Handling Precautions: Containers should be stored in a cool, dry, well-ventilated area. Store away from flammable materials, sources of heat, flame, sparks, and foodstuffs. Exercise due caution to prevent damage to or leakage from the container. Avoid any conditions that might tend to create a dust explosion. Maintain good housekeeping practices to minimize dust buildup. *Note:* This product has a 1-yr shelf life.

Other Precautions: Avoid breathing oil mist. Remove oil-soiled clothing and launder before reuse. Discard oil-soaked shoes. Wash skin thoroughly with soap and water after handling.

--- REGULATORY INFORMATION ---

Transportation Incident Information: This product is unregulated by DOT at present.

DOT Identification Number: Not applicable.

▼   ▼   ▼   ▼   ▼   ▼   ▼   ▼   ▼   ▼   ▼   ▼   ▼   ▼

--- IDENTIFIERS ---

Name: HEPTANE
Synonyms: Dipropyl Methane; Eptani (Italian); Heptan (Polish); *n*-Heptane; Heptane (DOT); Heptanen (Dutch); Heptyl Hydride; Dipropal Methane; Neptanen (Dutch); Normal Heptane
CAS: 142-82-5; **RTECS:** MI7700000
Formula: C_7H_{16}; **Mol Wt:** 100.23
WLN: 7H
Chemical Class: Paraffin

See other identifiers listed below under Regulations.

--- PROPERTIES ---

Physical Description: Colorless watery liquid with a gasolinelike odor
Boiling Point: 371.49 K; 98.3°C; 209°F
Melting Point: 182.04 K; -91.2°C; -132°F
Flash Point: 269.1 K; -4.1°C; 24.7°F
Autoignition: 495 K; 221.8°C; 431.3°F
Critical Temp: 540 K; 266.85°C; 512.33°F
Critical Press: 2.7 kN/m²; 26.6 atm; 391 psia

Heat of Vap: 136.1 Btu/lb; 75.58 cal/g; 3.162x E5 J/kg
Heat of Comb: -19,170 Btu/lb; -10,658 cal/g; -446x E5 J/kg
Vapor Pressure: 40 mm at 22.3
UEL: 6.7%
LEL: 1.1%
Ionization Potential (eV): 9.90
Vapor Density: No data
Evaporation Rate: 3.18 (*n*-butyl acetate = 1)
Specific Gravity: 0.6838 at 20°C
Density: 0.684
Water Solubility: 0.005%
Incompatibilities: Strong oxidizers

Reactivity with Water: No data on water reactivity
Reactivity with Common Materials: No data
Stability during Transport: No data
Neutralizing Agents: No data
Polymerization Possibilities: No data

Toxic Fire Gases: None reported other than possible unburned vapors
Odor Detected at (ppm): 220 ppm
Odor Description: Gasoline (Source: CHRIS)
100% Odor Detection: 200 ppm

--- REGULATIONS ---

DOT hazard class: 3 FLAMMABLE LIQUID
DOT guide: 27
Identification number: UN1206
DOT shipping name: Heptanes
Packing group: II
Label(s) required: FLAMMABLE LIQUID
Special Provisions: T2
Packaging exceptions: 173.150
Nonbulk packaging: 173.202
Bulk packaging: 173.242
Quantity limitations:
  Passenger air/rail: 5 L
  Cargo aircraft only: 60 L
  Vessel stowage: B

STCC Number: 4909190

Clean Water Act Sect. 307: No
Clean Water Act Sect. 311: No
Clean Air Act: Not listed
EPA Waste Number: D001
CERCLA Ref: N
RQ Designation: Not listed
SARA TPQ Value: Not listed
SARA Sect. 312 Categories:
  Fire hazard: flammable.

U.S. Postal Service Mailability:
  Hazard class: Flammable liquid; mailable as ORM-D
  Mailability: Domestic surface mail only
  Max per parcel: 1 qt metal; 1 pt other

NFPA Codes:
  Health Hazard (blue): (1) Slightly hazardous to health. As a precaution wear self-contained breathing apparatus.
  Flammability (red): (3) This material can be ignited under almost all temperature conditions.
  Reactivity (yellow): (0) Stable even under fire conditions.
  Special: Unspecified.

--- TOXICITY DATA ---

Short-term Toxicity: Lightheadedness, giddy, stupor, no appetite, nausea, dermatitis, chemical pneumonia and unconsciousness.

Long-term Toxicity: Paralysis by high concentrations.

Symptoms:
  Inhalation: Irritation of respiratory tract, coughing, depression, cardiac arrhythmias.
  Aspiration: Severe lung irritation, pulmonary edema, mild excitement followed by depression.
  Ingestion: Nausea, vomiting, swelling of abdomen, depression, headache.
  (Source: CHRIS)

Conc IDLH: 5000 ppm

NIOSH REL: 85 ppm time-weighted averages for 8-hr exposure; 350 mg/m^3 time-weighted averages for 8-hr exposure; 440 ppm ceiling exposures which shall at no time be exceeded or 1800 mg/m^3 ceiling exposures which shall at no time be exceeded.

ACGIH TLV: TLV = 400 ppm (1600 mg/m^3)
ACGIH STEL: STEL = 500 ppm (2000 mg/m^3)

OSHA PEL: Transitional Limits: PEL = 500 ppm (2000 mg/m^3); Final Rule Limits: TWA = 400 ppm (1600 mg/m^3); STEL = 500 ppm (2000 mg/m^3)

MAK Information: 50 ppm; 300 mg/m^3; local irritant: peak = 2xMAK for 5 min, 8 times per shift.

Carcinogen: N; **Status:** See below

Carcinogen Lists: *IARC*: Not listed; *MAK*: Not listed; *NIOSH*: Not listed; *NTP*: Not listed; *ACGIH*: Not listed; *OSHA*: Not listed.

Human Toxicity Data: (Source: NIOSH RTECS)
   ihl-hmn TCLo:1000 ppm/6M BMRII* 2979, - , 29
   Behavioral
       Hallucinations, distorted perceptions

LD50 Value: No LD50 in RTECS 1992

Other Species Toxicity Data: (Source: NIOSH RTECS 1992)
   ihl-mus LC50:75 g/m^3/2H
   ivn-mus LD50:222 mg/kg

Reproductive Toxicity (1992 RTECS): This chemical has no known mammalian reproductive toxicity.

--- PROTECTION AND FIRST AID ---

Protection Suggested from the CHRIS Manual: Safety glasses; gloves; similar to gasoline.

NIOSH Pocket Guide to Chemical Hazards:
   **Wear Appropriate Equipment to Prevent: Repeated or prolonged skin contact.

**Wear Eye Protection to Prevent: Reasonable probability of eye contact.

**Exposed Personnel Should Wash: Promptly when skin becomes wet.

**Remove Clothing: Immediately remove any clothing that becomes wet to avoid any flammability.

**Reference: NIOSH

Recommended Respiration Protection Source: NIOSH Pocket Guide (85-114) NIOSH (Heptane) -

850 ppm: Any chemical cartridge respirator with organic vapor cartridge(s). Any supplied-air respirator. Any self-contained breathing apparatus.

1000 ppm: Any powered air-purifying respirator with organic vapor cartridge(s). Any chemical cartridge respirator with a full facepiece and organic vapor cartridge(s).

2125 ppm: Any supplied-air respirator operated in a continuous-flow mode.

4250 ppm: Any air-purifying full facepiece respirator (gas mask) with a chin-style or front- or back-mounted organic vapor canister. Any supplied-air respirator with a full facepiece. Any supplied-air respirator with a tight-fitting facepiece operated in a continuous-flow mode.

Emergency or Planned Entry in Unknown Concentrations or IDLH Conditions: Any self-contained breathing apparatus with a full facepiece and operated in a pressure-demand or other positive-pressure mode. Any supplied-air respirator with a full facepiece and operated in a pressure-demand or other positive-pressure mode in combination with an auxiliary self-contained breathing apparatus operated in a pressure-demand or other positive-pressure mode.

Escape: Any air-purifying full facepiece respirator (gas mask) with a chin-style or front- or back-mounted organic vapor canister. Any appropriate escape-type self-contained breathing apparatus.

First Aid Source: CHRIS Manual 1991.

Inhalation: Maintain respiration; give oxygen if needed.

Aspiration: Enforce bed rest; administer oxygen.

Ingestion: Do *not* induce vomiting.

Skin or Eyes: Remove contaminated clothing, wipe and wash skin area with soap and water; wash eyes with plenty of water.

First Aid Source: DOT Emergency Response Guide 1993.

Move victim to fresh air and call emergency medical care; if not breathing, give artificial respiration; if breathing is difficult, give oxygen. In case of contact with material, immediately flush eyes with running water for at least 15 min. Wash skin with soap and water. Remove and isolate contaminated clothing and shoes at the site.

--- INITIAL INCIDENT RESPONSE ---

Fire Extinguishment: Foam, dry chemical, carbon dioxide. (CHRIS 91)

U.S. Department of Transportation Guide to Hazardous Materials Transport Information - Publication DOT 5800.5 (1990).
DOT Shipping Name: Heptanes
DOT ID Number: UN1206

ERG90   GUIDE 27
* POTENTIAL HAZARDS *

*Health Hazards
  May be poisonous if inhaled or absorbed through skin.
  Vapors may cause dizziness or suffocation.
  Contact may irritate or burn skin and eyes.
  Fire may produce irritating or poisonous gases.
  Runoff from fire control or dilution water may cause pollution.

*Fire or Explosion
  Flammable/combustible material; may be ignited by heat, sparks, or flames.
  Vapors may travel to a source of ignition and flash back.
  Container may explode in heat of fire.
  Vapor explosion hazard indoors, outdoors, or in sewers.
  Runoff to sewer may create fire or explosion hazard.

* EMERGENCY ACTION *

Keep unnecessary people away; isolate hazard area and deny entry.
Stay upwind; keep out of low areas.
Positive-pressure self-contained breathing apparatus (SCBA) and structural
  firefighters' protective clothing will provide limited protection.
Isolate for 1/2 mile in all directions if tank, rail car, or tank truck is involved
  in fire.
**CALL CHEMTREC AT 1-800-424-9300 FOR EMERGENCY ASSIS-
TANCE.** If water pollution occurs, notify the appropriate authorities.

*Fire
Small fires: Dry chemical, CO_2, water spray, or regular foam.
Large fires: Water spray, fog, or regular foam.
Move container from fire area if you can do it without risk.
Apply cooling water to sides of containers that are exposed to flames until well after fire is out.
For massive fire in cargo area, use unmanned hose holder or monitor nozzles; if this is impossible, withdraw from area and let fire burn.
Withdraw immediately in case of rising sound of venting safety device or any discoloration of tank due to fire.

*Spill or Leak
Shut off ignition sources; no flares, smoking, or flames in hazard area.
Stop leak if you can do it without risk.
Water spray may reduce vapor; but it may not prevent ignition in closed spaces.
Small spills: Take up with sand or other noncombustible absorbent material and place into containers for later disposal.
Large spills: Dike far ahead of liquid spill for later disposal.

*First Aid
Move victim to fresh air and call emergency medical care; if not breathing, give artificial respiration; if breathing is difficult, give oxygen.
In case of contact with material, immediately flush eyes with running water for at least 15 min. Wash skin with soap and water.
Remove and isolate contaminated clothing and shoes at the site.

▼ ▼ ▼ ▼ ▼ ▼ ▼ ▼ ▼ ▼ ▼ ▼ ▼ ▼

--- IDENTIFIERS ---

Name: *n*-HEXANE
Synonyms: Esani (Italian); Heksan (Polish); *n*-Hexane; *N*-Hexane; Hexane (DOT); Hexanen (Dutch); NCI-C60571; Hexyl Hydride; Normal Hexane; Hexanen (Dutch)
CAS: 110-54-3; **RTECS:** MN9275000
Formula: C_6H_{14}; **Mol Wt:** 86.17
WLN: 6H
Chemical Class: Paraffin

See other identifiers listed below under Regulations.

--- PROPERTIES ---

Physical Description: Colorless liquid with a mild gasolinelike odor
Boiling Point: 342.04 K; 68.8°C; 156°F
Melting Point: 178.16 K; -95°C; -139°F
Flash Point: 250 K; -23.2°C; -9.7°F
Autoignition: 498 K; 224.8°C; 436.7°F
Critical Temp: 507.4 K; 234.25°C; 453.65°F
Critical Press: 3.01 kN/m^2; 29.6 atm; 436 psia
Heat of Vap: 144 Btu/lb; 79.97 cal/g; 3.346x E5 J/kg
Heat of Comb: -19,246 Btu/lb; -10,700 cal/g; -447x E5 J/kg
Vapor Pressure: 124 mm
UEL: 7.5%
LEL: 1.1%
Ionization Potential (eV): 10.18
Vapor Density: 3.0 (air = 1)
Evaporation Rate: 6.82 (*n*-butyl acetate = 1)
Specific Gravity: 0.659 at 20°C
Density: 0.659
Water Solubility: 0.014%
Incompatibilities: Strong oxidizers

Reactivity with Water: No data on water reactivity
Reactivity with Common Materials: No data
Stability during Transport: No data
Neutralizing Agents: No data
Polymerization Possibilities: No data

Toxic Fire Gases: None reported other than possible unburned vapors
Odor Detected at (ppm): 130 ppm
Odor Description: Mild, gasolinelike (Source: NYDH)
100% Odor Detection: No data

--- REGULATIONS ---

DOT hazard class: 3 FLAMMABLE LIQUID
DOT guide: 27
Identification number: UN1208
DOT shipping name: HEXANES
Packing group: II
Label(s) required: FLAMMABLE LIQUID
Special Provisions: T8

Packaging exceptions: 173.150
Nonbulk packaging: 173.202
Bulk packaging: 173.242
Quantity limitations:
  Passenger air/rail: 5 L
  Cargo aircraft only: 60 L
  Vessel stowage: E

STCC Number: 4908183

Clean Water Act Sect. 307: No
Clean Water Act Sect. 311: No
Clean Air Act: CAA '90 listed
EPA Waste Number: D001
CERCLA Ref: Not listed
RQ Designation: Not listed
SARA TPQ Value: Not listed
SARA Sect. 312 Categories:
  Acute toxicity: irritant.
  Acute toxicity: adverse effect to target organs.
  Chronic toxicity: mutagen.
  Chronic toxicity: reproductive toxin.
  Fire hazard: flammable.

U.S. Postal Service Mailability:
  Hazard class: Flammable liquid - mailable as ORM-D
  Mailability: Domestic surface mail only
  Max per parcel: 1 qt metal; 1 pt other

NFPA Codes:
  Health Hazard (blue): (1) Slightly hazardous to health. As a precaution wear self-contained breathing apparatus.
  Flammability (red): (3) This material can be ignited under almost all temperature conditions.
  Reactivity (yellow): (0) Stable even under fire conditions.
  Special: Unspecified.

--- TOXICITY DATA ---

Short-term Toxicity: *Inhalation:* Exposure to levels above 500 ppm may cause headache, abdominal cramps, a burning feeling of the face, and numbness and weakness of the fingers and toes. Levels above 1300 ppm may cause the above plus nausea and irritation of the nose and throat. Levels above 1500 ppm may cause the above plus blurred vision, loss of appetite, and loss of weight. Most symptoms disappear within a few months if exposure ceases. Breathing liquid into the lungs may cause a chemical pneumonia. *Skin:* Contact may cause irritation, redness, swelling, blisters, and pain. Skin exposure may contribute to symptoms listed under inhalation. *Eyes:* Levels over 880 ppm may cause irritation. *Ingestion:* May contribute to symptoms listed under inhalation. Estimated lethal dose is 1 oz to 1 pt. (NYDH)

Long-term Toxicity: May cause symptoms listed under inhalation. Exposure to levels above 650 ppm for 2 to 4 months can result in weakness and numbness of the arms and legs. Symptoms go away within a few months if exposure stops. (NYDH)

Target Organs: Eyes, respiratory system

Symptoms:
Inhalation: Causes irritation of respiratory tract, cough, mild depression, and cardiac arrhythmias.
Aspiration: Causes severe lung irritation, coughing, pulmonary edema; excitement followed by depression.
Ingestion: Causes nausea, vomiting, swelling of abdomen, headache, depression. (Source: CHRIS)

Conc IDLH: 5000 ppm

NIOSH REL: 100 ppm time-weighted averages for 8-hr exposure; 350 mg/m^3 time-weighted averages for 8-hr exposure

ACGIH TLV: TLV = 500 ppm
ACGIH STEL: STEL = 1000 ppm

OSHA PEL: Final Rule Limits: TWA = 500 ppm (1800 mg/m^3); STEL = 1000 ppm (3600 mg/m^3)

MAK Information: 50 ppm; 180 mg/m^3
Substance with systemic effects, onset of effect less than or equal to 2 hr: Peak = 2xMAK for 30 min, 4 times per shift of 8 hr.

There is no reason to fear a risk of damage to the developing embryo or fetus when MAK values are adhered to.

Carcinogen: N; **Status:** See below

Carcinogen Lists: *IARC*: Not listed; *MAK*: Not listed; *NIOSH*: Not listed; *NTP*: Not listed; *ACGIH*: Not listed; *OSHA*: Not listed.

Human Toxicity Data: (Source: NIOSH RTECS)
ihl-hmn TCLo:190 ppm/8W AJIMD8 10, 111, 86
Peripheral Nerve and Sensation
Structural change in nerve or sheath

LD50 Value: orl-rat LD50:28710 mg/kg

Other Species Toxicity Data: (Source: NIOSH RTECS 1992)
orl-rat LD50:28710 mg/kg
ipr-rat LDLo:9100 mg/kg
ihl-mus LCLo:120 g/m^3
ivn-mus LDLo:831 mg/kg
ivn-rbt LDLo:132 mg/kg

Irritation Data: (Source: NIOSH RTECS 1992)
eye-hmn 5 ppm

Reproductive Toxicity (1992 RTECS): This chemical is a mammalian reproductive toxin.

Reproductive Toxicity Data (1992 RTECS):
ihl-rat TCLo:10000 ppm/7H (15D pre/1-18D preg) TOXID9 1, 152, 81
Effects on Newborn
Behavioral

ihl-rat TCLo:1000 ppm/6H (8-16D preg) TJADAB 19, 22A, 79
Effects on Newborn
Growth statistics (e.g., reduced weight gain)

ihl-rat TCLo:5000 ppm/20H (6-19D preg) NTIS** DE88-006812
Effects on Embryo or Fetus
Fetotoxicity (except death, e.g., stunted fetus)

ihl-rat TCLo:1 pph/6H (65D male) FAATDF 4, 191, 84
Paternal Effects
Testes, epididymis, sperm duct

ihl-rat TCLo:1 pph/6H (65D male) FAATDF 4, 191, 84
Paternal Effects
Testes, epididymis, sperm duct

--- PROTECTION AND FIRST AID ---

Protection Suggested from the CHRIS Manual: Eye protection (as for gasoline).

NIOSH Pocket Guide to Chemical Hazards:
 **Wear Appropriate Equipment to Prevent: Repeated or prolonged skin contact.

 **Wear Eye Protection to Prevent: Reasonable probability of eye contact.

 **Exposed Personnel Should Wash: Promptly when skin becomes wet.

 **Remove Clothing: Immediately remove any clothing that becomes wet to avoid any flammability.

 **Reference: NIOSH

Recommended Respiration Protection Source: NIOSH Pocket Guide (85-114)
NIOSH (*n*-Hexane) -
1000 ppm: Any supplied-air respirator. Any self-contained breathing apparatus. Substance reported to cause eye irritation or damage may require eye protection.
2500 ppm: Any supplied-air respirator operated in a continuous-flow mode. Substance reported to cause eye irritation or damage may require eye protection.
5000 ppm: Any self-contained breathing apparatus with a full facepiece. Any supplied-air respirator with a full facepiece.
Emergency or Planned Entry in Unknown Concentrations or IDLH Conditions: Any self-contained breathing apparatus with a full facepiece and operated in a pressure-demand or other positive-pressure mode. Any supplied-air respirator with a full facepiece and operated in a pressure-demand or other positive-pressure mode in combination with an auxiliary self-contained breathing apparatus operated in a pressure-demand or other positive-pressure mode.
Escape: Any air-purifying full facepiece respirator (gas mask) with a chin-style or front- or back-mounted organic vapor canister. Any appropriate escape-type self-contained breathing apparatus.

First Aid Source: CHRIS Manual 1991.
Call a doctor.
Inhalation: Maintain respiration; give oxygen if needed.
Aspiration: Enforce bed rest; give oxygen if needed.
Ingestion: Do *not* induce vomiting.
Skin or Eyes: Wipe off; wash skin with soap and water; wash eyes with copious amounts of water.

First Aid Source: DOT Emergency Response Guide 1993.
Move victim to fresh air and call emergency medical care; if not breathing, give artificial respiration; if breathing is difficult, give oxygen. In case of contact with material, immediately flush eyes with running water for at least 15 min. Wash skin with soap and water. Remove and isolate contaminated clothing and shoes at the site.

--- INITIAL INCIDENT RESPONSE ---

Fire Extinguishment: Foam, dry chemical, carbon dioxide. (CHRIS 91)

U.S. Department of Transportation Guide to Hazardous Materials Transport Information - Publication DOT 5800.5 (1990).
DOT Shipping Name: Hexanes
DOT ID Number: UN1208

ERG90   GUIDE 27
* POTENTIAL HAZARDS *

*Health Hazards
May be poisonous if inhaled or absorbed through skin.
Vapors may cause dizziness or suffocation.
Contact may irritate or burn skin and eyes.
Fire may produce irritating or poisonous gases.
Runoff from fire control or dilution water may cause pollution.

*Fire or Explosion
Flammable/combustible material; may be ignited by heat, sparks, or flames.
Vapors may travel to a source of ignition and flash back.
Container may explode in heat of fire.
Vapor explosion hazard indoors, outdoors, or in sewers.
Runoff to sewer may create fire or explosion hazard.

* EMERGENCY ACTION *

Keep unnecessary people away; isolate hazard area and deny entry.

Stay upwind; keep out of low areas.

Positive-pressure self-contained breathing apparatus (SCBA) and structural firefighters' protective clothing will provide limited protection.

Isolate for 1/2 mile in all directions if tank, rail car, or tank truck is involved in fire.

CALL CHEMTREC AT 1-800-424-9300 FOR EMERGENCY ASSISTANCE. If water pollution occurs, notify the appropriate authorities.

*Fire

Small fires: Dry chemical, CO_2, water spray, or regular foam.

Large fires: Water spray, fog, or regular foam.

Move container from fire area if you can do it without risk.

Apply cooling water to sides of containers that are exposed to flames until well after fire is out. Stay away from ends of tanks.

For massive fire in cargo area, use unmanned hose holder or monitor nozzles; if this is impossible, withdraw from area and let fire burn.

Withdraw immediately in case of rising sound of venting safety device or any discoloration of tank due to fire.

*Spill or Leak

Shut off ignition sources; no flares, smoking, or flames in hazard area.

Stop leak if you can do it without risk.

Water spray may reduce vapor; but it may not prevent ignition in closed spaces.

Small spills: Take up with sand or other noncombustible absorbent material and place into containers for later disposal.

Large spills: Dike far ahead of liquid spill for later disposal.

*First Aid

Move victim to fresh air and call emergency medical care; if not breathing, give artificial respiration; if breathing is difficult, give oxygen.

In case of contact with material, immediately flush eyes with running water for at least 15 min. Wash skin with soap and water.

Remove and isolate contaminated clothing and shoes at the site.

▼   ▼   ▼   ▼   ▼   ▼   ▼   ▼   ▼   ▼   ▼   ▼   ▼   ▼

--- IDENTIFIERS ---

Name: ISOPROPYL ALCOHOL
Synonyms: Alcool Isopropilico (Italian); Alcool Isopropylique (French); Avantine; Dimethylcarbinol; Isohol; Isopropanol; Iso-Propylalkohol (German); Lutosol; Petrohol; Propropan-2-Ol; 2-Propanol; *i*-Propanol (German); sec-Propyl Alcohol; *i*-Propylalkohol
CAS: 67-63-0; **RTECS:** NT8050000
Formula: C_3H_8O; **Mol Wt:** 60.11
WLN: QY1
Chemical Class: Alcohol

See other identifiers listed below under Regulations.

--- PROPERTIES ---

Physical Description: Colorless liquid with a sweet odor
Boiling Point: 355.65 K; 82.5°C; 180.5°F
Melting Point: 184.27 K; -88.9°C; -128°F
Flash Point: 294.15 K; 21°C; 69.8°F
Autoignition: 672.15 K; 399°C; 750.2°F
Vapor Pressure: 89.26 mm at 56.77°C
UEL: 12.7%
LEL: 2.3%
Ionization Potential (eV): 10.15
Vapor Density: 2.07 (air = 1)
Evaporation Rate: 1.70 (*n*-butyl acetate = 1)
Specific Gravity: 0.785 at 20°C
Density: 0.785
Water Solubility: Miscible
Incompatibilities: Keep away from heat and open flame; reacts vigorously with oxidizing materials.

Reactivity with Water: No data on water reactivity
Reactivity with Common Materials: Reacts violently with Nitroform, Oleum, Phosgene, Potassium, Tert-Butoxide, Aluminum, Al Triisopropoxide, Crotonaldehyde, Oxidants (Source: SAX)
Stability during Transport: No data
Neutralizing Agents: No data
Polymerization Possibilities: No data

Toxic Fire Gases: NA
Odor Detected at (ppm): 90 mg/m#L3
Odor Description: Nonresidual (Source: CHRIS)
100% Odor Detection: No data

--- REGULATIONS ---

DOT hazard class: 3 FLAMMABLE LIQUID
DOT guide: 26
Identification number: UN1219
DOT shipping name: Isopropanol or isopropyl alcohol
Packing group: II
Label(s) required: FLAMMABLE LIQUID
Special Provisions: T1
Packaging exceptions: 173.150
Nonbulk packaging: 173.202
Bulk packaging: 173.242
Quantity limitations:
  Passenger air/rail: 5 L
  Cargo aircraft only: 60 L
  Vessel stowage: B

STCC Number: 4909205

Clean Water Act Sect. 307: No
Clean Water Act Sect. 311: No
Clean Air Act: Not listed
EPA Waste Number: None
CERCLA Ref: N
RQ Designation: Not listed
SARA TPQ Value: Not listed
SARA Sect. 312 Categories:
  Acute toxicity: irritant.
  Acute toxicity: adverse effect to target organs.
  Chronic toxicity: mutagen.
  Chronic toxicity: reproductive toxin.
  Fire hazard: flammable.
Listed in SARA Sect 313: Yes
De Minimus Concentration: 0.1%

U.S. Postal Service Mailability:
  Hazard class: Flammable liquid - mailable as ORM-D
  Mailability: Domestic surface mail only
  Max per parcel: 1 qt metal; 1 pt other

NFPA Codes:
  Health Hazard (blue): (1) Slightly hazardous to health. As a precaution wear self-contained breathing apparatus.
  Flammability (red): (3) This material can be ignited under almost all temperature conditions.
  Reactivity (yellow): (0) Stable even under fire conditions.
  Special: Unspecified.

--- TOXICITY DATA ---

Short-term Toxicity: *Inhalation:* Irritation of the nose and throat may occur at 400 ppm and above. *Skin:* 5% solution may cause irritation and dryness. *Eyes:* Vapor levels of 20 ppm or above may result in irritation. Liquid may cause corneal burns and eye damage. *Ingestion:* 22.5 ml (2/3 oz) has caused salivation, reddening of face, stomach pain, depression, dizziness, headache, vomiting, and unconsciousness. Ingestion of 100 ml (3 oz) has caused death. (NYDH)

Long-term Toxicity: No reported long-term exposure effects. (NYDH)

Target Organs: Eyes, skin, respiratory system

Symptoms: Vapors cause mild irritation of eyes and upper respiratory tract; high concentrations may be anesthetic. Liquid irritates eyes and may cause injury; harmless to skin; if ingested, causes drunkenness and vomiting. (Source: CHRIS)

Conc IDLH: 12,000 ppm

NIOSH REL: 400 ppm time-weighted averages for 8-hr exposure; 984 mg/m^3 time-weighted averages for 8-hr exposure; 800 ppm ceiling exposures, which shall at no time be exceeded or 1968 mg/m^3 ceiling exposures, which shall at no time be exceeded

ACGIH TLV: TLV = 400 ppm (980 mg/m^3)
ACGIH STEL: STEL = 500 ppm (1225 mg/m^3)

OSHA PEL: Transitional Limits: PEL = 400 ppm (980 mg/m³; Final Rule Limits: TWA = 400 ppm (980 mg/m³); STEL = 500 ppm (1225 mg/m³)

MAK Information: 400 ppm; 980 mg/m³
Substance with systemic effects, onset of effect less than or equal to 2 hr: Peak = 2xMAK for 30 min, 4 times per shift of 8 hr.

Carcinogen: N; **Status:** See below
References: Human suspected IARC** 15, 223, 77; Animal indefinite IARC** 15, 223, 77

Carcinogen Lists: *IARC:* Not classified as to human carcinogenicity or probably not carcinogenic to humans; *MAK:* Not listed; *NIOSH:* Not listed; *NTP:* Not listed; *ACGIH:* Not listed; *OSHA:* Not listed.

Human Toxicity Data: (Source: NIOSH RTECS)
  orl-man TDLo:14432 mg/kg NEJMAG 277, 699, 67
      Behavioral
        Coma
      Vascular
        BP lowering not characterized in autonomic section
      Lungs, Thorax, or Respiration
        Dyspnea

  orl-hmn TDLo:223 mg/kg JLCMAK 12, 326, 27
      Behavioral
        Hallucinations, distorted perceptions
      Cardiac
        Pulse rate decreased with fall in BP
      Vascular
        BP lowering not characterized in autonomic section

  orl-man LDLo:5272 mg/kg AJCPAI 38, 144, 62
      Behavioral
        Coma
      Vascular
        BP lowering not characterized in autonomic section.
      Lungs, Thorax, or Respiration
        Chronic pulmonary edema or congestion

orl-hmn LDLo:3570 mg/kg 34ZIAG -, 339, 69
  Behavioral
    Coma
  Lungs, Thorax, or Respiration
    Respiratory depression
  Gastrointestinal
    Nausea or vomiting

LD50 Value: orl-rat LD50:5045 mg/kg

Other Species Toxicity Data: (Source: NIOSH RTECS 1992)
  orl-rat LD50:5045 mg/kg
  ihl-rat LCLo:16000 ppm/4H
  ipr-rat LD50:2735 mg/kg
  ivn-rat LD50:1088 mg/kg
  orl-mus LD50:3600 mg/kg
  ihl-mus LCLo:12800 ppm/3H
  ipr-mus LD50:4477 mg/kg
  scu-mus LDLo:6 g/kg
  ivn-mus LD50:1509 mg/kg
  orl-dog LDLo:1537 mg/kg
  ivn-dog LDLo:1024 mg/kg
  ivn-cat LDLo:1963 mg/kg
  orl-rbt LD50:6410 mg/kg
  skn-rbt LD50:12800 mg/kg
  ipr-rbt LD50:667 mg/kg
  ivn-rbt LD50:1184 mg/kg
  ipr-gpg LD50:2560 mg/kg
  ipr-ham LD50:3444 mg/kg
  par-frg LDLo:20 g/kg
  scu-mam LDLo:6 mg/kg

Reproductive Toxicity (1992 RTECS): This chemical is a mammalian reproductive toxin.

Reproductive Toxicity Data (1992 RTECS):
  orl-rat TDLo:11340 mg/kg (45D pre) GISAAA 43(1), 8, 78
    Maternal Effects
      Menstrual cycle changes or disorders

orl-rat TDLo:5040 mg/kg (1-20D preg) GISAAA 43(1), 8, 78
Effects on Fertility
Litter size (number of fetuses per litter; measured before birth)

orl-rat TDLo:20160 mg/kg (1-20D preg) GISAAA 43(1), 8, 78
Effects on Fertility
Preimplantation mortality

orl-rat TDLo:32400 mg/kg (26W pre) GISAAA 43(1), 8, 78
Effects on Embryo or Fetus
Fetal death

orl-rat TDLo:6480 mg/kg (26W male/26W pre) GISAAA 43(1), 8, 78
Effects on Newborn
Growth statistics (e.g., reduced weight gain)

ihl-rat TCLo:3500 ppm/7H (1-19D preg) FCTOD7 26, 247, 88
Effects on Embryo or Fetus
Fetotoxicity (except death, e.g., stunted fetus)

ihl-rat TCLo:10000 ppm/7H (1-19D preg) FCTOD7 26, 247, 88
Effects on Fertility
Preimplantation mortality
Postimplantation mortality
Effects on Embryo or Fetus
Fetal death

--- PROTECTION AND FIRST AID ---

Protection Suggested from the CHRIS Manual: Organic vapor canister or air-supplied mask; chemical goggles or face splash shield.

NIOSH Pocket Guide to Chemical Hazards:
**Wear Appropriate Equipment to Prevent: Repeated or prolonged skin contact.

**Wear Eye Protection to Prevent: Reasonable probability of eye contact.

**Exposed Personnel Should Wash: Promptly when skin becomes wet.

**Remove Clothing: Immediately remove any clothing that becomes wet to avoid any flammability.

**Reference: NIOSH

Recommended Respiration Protection Source: NIOSH Pocket Guide (85-114) NIOSH (Isopropyl Alcohol)
1000 ppm: Any powered air-purifying respirator with organic vapor cartridge(s). Substance causes eye irritation or damage; eye protection needed. Any chemical cartridge respirator with a full facepiece and organic vapor cartridge(s).
10,000 ppm: Any supplied-air respirator operated in a continuous-flow mode. Substance causes eye irritation or damage; eye protection needed.
20,000 ppm: Any air-purifying full facepiece respirator (gas mask) with a chin-style or front- or back-mounted organic vapor canister. Any self-contained breathing apparatus with a full facepiece. Any supplied-air respirator with a full facepiece.
Emergency or Planned Entry in Unknown Concentrations or IDLH Conditions: Any self-contained breathing apparatus with a full facepiece and operated in a pressure-demand or other positive-pressure mode. Any supplied-air respirator with a full facepiece and operated in a pressure-demand or other positive-pressure mode in combination with an auxiliary self-contained breathing apparatus operated in a pressure-demand or other positive-pressure mode.
Escape: Any air-purifying full facepiece respirator (gas mask) with a chin-style or front- or back-mounted organic vapor canister. Any appropriate escape-type self-contained breathing apparatus.

First Aid Source: CHRIS Manual 1991.
Inhalation: If victim is overcome by vapors, remove from exposure immediately; call a physician; if breathing is irregular or has stopped, start resuscitation and administer oxygen.
Eyes: Flush with water for at least 15 min.

First Aid Source: DOT Emergency Response Guide 1993.
Move victim to fresh air and call emergency medical care; if not breathing, give artificial respiration; if breathing is difficult, give oxygen. In case of contact with material, immediately flush eyes with running water for at least 15 min. Wash skin with soap and water. Remove and isolate contaminated clothing and shoes at the site.

--- INITIAL INCIDENT RESPONSE ---

Fire Extinguishment: Alcohol foam, dry chemical, carbon dioxide. *Note:* Water may be ineffective. (CHRIS 91)

U.S. Department of Transportation Guide to Hazardous Materials Transport Information - Publication DOT 5800.5 (1990).
DOT Shipping Name: Isopropanol or isopropyl alcohol
DOT ID Number: UN1219

ERG90   GUIDE 26
* POTENTIAL HAZARDS *

*Health Hazards
May be poisonous if inhaled or absorbed through skin.
Vapors may cause dizziness or suffocation.
Fire may produce irritating or poisonous gases.
Runoff from fire control or dilution water may cause pollution.

*Fire or Explosion
Flammable/combustible material; may be ignited by heat, sparks, or flames.
Vapors may travel to a source of ignition and flash back.
Container may explode in heat of fire.
Vapor explosion hazard indoors, outdoors, or in sewers.
Runoff to sewer may create fire or explosion hazard.

* EMERGENCY ACTION *

Keep unnecessary people away; isolate hazard area and deny entry.
Stay upwind; keep out of low areas.
Positive-pressure self-contained breathing apparatus (SCBA) and structural
  firefighters' protective clothing will provide limited protection.
Isolate for 1/2 mile in all directions if tank, rail car, or tank truck is involved
in   fire.
**CALL CHEMTREC AT 1-800-424-9300 FOR EMERGENCY ASSIS-
TANCE.** If water pollution occurs, notify the appropriate authorities.

*Fire
  Small fires: Dry chemical, CO_2 or Halon, water spray, or alcohol-resistant
    foam.

Large fires: Water spray, fog, or alcohol-resistant foam.

Move container from fire area if you can do it without risk.

Apply cooling water to sides of containers that are exposed to flames until well after fire is out. Stay away from ends of tanks.

For massive fire in cargo area, use unmanned hose holder or monitor nozzles; if this is impossible, withdraw from area and let fire burn.

Withdraw immediately in case of rising sound of venting safety device or any discoloration of tank due to fire.

*Spill or Leak

Shut off ignition sources; no flares, smoking, or flames in hazard area.

Stop leak if you can do it without risk.

Water spray may reduce vapor; but it may not prevent ignition in closed spaces.

Small spills: Take up with sand or other noncombustible absorbent material and place into containers for later disposal.

Large spills: Dike far ahead of liquid spill for later disposal.

*First Aid

Move victim to fresh air and call emergency medical care; if not breathing, give artificial respiration; if breathing is difficult, give oxygen.

In case of contact with material, immediately flush eyes with running water for at least 15 min. Wash skin with soap and water.

Remove and isolate contaminated clothing and shoes at the site.

▼   ▼   ▼   ▼   ▼   ▼   ▼   ▼   ▼   ▼   ▼   ▼   ▼

--- IDENTIFIERS ---

Name: METHANOL

Synonyms: Alcool Methylique (French); Alcool Metilico (Italian); Carbinol; Colonial Spirit; Columbian Spirit; Columbian Spirits (DOT); Methanol (DOT); Metanolo (Italian); Methyl Alcohol; Methyl Alcohol (DOT); Methylol; Methylalkohol (German); Methyl Hydroxide; Metylowy Alkohol (Polish); Monohydroxymethane; Pyroxylic Spirit; Wood Alcohol; Wood Naphtha; Wood Spirit

CAS: 67-56-1; **RTECS:** PC1400000

Formula: CH_4O; **Mol Wt:** 32.05

WLN: Q1

Chemical Class: Alcohol

See other identifiers listed below under Regulations.

--- PROPERTIES ---

Physical Description: Clear, colorless, flammable, poisonous liquid with slight alcoholic odor
Boiling Point: 337.65 K; 64.5°C; 148.1°F
Melting Point: 175.38 K; -97.8°C; -144°F
Flash Point: 288.65 K; 15.5°C; 59.9°F
Autoignition: 737.15 K; 464°C; 867.2°F
Vapor Pressure: 92 mm at 20°C
UEL: 36.5%
LEL: 6.0%
Ionization Potential (eV): 10.84
Vapor Density: 1.11 (air = 1)
Evaporation Rate: 2.10 (*n*-butyl acetate = 1)
Specific Gravity: 0.792 at 20°C
Density: 0.7915 g/ml at 20°C
Water Solubility: Miscible
Incompatibilities: Strong oxidizers

Reactivity with Water: No data on water reactivity
Reactivity with Common Materials: Oxidizing materials (Source: SAX)
Stability during Transport: No data
Neutralizing Agents: No data
Polymerization Possibilities: No data

Toxic Fire Gases: None reported other than possible unburned vapors
Odor Detected at (ppm): 100 ppm
Odor Description: Faintly sweet; characteristic, pungent (Source: CHRIS)
100% Odor Detection: No data

--- REGULATIONS ---

DOT hazard class: 3 FLAMMABLE LIQUID
DOT guide: 28
Identification number: UN1230
DOT shipping name: Methanol or methyl alcohol
Packing group: II
Label(s) required: FLAMMABLE LIQUID, POISON
Special Provisions: T8
Packaging exceptions: 173.None
Nonbulk packaging: 173.202
Bulk packaging: 173.243

Quantity limitations:
  Passenger air/rail: 1 L
  Cargo aircraft only: 60 L
  Vessel stowage: B
  Other stowage provisions: 40

STCC Number: 4909230

Clean Water Act Sect. 307: No
Clean Water Act Sect. 311: No
Clean Air Act: CAA '90 listed
EPA Waste Number: U154
CERCLA Ref: Y
RQ Designation: D   5000 lb (2270 kg) CERCLA
SARA TPQ Value: Not listed
SARA Sect. 312 Categories:
  Acute toxicity: irritant.
  Acute toxicity: adverse effect to target organs.
  Chronic toxicity: mutagen.
  Chronic toxicity: reproductive toxin.
  Fire hazard: flammable.
Listed in SARA Sect 313: Yes
De Minimus Concentration: 1.0%

U.S. Postal Service Mailability:
  Hazard class: Flammable liquid - mailable as ORM-D
  Mailability: Domestic service and air transportation
  Max per parcel: 1 qt metal; 1 pt other

NFPA Codes:
  Health Hazard (blue): (1) Slightly hazardous to health. As a precaution wear self-contained breathing apparatus.
  Flammability (red): (3) This material can be ignited under almost all temperature conditions.
  Reactivity (yellow): (0) Stable even under fire conditions.
  Special: Unspecified.

--- TOXICITY DATA ---

Short-term Toxicity: *Inhalation:* Below 500 ppm, symptoms are rarely felt. Can cause headache, vomiting, irritation of the nose and throat, dilation of the pupils, feeling of intoxication, loss of muscle coordination, excessive sweating, bronchitis, and convulsions. Very high exposures may result in stupor, cramps and visual difficulties such as spotted vision, sensitivity to light, eye tenderness, and blindness. Recovery is not always complete and symptoms may recur without additional exposure. *Skin:* Can cause dry and cracked skin irritation and reddening. Skin absorption can be enough to contribute to symptoms described under inhalation. *Eyes:* Can cause irritation of eye. *Ingestion:* Symptoms are similar to those under inhalation, plus damage to liver, kidneys, and heart. Nerve damage may occur causing loss of coordination and blindness. Recovery is not always complete. Death may occur. Usual fatal dose is about 100 to 250 ml, but death from ingestion has occurred from as little as 30 ml (about 1 oz). (NYDH)

Long-term Toxicity: Exposure to low levels may cause many of the symptoms listed above. Because methyl alcohol is slowly eliminated from the body, repeated low exposures may build up to high levels, causing severe symptoms. Recovery is not always complete. Methanol has been found to cause changes in the genetic material of some test animals. Whether it does in humans is unknown. (NYDH)

Target Organs: Eyes, skin, central nervous system, gastrointestinal.

Symptoms: Exposure to excessive vapor causes eye irritation, headache, fatigue, and drowsiness. High concentrations can produce central nervous system depression and optic nerve damage. 50,000 ppm will probably cause death in 1 to 2 hr. Can be absorbed through skin. Swallowing may cause death or eye damage. (Source: CHRIS)

Conc IDLH: 25,000 ppm

NIOSH REL: 200 ppm time-weighted averages for 8-hr exposure; 262 mg/m^3 time-weighted averages for 8-hr exposure; 800 ppm ceiling exposures, which shall at no time be exceeded; 1048 mg/m^3 ceiling exposures, which shall at no time be exceeded.

ACGIH TLV: TLV = 200 ppm SKIN
ACGIH STEL: STEL = 250 ppm > > SKIN

OSHA PEL: Transitional Limits: PEL = 200 ppm (260 mg/m^3) (skin); Final Rule Limits: TWA = 200 ppm (260 mg/m^3) (skin); STEL = 250 ppm (325 mg/m^3) (skin)

MAK Information: 200 ppm; 260 mg/m^3
   Substance with systemic effects, onset of effect less than or equal to 2 hr:
   Peak = 2xMAK for 30 min, 4 times per shift of 8 hr.
   Danger of cutaneous absorption

Carcinogen: N; **Status:** See below

Carcinogen Lists: *IARC*: Not listed; *MAK*: Not listed; *NIOSH*: Not listed;
   NTP: Not listed; *ACGIH*: Not listed; *OSHA*: Not listed.

Human Toxicity Data: (Source: NIOSH RTECS)
   orl-man LDLo:6422 mg/kg CMAJAX 128, 14, 83
      Brain and Coverings
         Changes in circulation (hemorrhage, thrombosis, etc.)
      Lungs, Thorax, or Respiration
         Dyspnea
      Gastrointestinal
         Nausea or vomiting

   orl-man TDLo:3429 mg/kg AMSVAZ 212, 5, 82
      Sense Organs
         Eye
            Visual field changes

   orl-hmn LDLo:428 mg/kg NPIRI* 1, 74, 74
      Behavioral
         Headache
      Lungs, Thorax, or Respiration
         Other changes

   orl-hmn LDLo:143 mg/kg 34ZIAG -, 382, 69
      Sense Organs
         Eye
            Optic nerve neuropathy
      Lungs, Thorax, or Respiration
         Dyspnea
      Gastrointestinal
         Nausea or vomiting

orl-wmn TDLo:4 g/kg AMSVAZ 212, 5, 82
  Sense Organs
    Eye
      Visual field changes
  Lungs, Thorax, or Respiration
    Dyspnea
  Gastrointestinal
    Nausea or vomiting

ihl-hmn TCLo:86000 mg/m³ AGGHAR 5, 1, 33
  Sense Organs
    Eye
      Lacrimation
  Lungs, Thorax, or Respiration
    Cough
    Other changes

ihl-hmn TCLo:300 ppm NPIRI* 1, 74, 74
  Sense Organs
    Eye
      Visual field changes
  Behavioral
    Headache
  Lungs, Thorax, or Respiration
    Other changes

LD50 Value: orl-rat LD50:5628 mg/kg

Other Species Toxicity Data: (Source: NIOSH RTECS 1992)
  orl-rat LD50:5628 mg/kg
  ihl-rat LC50:64000 ppm/4H
  ipr-rat LD50:7529 mg/kg
  ivn-rat LD50:2131 mg/kg
  orl-mus LD50:7300 mg/kg
  ihl-mus LCLo:50 g/m³/2H
  ipr-mus LD50:10765 mg/kg
  scu-mus LD50:9800 mg/kg
  ivn-mus LD50:4710 mg/kg
  orl-dog LDLo:7500 mg/kg
  orl-mky LD50:7 g/kg
  ihl-mky LCLo:1000 ppm
  skn-mky LDLo:393 mg/kg

ihl-cat LCLo:44000 mg/m^3/6H
ivn-cat LDLo:4641 mg/kg
orl-rbt LD50:14200 mg/kg
skn-rbt LD50:15800 mg/kg
ipr-rbt LD50:1826 mg/kg
ivn-rbt LD50:8907 mg/kg
ipr-gpg LD50:3556 mg/kg
ipr-ham LD50:8555 mg/kg
par-frg LDLo:59 g/kg

Irritation Data: (Source: NIOSH RTECS 1992)
eye-hmn 5 ppm
skn-rbt 500 mg/24H MOD
eye-rbt 40 mg MOD

Reproductive Toxicity (1992 RTECS): This chemical is a mammalian reproductive toxin.

Reproductive Toxicity Data (1992 RTECS):
orl-rat TDLo:7500 mg/kg (17-19D preg) TJADAB 33, 259, 86
Effects on Newborn
Behavioral

orl-rat TDLo:35295 mg/kg (1-15D preg) ONGZAC 22(1), 71, 91
Effects on Fertility
Female fertility index
Preimplantation mortality
Postimplantation mortality

orl-rat TDLo:35295 mg/kg (1-15D preg) ONGZAC 22(1), 71, 91
Effects on Embryo or Fetus
Fetotoxicity (except death, e.g., stunted fetus)

ihl-rat TCLo:20000 ppm/7H (1-22D preg) TJADAB 29(2), 48A, 84
Specific Developmental Abnormalities
Musculoskeletal system
Cardiovascular (circulatory) system
Urogenital system

ihl-rat TCLo:20,000 ppm/7H (7-15D preg) FAATDF 5, 727, 85
　　Specific Developmental Abnormalities
　　　　Musculoskeletal system
　　　　Endocrine system

ihl-rat TCLo:10,000 ppm/7H (7-15D preg) FAATDF 5, 727, 85
　　Effects on Embryo or Fetus
　　　　Fetotoxicity (except death, e.g., stunted fetus)

--- PROTECTION AND FIRST AID ---

Protection Suggested from the CHRIS Manual: Approved canister mask for high vapor concentrations; safety goggles; rubber gloves.

NIOSH Pocket Guide to Chemical Hazards:
　　**Wear Appropriate Equipment to Prevent: Repeated or prolonged skin contact.

　　**Wear Eye Protection to Prevent: Reasonable probability of eye contact.

　　**Exposed Personnel Should Wash: Promptly when skin becomes wet.

　　**Remove Clothing: Immediately remove any clothing that becomes wet to avoid any flammability.

　　**Reference: NIOSH

First Aid Source: CHRIS Manual 1991.
Remove victim from exposure and apply artificial respiration if breathing has ceased.
Ingestion: Induce vomiting; then give 2 teaspoons of baking soda in glass of water. Call a physician.
Skin or Eyes: Flush with water for 15 min.

First Aid Source: DOT Emergency Response Guide 1993.
Move victim to fresh air and call emergency medical care; if not breathing, give artificial respiration; if breathing is difficult, give oxygen. In case of contact with material, immediately flush eyes with running water for at least 15 min. Wash skin with soap and water. Remove and isolate contaminated clothing and shoes at the site. Keep victim quiet and maintain normal body temperature. Effects may be delayed; keep victim under observation.

--- INITIAL INCIDENT RESPONSE ---

Fire Extinguishment: Alcohol foam, dry chemical, carbon dioxide. *Note:* Water may be ineffective. (CHRIS 91)

U.S. Department of Transportation Guide to Hazardous Materials Transport Information - Publication DOT 5800.5 (1990).
DOT Shipping Name: Methanol or methyl alcohol
DOT ID Number: UN1230

ERG90   GUIDE 28
* POTENTIAL HAZARDS *

*Health Hazards
   Poisonous; may be fatal if inhaled, swallowed, or absorbed through skin.
   Contact may irritate or burn skin and eyes.
   Runoff from fire control or dilution water may cause pollution.

*Fire or Explosion
   Flammable/combustible material; may be ignited by heat, sparks, or flames.
   Vapors may travel to a source of ignition and flash back.
   Container may explode in heat of fire.
   Vapor explosion and poison hazard indoors, outdoors, or in sewers.
   Runoff to sewer may create fire or explosion hazard.

* EMERGENCY ACTION *

Keep unnecessary people away; isolate hazard area and deny entry.
Stay upwind; keep out of low areas.
Positive-pressure self-contained breathing apparatus (SCBA) and chemical
   protective clothing that is specifically recommended by the shipper or
   manufacturer may be worn. It may provide little or no thermal protection.
Structural firefighter's protective clothing is not effective for these materials.
Isolate for 1/2 mile in all directions if tank, rail car, or tank truck is involved
in   fire.
**CALL CHEMTREC AT 1-800-424-9300 FOR EMERGENCY ASSIS-
TANCE.**

*Fire
   Small fires: Dry chemical, CO_2, water spray, or alcohol-resistant foam.
   Large fires: Water spray, fog, or alcohol-resistant foam.
   Move container from fire area if you can do it without risk.

Dike fire-control water for later disposal; do not scatter the material.

Apply cooling water to sides of containers that are exposed to flames until well after fire is out. Stay away from ends of tanks.

Withdraw immediately in case of rising sound of venting safety device or any discoloration of tank due to fire.

*Spill or Leak

Shut off ignition sources; no flares, smoking, or flames in hazard area.

Fully encapsulating, vapor-protective clothing should be worn for spills and leaks with no fire.

Do not touch or walk through spilled material; stop leak if you can do it without risk.

Water spray may reduce vapor; but it may not prevent ignition in closed spaces.

Small spills: Take up with sand or other noncombustible absorbent material and place into containers for later disposal.

Large spills: Dike far ahead of liquid spill for later disposal.

*First Aid

Move victim to fresh air and call emergency medical care; if not breathing, give artificial respiration; if breathing is difficult, give oxygen.

In case of contact with material, immediately flush eyes with running water for at least 15 min. Wash skin with soap and water.

Remove and isolate contaminated clothing and shoes at the site.

Keep victim quiet and maintain normal body temperature.

Effects may be delayed; keep victim under observation.

▼   ▼   ▼   ▼   ▼   ▼   ▼   ▼   ▼   ▼   ▼   ▼   ▼   ▼

--- IDENTIFIERS ---

Name: 2-BUTANONE

Synonyms: Acetone, Methyl-; Aethylmethylketon (German); Butanone; Butanone 2 (French); Ethyl Methyl Cetone (French); Ethylmethylketon (Dutch); Ethyl Methyl Ketone; Ethyl Methyl Ketone (DOT); Ketone, Ethyl Methyl; Meetco; Methyl Acetone; Methyl Acetone (DOT); Methyl Ethyl Ketone; Methyl Ethyl Ketone (DOT); Metiletilchetone (Italian); Methyloetyloketon (Polish); 2-Butanone; MEK

CAS: 78-93-3; **RTECS:** EL6475000

Formula: C_4H_8O; **Mol Wt:** 72.12

WLN: 2V1
Chemical Class: Ketone

See other identifiers listed below under Regulations.

--- PROPERTIES ---

Physical Description: Clear colorless liquid with a fragrant, mintlike, moderately sharp odor
Boiling Point: 352.72 K; 79.5°C; 175.2°F
Melting Point: 187.04 K; -86.2°C; -123°F
Flash Point: 267.4 K; -5.8°C; 21.6°F
Autoignition: 788.7 K; 515.5°C; 959.9°F
Critical Temp: 535.7 K; 262.55°C; 504.59°F
Critical Press: 4.15 kN/m^2; 40.9 atm; 601 psia
Heat of Vap: 191 Btu/lb; 106.07 cal/g; 4.438x E5 J/kg
Heat of Comb: -13,480 Btu/lb; -7494 cal/g; -313x E5 J/kg
Vapor Pressure: 71.2 mm at 20°C
UEL: 11.5%
LEL: 1.8%
Ionization Potential (eV): 6.7
Vapor Density: 2.42 (air = 1)
Specific Gravity: 0.806 at 20°C
Density: 0.805 g/ml at 20°C
Water Solubility: 27%
Incompatibilities: Very strong oxidizers, chlorosulfonic acid, oleum, potassium-tert-butoxide, heat or flame, chloroform, hydrogen peroxide, nitric acid

Reactivity with Water: No data on water reactivity
Reactivity with Common Materials: No reaction (Source: SAX)
Stability during Transport: No data
Neutralizing Agents: Not pertinent (Source: SAX)
Polymerization Possibilities: Not pertinent (Source: SAX)

Toxic Fire Gases: Unburned vapors
Odor Detected at (ppm): 10 ppm
Odor Description: Like acetone; pleasant; pungent. (Source: CHRIS)
100% Odor Detection: 6.0 ppm

--- REGULATIONS ---

DOT hazard class: 3 FLAMMABLE LIQUID
DOT guide: 26
Identification number: UN1193
DOT shipping name: Ethyl methyl ketone or methyl ethyl ketone
Packing group: II
Label(s) required: FLAMMABLE LIQUID
Special Provisions: T8
Packaging exceptions: 173.150
Nonbulk packaging: 173.202
Bulk packaging: 173.242
Quantity limitations:
  Passenger air/rail: 5 L
  Cargo aircraft only: 60 L
  Vessel stowage: B

STCC Number: 4909243

Clean Water Act Sect. 307: No
Clean Water Act Sect. 311: No
Clean Air Act: CAA '90 listed
EPA Waste Number: U159, D035
CERCLA Ref: Y
RQ Designation: D   5000 lb (2270 kg) CERCLA
SARA TPQ Value: Not listed
SARA Sect. 312 Categories:
  Acute toxicity: irritant.
  Acute toxicity: adverse effect to target organs.
  Chronic toxicity: mutagen.
  Chronic toxicity: reproductive toxin.
  Fire hazard: flammable.
Listed in SARA Sect 313: Yes
De Minimus Concentration: 1.0%

U.S. Postal Service Mailability:
  Hazard class: Flammable liquid - mailable as ORM-D
  Mailability: Domestic surface mail only
  Max per parcel: 1 qt metal; 1 pt other

NFPA Codes:
Health Hazard (blue): (1) Slightly hazardous to health. As a precaution, wear self-contained breathing apparatus.
Flammability (red): (3) This material can be ignited under almost all temperature conditions.
Reactivity (yellow): (0) Stable even under fire conditions.
Special: Unspecified.

--- TOXICITY DATA ---

Short-term Toxicity: *Inhalation:* Human exposure to levels above 350 ppm caused irritation of the nose and throat. Numbness in fingers, arms, and legs accompanied by headache, nausea, vomiting, and fainting have occurred after exposure to levels of 300 to 600 ppm. *Skin:* Contact with liquid or vapor at levels of 300 to 600 ppm caused severe irritation. Liquid is absorbed readily and may cause numbing of fingers and arms. *Eyes:* Exposure to levels of 200 ppm produced irritation. *Ingestion:* Can cause irritation of the mouth, throat, and stomach, the severity of which will depend on amount swallowed. Symptoms of poisoning include nausea, vomiting, stomach pain, and diarrhea. Death can occur from ingestion of as little as 1 oz. (NYDH)

Long-term Toxicity: Has been implicated in certain nervous disorders characterized by weakness, fatigue, heaviness in chest, and numbness of hands and feet. These symptoms may develop after 1 yr of exposure to vapor concentrations of 50 to 200 ppm. Improvement is gradual and may take years after exposure is discontinued. (NYDH)

Target Organs: Central nervous system, lungs. Peripheral nervous system. Eye irritation at 350 ppm.

Symptoms: Liquid causes eye burn. Vapor irritates eyes, nose, and throat; can cause headache, dizziness, nausea, weakness, and loss of consciousness. (Source: CHRIS)

Conc IDLH: 3000 ppm

NIOSH REL: 200 ppm time-weighted averages for 8-hr exposure; 590 mg/m^3 time-weighted averages for 8-hr exposure.

ACGIH TLV: TLV = 200 ppm (590 mg/m^3)
ACGIH STEL: STEL = 300 ppm

OSHA PEL: Transitional Limits: PEL = 200 ppm (590 mg/m³); Final Rule Limits: TWA = 200 ppm (590 mg/m³); STEL = 300 ppm (885 mg/m³)

MAK Information: 200 ppm; 590 mg/m³
Substance with systemic effects, onset of effect less than or equal to 2 hr: Peak = 2xMAK for 30 min, 4 times per shift of 8 hr.

Carcinogen: N; **Status:** See below

Carcinogen Lists: *IARC*: Not listed; *MAK*: Not listed; *NIOSH*: Not listed; *NTP*: Not listed; *ACGIH*: Not listed; *OSHA*: Not listed.

Human Toxicity Data: (Source: NIOSH RTECS)
ihl-hmn TCLo:100 ppm/5M JIHTAB 25, 282, 43
  Sense Organs
    Nose
      Other
    Eye
      Conjunctive irritation
  Lungs, Thorax, or Respiration
    Other changes

LD50 Value: orl-rat LD50:2737 mg/kg

Other Species Toxicity Data: (Source: NIOSH RTECS 1992)
orl-rat LD50:2737 mg/kg
ihl-rat LC50:23500 mg/m³/8H
ipr-rat LD50:607 mg/kg
orl-mus LD50:4050 mg/kg
ihl-mus LC50:40 g/m³/2H
ipr-mus LD50:616 mg/kg
skn-rbt LD50:6480 mg/kg
ipr-gpg LDLo:2000 mg/kg
ihl-mam LC50:38 g/m³

Irritation Data: (Source: NIOSH RTECS 1992)
eye-hmn 350 ppm
skn-rbt 500 mg/24H MOD
skn-rbt 402 mg/24H MLD
skn-rbt 13780 mg/24H open MLD
eye-rbt 80 mg

Reproductive Toxicity (1992 RTECS):   This chemical is a mammalian reproductive toxin.

Reproductive Toxicity Data (1992 RTECS):
ihl-rat TCLo:3000 ppm/7H (6-15D preg) TXAPA9 28, 452, 74
   Specific Developmental Abnormalities
    Craniofacial (including nose and tongue)
    Urogenital system
    Homeostasis

--- PROTECTION AND FIRST AID ---

Protection Suggested from the CHRIS Manual: Organic canister or air pack; plastic gloves; goggles or face shield.

NIOSH Pocket Guide to Chemical Hazards:
  **Wear Appropriate Equipment to Prevent: Repeated or prolonged skin contact.

  **Wear Eye Protection to Prevent: Reasonable probability of eye contact.

  **Remove Clothing: Promptly remove nonimpervious clothing that becomes contaminated.

  **Reference: NIOSH

Recommended Respiration Protection Source: NIOSH Pocket Guide (85-114) NIOSH (2-Butanone)
1000 ppm: Any powered air-purifying respirator with organic vapor cartridge(s).   Substance causes eye irritation or damage; eye protection needed.   Any chemical cartridge respirator with a full facepiece and organic vapor cartridge(s).
3000 ppm: Any air-purifying full facepiece respirator (gas mask) with a chin-style or front- or back-mounted organic vapor canister.   Any supplied-air respirator operated in a continuous-flow mode.   Substance causes eye irritation or damage; eye protection needed.   Any self-contained breathing apparatus with a full facepiece.   Any supplied-air respirator with a full facepiece.
Emergency or Planned Entry in Unknown Concentrations or IDLH Conditions:
Any self-contained breathing apparatus with a full facepiece and operated in a pressure-demand or other positive-pressure mode.   Any supplied-air

respirator with a full facepiece and operated in a pressure-demand or other positive-pressure mode in combination with an auxiliary self-contained breathing apparatus operated in a pressure-demand or other positive-pressure mode.

Escape: Any air-purifying full facepiece respirator (gas mask) with a chin-style or front- or back-mounted organic vapor canister. Any appropriate escape-type self-contained breathing apparatus.

First Aid Source: CHRIS Manual 1991.

Inhalation: Remove victim to fresh air; if breathing is irregular or has stopped, start resuscitation and administer oxygen.

Eyes: Wash with plenty of water for at least 15 min and call physician.

First Aid Source: DOT Emergency Response Guide 1993.

Move victim to fresh air and call emergency medical care; if not breathing, give artificial respiration; if breathing is difficult, give oxygen. In case of contact with material, immediately flush eyes with running water for at least 15 min. Wash skin with soap and water. Remove and isolate contaminated clothing and shoes at the site.

--- INITIAL INCIDENT RESPONSE ---

Fire Extinguishment: Alcohol foam, dry chemical, carbon dioxide. *Note:* Water may be ineffective. (CHRIS 91)

U.S. Department of Transportation Guide to Hazardous Materials Transport Information - Publication DOT 5800.5 (1990).

DOT Shipping Name: Ethyl methyl ketone or methyl ethyl ketone
DOT ID Number: UN1193

ERG90   GUIDE 26
* POTENTIAL HAZARDS *

*Health Hazards

May be poisonous if inhaled or absorbed through skin.
Vapors may cause dizziness or suffocation.
Fire may produce irritating or poisonous gases.
Runoff from fire control or dilution water may cause pollution.

*Fire or Explosion

Flammable/combustible material; may be ignited by heat, sparks, or flames.
Vapors may travel to a source of ignition and flash back.

Container may explode in heat of fire.

Vapor explosion hazard indoors, outdoors, or in sewers.

Runoff to sewer may create fire or explosion hazard.

* EMERGENCY ACTION *

Keep unnecessary people away; isolate hazard area and deny entry.

Stay upwind; keep out of low areas.

Positive-pressure self-contained breathing apparatus (SCBA) and structural firefighters' protective clothing will provide limited protection.

Isolate for 1/2 mile in all directions if tank, rail car, or tank truck is involved in fire.

CALL CHEMTREC AT 1-800-424-9300 FOR EMERGENCY ASSIS-TANCE. If water pollution occurs, notify the appropriate authorities.

*Fire

Small fires: Dry chemical, CO_2 or Halon, water spray, or alcohol-resistant foam.

Large fires: Water spray, fog, or alcohol-resistant foam.

Move container from fire area if you can do it without risk.

Apply cooling water to sides of containers that are exposed to flames until well after fire is out. Stay away from ends of tanks.

For massive fire in cargo area, use unmanned hose holder or monitor nozzles; if this is impossible, withdraw from area and let fire burn.

Withdraw immediately in case of rising sound of venting safety device or any discoloration of tank due to fire.

*Spill or Leak

Shut off ignition sources; no flares, smoking, or flames in hazard area.

Stop leak if you can do it without risk.

Water spray may reduce vapor; but it may not prevent ignition in closed spaces.

Small spills: Take up with sand or other noncombustible absorbent material and place into containers for later disposal.

Large spills: Dike far ahead of liquid spill for later disposal.

*First Aid

Move victim to fresh air and call emergency medical care; if not breathing, give artificial respiration; if breathing is difficult, give oxygen.

In case of contact with material, immediately flush eyes with running water for at least 15 min. Wash skin with soap and water.

Remove and isolate contaminated clothing and shoes at the site.

▼   ▼   ▼   ▼   ▼   ▼   ▼   ▼   ▼   ▼   ▼   ▼   ▼   ▼

--- IDENTIFIERS ---

Name: ASPHALT
CAS: 8052-42-4; **RTECS:** CI9900000

See other identifiers listed below under Regulations.

--- PROPERTIES ---

Physical Description: Dark brown to black thick liquid with a tar odor; softens
to viscous liquid >90°C.
Boiling Point: >973.15 K
Melting Point: NA
Flash Point: 422.03 K to .8°C to 560.92 .9°F to 560.92
Autoignition: NA
UEL: NA
LEL: NA
Vapor Density: No data
Specific Gravity: 1 at 20°C
Density: 1 g/cc or 9.3 lb/gal
Water Solubility: Unknown
Incompatibilities: Unknown

Reactivity with Water: No data on water reactivity
Reactivity with Common Materials: No data
Stability during Transport: No data
Neutralizing Agents: No data
Polymerization Possibilities: No data

Toxic Fire Gases: None reported other than possible unburned vapors
Odor Detected at (ppm): Data not available
Odor Description: Tarry (Source: CHRIS)
100% Odor Detection: No data

--- REGULATIONS ---

DOT hazard class: 3 FLAMMABLE LIQUID
DOT guide: 27
Identification number: NA1999
DOT shipping name: Asphalt (at or above its flashpoint)
Packing group: III

Label(s) required: None
Packaging exceptions: 173.150
Nonbulk packaging: 173.203
Bulk packaging: 173.242
Quantity limitations:
  Passenger air/rail: Forbidden
  Cargo aircraft only: Forbidden
  Vessel stowage: D
  Other stowage provisions: M4

STCC Number: Not listed

Clean Water Act Sect. 307: No
Clean Water Act Sect. 311: No
Clean Air Act: Not listed
EPA Waste Number: None
CERCLA Ref: Not listed
RQ Designation: Not listed
SARA TPQ Value: Not listed
SARA Sect. 312 Categories:
  Acute toxicity: adverse effect to target organs.
  Chronic toxicity: carcinogen.
  Chronic toxicity: mutagen.

U.S. Postal Service Mailability:
  Hazard class: Not given
  Mailability: Nonmailable
  Max per parcel: 0

NFPA Codes:
  Health Hazard (blue): (0) No unusual health hazard.
  Flammability (red): (1) This material must be preheated before ignition can occur.
  Reactivity (yellow): (0) Stable even under fire conditions.
  Special: Unspecified.

--- TOXICITY DATA ---

Short-term Toxicity: *Inhalation:* Symptoms listed here are from reports of industrial accidents. No levels were reported. Can cause irritation of the nose, throat, and lungs; increase in coughing and spitting; burning sensation in throat and chest; hoarseness, headache and runny nose. *Skin:* Can cause heat burns and irritation. *Eyes:* Can cause irritation. *Ingestion:* No information available. (NYDH)

Long-term Toxicity: Can cause acnelike sores, thickening, and yellow discoloration of the skin. Symptoms similar to those under inhalation may be seen. (NYDH)

Target Organs: Skin, eyes, respiratory tract.

Symptoms: Contact with skin may cause dermatitis. Inhalation of vapors may cause moderate irritation of nose and throat. Hot liquid burns skin. (Source: CHRIS)

Conc IDLH: Unknown

NIOSH REL: Potential occupational carcinogen, 5 mg/m^3 ceiling exposures, which shall at no time be exceeded.

ACGIH TLV: TLV = 5 mg/m^3
ACGIH STEL: Not listed

OSHA PEL: Not in Table Z-1-A

MAK Information: Carcinogenic working material without MAK. A compound that is justifiably suspected of having carcinogenic potential.

Carcinogen: Y; **Status:** See below

Carcinogen Lists: *IARC:* Not listed; *MAK:* A compound that is justifiably suspected of having carcinogenic potential; *NIOSH:* Carcinogen defined by NIOSH with no further categorization; *NTP:* Not listed; *ACGIH:* Not listed; *OSHA:* Not listed.

LD50 Value: No LD50 in RTECS 1992

Reproductive Toxicity (1992 RTECS): This chemical has no known mammalian reproductive toxicity.

--- PROTECTION AND FIRST AID ---

Protection Suggested from the CHRIS Manual: Protective clothing; face and eye protection when handling hot material.

First Aid Source: NIOSH
Eye: None given
Skin: None given
Inhalation: None given
Ingestion: None given

First Aid Source: CHRIS Manual 1991.
Severe burns may result from contact with hot asphalt. If molten asphalt strikes the exposed skin, cool the skin immediately by quenching with cold water. A burn should be covered with a sterile dressing, and the patient should be taken immediately to a hospital.

First Aid Source: DOT Emergency Response Guide 1993.
Move victim to fresh air and call emergency medical care; if not breathing, give artificial respiration; if breathing is difficult, give oxygen. In case of contact with material, immediately flush eyes with running water for at least 15 min. Wash skin with soap and water. Remove and isolate contaminated clothing and shoes at the site.

--- INITIAL INCIDENT RESPONSE ---

Fire Extinguishment: Water spray, dry chemical, foam or carbon dioxide.
Note: Water or foam may cause frothing. (CHRIS 91)

U.S. Department of Transportation Guide to Hazardous Materials Transport Information - Publication DOT 5800.5 (1990).
DOT Shipping Name: Asphalt (at or above its flashpoint)
DOT ID Number: NA1999

ERG90   GUIDE 27
* POTENTIAL HAZARDS *

*Health Hazards
May be poisonous if inhaled or absorbed through skin.
Vapors may cause dizziness or suffocation.
Contact may irritate or burn skin and eyes.
Fire may produce irritating or poisonous gases.
Runoff from fire control or dilution water may cause pollution.

*Fire or Explosion
Flammable/combustible material; may be ignited by heat, sparks, or flames.
Vapors may travel to a source of ignition and flash back.

Container may explode in heat of fire.

Vapor explosion hazard indoors, outdoors, or in sewers.

Runoff to sewer may create fire or explosion hazard.

* EMERGENCY ACTION *

Keep unnecessary people away; isolate hazard area and deny entry.

Stay upwind; keep out of low areas.

Positive-pressure self-contained breathing apparatus (SCBA) and structural firefighters' protective clothing will provide limited protection.

Isolate for 1/2 mile in all directions if tank, rail car, or tank truck is involved in fire.

CALL CHEMTREC AT 1-800-424-9300 FOR EMERGENCY ASSISTANCE. If water pollution occurs, notify the appropriate authorities.

*Fire

Small fires: Dry chemical, CO_2, water spray, or regular foam.

Large fires: Water spray, fog, or regular foam.

Move container from fire area if you can do it without risk.

Apply cooling water to sides of containers that are exposed to flames until well after fire is out. Stay away from ends of tanks.

For massive fire in cargo area, use unmanned hose holder or monitor nozzles; if this is impossible, withdraw from area and let fire burn.

Withdraw immediately in case of rising sound of venting safety device or any discoloration of tank due to fire.

*Spill or Leak

Shut off ignition sources; no flares, smoking, or flames in hazard area.

Stop leak if you can do it without risk.

Water spray may reduce vapor; but it may not prevent ignition in closed spaces.

Small spills: Take up with sand or other noncombustible absorbent material and place into containers for later disposal.

Large spills: Dike far ahead of liquid spill for later disposal.

*First Aid

Move victim to fresh air and call emergency medical care; if not breathing, give artificial respiration; if breathing is difficult, give oxygen.

In case of contact with material, immediately flush eyes with running water for at least 15 min. Wash skin with soap and water.

Remove and isolate contaminated clothing and shoes at the site.

▼   ▼   ▼   ▼   ▼   ▼   ▼   ▼   ▼   ▼   ▼   ▼   ▼   ▼

--- IDENTIFIERS ---

Chemical Name and Synonyms: **MIXTURE OF PETROLEUM HYDROCARBONS**

Trade Name and Synonyms: Mobilsol K

Chemical Family: Aromatic Oil

Manufacturer: Mobil Oil Company, New York, New York

UN ID Number: NA

--- INGREDIENTS AND COMPOSITION DATA ---

Material	Cas No.	%	TLV
High-boiling aromatic oil	NA	100	5 mg/m^3

--- PHYSICAL DATA ---

Boiling Point (°F): 90% at 660 to 925

Specific Gravity (H$_2$O = 1): 0.995

Vapor Pressure (mm Hg): <0.00002 at 80°F

Percent Volatile by Volume (%): Nil

Vapor Density (air = 1): Nil

Freezing Point: NA

Evaporation Rate: 0

pH: NA

Solubility in Water: Negligible

Melting Point/Pour or Congealing Point: NA

Viscosity: NA

Appearance/Color/Odor: Light mineral oil

--- FIRE AND EXPLOSION HAZARDS ---

Flash Point: 410°F (COC)

Flammability Limits: NA

Autoignition Temperature: NA

Extinguishing Media: Foam, dry chemical, carbon dioxide, or water fog or spray.

Special Fire Fighting Procedures: No special procedures are recommended by the manufacturer.

Unusual Fire and Explosion Hazards: None reported.

Hazardous Products of Combustion and Decomposition: None reported by manufacturer.

Hazardous Polymerization: Will not occur.

--- HEALTH HAZARD DATA ---

Effects of Overexposure: Primary skin irritation index, 0.25. This product is a high-boiling aromatic oil. After prolonged and repeated skin contact, irritation or more serious skin disorders may result.

Emergency Response and First Aid:
Skin: Flush all affected areas with plenty of water for 15 min. Remove and clean any contaminated clothing and shoes. Seek medical attention if skin irritation occurs.
Eyes: Flush the eyes with plenty of running water for at least 15 min. Hold the eyelids apart during the flushing to ensure rinsing of the surface of the eye and lids with water. Seek medical attention if eye irritation occurs.
Inhalation: There are no effects or recommendations reported.
Ingestion: Consult physician immediately.

Cancer Information: None listed under OSHA/NTP/IARC.

--- TOXICITY DATA ---

No information reported.

--- REACTIVITY DATA ---

Stability and Conditions to Avoid: Avoid open flames and hot surfaces.

Materials to Avoid (Incompatibility): Strong oxidizing agents and excessive heat.

Hazard Decomposition Products: Fumes, smoke, carbon monoxide, and carbon dioxide, in the case of incomplete combustion.

Hazardous Polymerization: Will not occur.

--- SPILL AND EMERGENCY RESPONSE ---

For Spills: Use any inert absorbent such as sand, earth, or vermiculite. The product is described as being slippery. Use proper safety equipment during cleanup. Dispose of in a manner that is consistent with local, state, and federal regulations.

Waste Disposal Method: Assure conformity with applicable disposal regulations. Dispose of absorbed material at an approved waste disposal site or facility. Waste material may be incinerated under conditions that meet federal, state, and local environmental control regulations. At present this is not classified as a RCRA hazardous waste. The manufacturer recommends mixing with more flammable solvent and incinerating.

Empty Containers: Keep containers closed when not in use. Do not handle or store near heat, sparks, flame, or strong oxidants.

--- SPECIAL PROTECTION AND PRECAUTIONS ---

Respiratory Protection: Normally not needed. Use supplied-air respiratory protection in confined or enclosed spaces.

Type of Ventilation Required: Use local exhaust to capture fumes and vapors. Provide greater than 60-fpm hood or face velocity for confined spaces.

Eye Protection Requirements: Use splash goggles or face shield when eye contact may occur.

Protective Gloves and Other: Use neoprene gloves, neoprene apron, and chemical goggles.

Storage and Handling Precautions: Containers should be stored in a cool, dry, well-ventilated area. Store away from flammable materials, sources of heat, flame, sparks, and foodstuffs. Exercise due caution to prevent damage to or leakage from the container. Avoid any conditions that might tend to create a dust explosion. Maintain good housekeeping practices to minimize dust buildup. *Note:* This product has a shelf life of 1 yr.

Other Precautions: Avoid breathing oil mist. Remove oil-soiled clothing and launder before reuse. Discard oil-soaked shoes. Wash skin thoroughly with soap and water after handling.

--- REGULATORY INFORMATION ---

Transportation Incident Information: This product is unregulated by DOT at present.

DOT Identification Number: Not applicable.

▼   ▼   ▼   ▼   ▼   ▼   ▼   ▼   ▼   ▼   ▼   ▼   ▼   ▼

--- IDENTIFIERS ---

Chemical Name and Synonyms: INTERNAL LUBRICANT, MOLD RELEASE AGENT

Trade Name and Synonyms: Mold Wiz

Chemical Family: Condensation product of synthetic resins, glycerides, and organic acid derivatives

Manufacturer: Axel Plastics Research Laboratories, Inc., Woodside, New York

UN ID Number: NA

--- INGREDIENTS AND COMPOSITION DATA ---

Material	Cas No.	%	TLV
Product mixture	NA	100	Considered nontoxic

--- PHYSICAL DATA ---

Boiling Point (°F): 220

Specific Gravity (H₂O = 1): 0.956 at 25°C

Vapor Pressure (mm Hg): NA

Percent Volatile by Volume (%): None at 25°C

Vapor Density (air = 1): NA

Freezing Point: NA

Evaporation Rate: None at 25°C

pH: NA

Solubility in Water: Insoluble

Melting Point/Pour or Congealing Point: NA

Viscosity: NA

Appearance/Color/Odor: Amber color, with aliphatic odor

--- FIRE AND EXPLOSION HAZARDS ---

Flash Point: 280°F (COC)

Flammability Limits: NA

Autoignition Temperature: NA

Extinguishing Media: Foam, dry chemical, carbon dioxide, or water fog or spray.

Special Fire Fighting Procedures: No special procedures are recommended by the manufacturer.

Unusual Fire and Explosion Hazards: None reported.

Hazardous Products of Combustion and Decomposition: None reported by manufacturer.

Hazardous Polymerization: Will not occur.

--- HEALTH HAZARD DATA ---

Effects of Overexposure: None reported. The manufacturer states that the product is totally inoffensive and nontoxic. There is no regulated TLV value for this material.

Emergency Response and First Aid:
Skin: Flush all affected areas with plenty of water for 15 min. Remove and clean any contaminated clothing and shoes. Seek medical attention if skin irritation occurs.
Eyes: Flush the eyes with plenty of running water for at least 15 min. Hold the eyelids apart during the flushing to ensure rinsing of the surface of the eye and lids with water. Seek medical attention if eye irritation occurs.
Inhalation: There are no effects or recommendations reported.
Ingestion: Consult physician immediately.

Cancer Information: None listed under OSHA/NTP/IARC.

--- TOXICITY DATA ---

No information reported.

--- REACTIVITY DATA ---

Stability and Conditions to Avoid: This product is stable but should be stored away from open flames.

Materials to Avoid (Incompatibility): Avoid contact with water.

Hazard Decomposition Products: Fumes, smoke, carbon monoxide, and carbon dioxide, in the case of incomplete combustion. This material has similar properties to that of any organic chemical.

Hazardous Polymerization: Will not occur.

--- SPILL AND EMERGENCY RESPONSE ---

For Spills: Use any inert absorbent such as sand, earth, or vermiculite. The product is described as being slippery. Use proper safety equipment during cleanup. Dispose of in a manner that is consistent with local, state, and federal regulations. After absorbent is applied, the area can be washed off with detergent.

Waste Disposal Method: Assure conformity with applicable disposal regulations. Dispose of absorbed material at an approved waste disposal site or facility. Waste material may be incinerated under conditions that meet federal, state, and local environmental control regulations. At present this is not classified as a RCRA hazardous waste. The manufacturer recommends mixing with more flammable solvent and incinerating.

Empty Containers: Keep containers closed when not in use. Do not handle or store near heat, sparks, flame, or strong oxidants.

--- SPECIAL PROTECTION AND PRECAUTIONS ---

Respiratory Protection: Normally not needed.

Type of Ventilation Required: Use local exhaust to capture fumes and vapors. Provide greater than 60-fpm hood or face velocity for confined spaces.

Eye Protection Requirements: Use splash goggles or face shield when eye contact may occur.

Protective Gloves and Other: Use polysulfide or polyethylene protective gloves.

Storage and Handling Precautions: Containers should be stored in a cool, dry, well-ventilated area. Store away from flammable materials, sources of heat, flame, sparks, and foodstuffs. Exercise due caution to prevent damage to or leakage from the container. Avoid any conditions that might tend to create a dust explosion. Maintain good housekeeping practices to minimize dust buildup. *Note:* This product has a shelf life of 1 yr.

Other Precautions: Avoid breathing oil mist. Remove oil-soiled clothing and launder before reuse. Discard oil-soaked shoes. Wash skin thoroughly with soap and water after handling.

--- REGULATORY INFORMATION ---

Transportation Incident Information: This product is unregulated by DOT at present.

DOT Identification Number: Not applicable.

▼   ▼   ▼   ▼   ▼   ▼   ▼   ▼   ▼   ▼   ▼   ▼   ▼   ▼

--- IDENTIFIERS ---

Chemical Name and Synonyms: RETENE, ROSIN OIL

Trade Name and Synonyms: "N" Rosin Oil

Chemical Family: Decarboxylated Abietic Acid

Manufacturer: Natrochem, Inc. Savannah, Georgia

UN ID Number: NA

--- INGREDIENTS AND COMPOSITION DATA ---

Material	Cas No.	%	TLV
Decarboxylated abietic acid	NA	100	Not established

--- PHYSICAL DATA ---

Boiling Point (°F): NA

Specific Gravity (H$_2$O = 1): 1.00

Vapor Pressure (mm Hg): NA

Percent Volatile by Volume (%): 3

Vapor Density (air = 1): NA

Freezing Point: NA

Evaporation Rate: NA

pH: NA

Solubility in Water: Insoluble

Melting Point/Pour or Congealing Point: NA

Viscosity: NA

Appearance/Color/Odor: Amber, terpenic oil

--- FIRE AND EXPLOSION HAZARDS ---

Flash Point: 300°F (COC)

Flammability Limits: NA

Autoignition Temperature: NA

Extinguishing Media: Steam, water fog, dry chemical, carbon dioxide.

Special Fire Fighting Procedures: No special procedures are recommended by the manufacturer.

Unusual Fire and Explosion Hazards: None reported.

Hazardous Products of Combustion and Decomposition: None reported by manufacturer.

Hazardous Polymerization: Will not occur.

--- HEALTH HAZARD DATA ---

Effects of Overexposure: None reported by manufacturer; however, this product should be handled as any hydrocarbon vapor, i.e., limit excessive exposure to fumes and skin contact.

Emergency Response and First Aid:
Skin: Flush all affected areas with plenty of water for 15 min. Remove and clean any contaminated clothing and shoes. Seek medical attention if skin irritation occurs.
Eyes: Flush the eyes with plenty of running water for at least 15 min. Hold the eyelids apart during the flushing to ensure rinsing of the surface of the eye and lids with water. Seek medical attention if eye irritation occurs.
Inhalation: There are no effects or recommendations reported.
Ingestion: Consult physician immediately.

Cancer Information: None listed under OSHA/NTP/IARC.

--- TOXICITY DATA ---

No information reported.

--- REACTIVITY DATA ---

Stability and Conditions to Avoid: This product is stable but should be stored away from open flames.

Materials to Avoid (Incompatibility): None reported by manufacturer.

Hazard Decomposition Products: Fumes, smoke, carbon monoxide, and carbon dioxide, in the case of incomplete combustion. This material has similar properties to that of any organic chemical.

Hazardous Polymerization: Will not occur.

--- SPILL AND EMERGENCY RESPONSE ---

For Spills: Use any inert absorbent such as sand, earth, or vermiculite. The product is described as being slippery. Use proper safety equipment during cleanup. Dispose of in a manner that is consistent with local, state, and federal regulations. After absorbent is applied, the area can be washed off with detergent.

Waste Disposal Method: Assure conformity with applicable disposal regulations. Dispose of absorbed material at an approved waste disposal site or facility. Waste material may be incinerated under conditions that meet federal, state, and local environmental control regulations. At present this is not classified as a RCRA hazardous waste. The manufacturer recommends mixing with more flammable solvent and incinerating.

Empty Containers: Keep containers closed when not in use. Do not handle or store near heat, sparks, flame, or strong oxidants.

--- SPECIAL PROTECTION AND PRECAUTIONS ---

Respiratory Protection: Normally not needed.

Type of Ventilation Required: No recommendations are given by manufacturer.

Eye Protection Requirements: Use splash goggles or face shield when eye contact may occur.

Protective Gloves and Other: Use polyethylene protective gloves.

Storage and Handling Precautions: Containers should be stored in a cool, dry, well-ventilated area. Store away from flammable materials, sources of heat, flame, sparks, and foodstuffs. Exercise due caution to prevent damage to or leakage from the container. Avoid any conditions that might tend to create a dust explosion. Maintain good housekeeping practices to minimize dust buildup. *Note:* This product has a shelf life of 1 yr.

Other Precautions: Avoid breathing oil mist. Remove oil-soiled clothing and launder before reuse. Discard oil-soaked shoes. Wash skin thoroughly with soap and water after handling.

--- REGULATORY INFORMATION ---

Transportation Incident Information: This product is unregulated by DOT at present.

DOT Identification Number: Not applicable.

10 AMINES

Name: *N*-NITROSODIPHENYLAMINE

Synonyms: Benzenamine, *N*-Nitroso-*N*-Phenyl- (9CI); Curetard A; Delac J; Diphenylnitrosamin (German); Diphenylnitrosamine; Diphenyl *N*-Nitrosoamine; *N,N*-Diphenylnitrosamine; Naugard TJB; NCI-C02880; NDPA; NDPhA; *N*-Nitrosodifenylamin (Czech); Nitrosodiphenylamine; *N*-Nitrosodiphenylamine; *N*-Nitroso-*N*-Phenylaniline; Nitrous Diphenylamide; Redax; Retarder J; TJB; Vulcalent A; Vulcatard; Vulcatard A; Vulkalent A (Czech); Vultrol

CAS: 86-30-6; **RTECS:** JJ9800000

Formula: $C_{12}H_{10}N_2O$; **Mol Wt:** 198.24

WLN: ONNR

Chemical Class: Nitro compound; aromatic amine

See other identifiers listed below under Regulations.

--- PROPERTIES ---

Boiling Point: NA

Melting Point: 418 K; 144.8°C; 292.7°F

Flash Point: NA

Autoignition: NA

UEL: NA

LEL: NA

Vapor Density: No data

Specific Gravity: No data

Reactivity with Water: No data on water reactivity

Reactivity with Common Materials: No data

Stability during Transport: No data

Neutralizing Agents: No data

Polymerization Possibilities: No data

Toxic Fire Gases: None reported other than possible unburned vapors
Odor Detected at (ppm): Unknown
Odor Description: No data
100% Odor Detection: No data

--- REGULATIONS ---

DOT hazard class: Not given
Packaging exceptions: 173
Nonbulk packaging: 173
Bulk packaging: 173

STCC Number: Not listed

Clean Water Act Sect. 307: Yes
Clean Water Act Sect. 311: No
Clean Air Act: Not listed
EPA Waste Number: None
CERCLA Ref: Y
RQ Designation: B, 100 lb (45.4 kg) CERCLA
SARA TPQ Value: Not listed
SARA Sect. 312 Categories:
  Acute toxicity: adverse effect to target organs.
Listed in SARA Sect. 313: Yes
De Minimus Concentration: 1.0%

U.S. Postal Service Mailability: Not given

NFPA Codes:
  Health Hazard (blue): Unspecified
  Flammability (red): Unspecified
  Reactivity (yellow): Unspecified
  Special: Unspecified

--- TOXICITY DATA ---

Short-term Toxicity: Unknown

Long-term Toxicity: Unknown

Target Organs: Eyes

Symptoms: Eye irritant (Source: SAX)

Conc IDLH: Unknown

NIOSH REL: Not given

ACGIH TLV: Not listed
ACGIH STEL: Not listed

OSHA PEL: Not in Table Z-1-A

MAK Information: Not listed

Carcinogen: N; **Status:** See below
References: Animal suspected IARC** 27, 213, 82; Animal positive IARC** 28, 151, 82; Human indefinite IARC** 27, 213, 82

Carcinogen Lists: *IARC*: Not classified as to human carcinogenicity or probably not carcinogenic to humans; *MAK*: Not listed; *NIOSH*: Not listed; *NTP*: Not listed; *ACGIH*: Not listed; *OSHA*: Not listed.

LD50 Value: orl-rat LD50:2500 mg/kg

Other Species Toxicity Data: (Source: NIOSH RTECS 1992)
orl-rat LD50:2500 mg/kg
unr-rat LD50:3000 mg/kg
orl-mus LD50:3850 mg/kg
ipr-mus LD50:1000 mg/kg

Irritation Data: (Source: NIOSH RTECS 1992)
eye-rbt 500 mg/24H SEV

Reproductive Toxicity (1992 RTECS): This chemical has no known mammalian reproductive toxicity.

No Significant Risk Level (Ca P65): N80 mg/day

▼ ▼ ▼ ▼ ▼ ▼ ▼ ▼ ▼ ▼ ▼ ▼ ▼ ▼

--- IDENTIFIERS ---

Name: *N*-**NITROSODI-*N*-PROPYLAMINE**
Synonyms: Dipropylnitrosamine; Di-*n*-Propylnitrosamine; DPN; DPNA;
NDPA;   *N*-Nitrosodi-*n*-Propylamine;   *N*-Nitroso-*N*-Propyl-1-Propanamine;
Propanamine, *N*-Nitroso-*N*-Propyl-; Propylamine, *N*-Nitroso-*N*-Di-; Di-*N*-
Propylnitrosamine
CAS: 621-64-7; **RTECS:** JL9700000
Formula: $C_6H_{14}N_2O$; **Mol Wt:** 130.22
WLN: ONN3
Chemical Class: Nitro compound; aliphatic amine

See other identifiers listed below under Regulations.

--- PROPERTIES ---

Boiling Point: 479 K; 205.8°C; 402.5°F
Melting Point: NA
Flash Point: NA
Autoignition: NA
UEL: NA
LEL: NA
Vapor Density: No data
Specific Gravity: No data
Density: 0.9163, 20/4

Reactivity with Water: No data on water reactivity
Reactivity with Common Materials: No data
Stability during Transport: No data
Neutralizing Agents: No data
Polymerization Possibilities: No data

Toxic Fire Gases: Toxic fumes of oxides of nitrogen
Odor Detected at (ppm): Unknown
Odor Description: No data
100% Odor Detection: No data

--- REGULATIONS ---

DOT hazard class: 9 CLASS 9
DOT guide: 31
Identification number: UN3077

DOT shipping name: Environmentally hazardous substances, solid, N.O.S.
Packing group: III
Label(s) required: CLASS 9
Special Provisions: 8, B54
Packaging exceptions: 173.155
Nonbulk packaging: 173.213
Bulk packaging: 173.240
Quantity limitations:
  Passenger air/rail: None
  Cargo aircraft only: None
  Vessel stowage: A

STCC Number: Not listed

Clean Water Act Sect. 307: Yes
Clean Water Act Sect. 311: No
Clean Air Act: Not listed
EPA Waste Number: U111
CERCLA Ref: Y
RQ Designation: A, 10 lb (4.54 kg) CERCLA
SARA TPQ Value: Not listed
SARA Sect. 312 Categories:
  Acute toxicity: toxic. LD50 > 50 and $< = 500$ mg/kg (oral rat).
  Chronic toxicity: carcinogen.
  Chronic toxicity: mutagen.
Listed in SARA Sect. 313: Yes
De Minimus Concentration: 0.1%

U.S. Postal Service Mailability: Not given

NFPA Codes:
  Health Hazard (blue): Unspecified
  Flammability (red): Unspecified
  Reactivity (yellow): Unspecified
  Special: Unspecified

--- TOXICITY DATA ---

Short-term Toxicity: Unknown

Long-term Toxicity: Unknown

Conc IDLH: Unknown

NIOSH REL: Not given

ACGIH TLV: Not listed
ACGIH STEL: Not listed

OSHA PEL: Not in Table Z-1-A

MAK Information: Carcinogenic working material without MAK. In the Commission's view, an animal carcinogen.

Carcinogen: Y; **Status:** See below

Carcinogen Lists: *IARC:* Carcinogen defined by IARC to be possibly carcinogenic to humans, but having (usually) no human evidence; *MAK:* An animal carcinogen; *NIOSH:* Not listed; *NTP:* Carcinogen defined by NTP as reasonably anticipated to be carcinogenic, with limited evidence in humans or sufficient evidence in experimental animals. *ACGIH:* Not listed; *OSHA:* Not listed.

LD50 Value: orl-rat LD50:480 mg/kg

Other Species Toxicity Data: (Source: NIOSH RTECS 1992)
    orl-rat LD50:480 mg/kg
    scu-rat LD50:487 mg/kg
    scu-mus LD50:689 mg/kg
    orl-ham LD50: >400 mg/kg
    scu-ham LD50:600 mg/kg

Reproductive Toxicity (1992 RTECS): This chemical has no known mammalian reproductive toxicity.

No Significant Risk Level (Ca P65): 0.1 mg/day

--- PROTECTION AND FIRST AID ---

First Aid Source: DOT Emergency Response Guide 1993.
    In case of contact with material, immediately flush eyes with running water for at least 15 min. Wash skin with soap and water. Remove and isolate contaminated clothing and shoes at the site.

--- **INITIAL INCIDENT RESPONSE** ---

U.S. Department of Transportation Guide to Hazardous Materials Transport Information - Publication DOT 5800.5 (1990).
DOT Shipping Name: Environmentally hazardous substances, solid, N.O.S.
DOT ID Number: UN3077

ERG90   GUIDE 31
* POTENTIAL HAZARDS *

*Health Hazards
Contact may cause burns to skin and eyes.
Fire may produce irritating or poisonous gases.
Runoff from fire control or dilution water may cause pollution.

*Fire or Explosion
Some of these materials may burn, but none of them ignites readily.

* EMERGENCY ACTION *

Keep unnecessary people away; isolate hazard area and deny entry.
Positive-pressure self-contained breathing apparatus (SCBA) and structural
    firefighters' protective clothing will provide limited protection.
**CALL CHEMTREC AT 1-800-424-9300 FOR EMERGENCY ASSIS-
TANCE.** If water pollution occurs, notify the appropriate authorities.

*Fire
Small fires: Dry chemical, CO_2, water spray, or regular foam.
Large fires: Water spray, fog, or regular foam.
Move container from fire area if you can do it without risk.
Do not scatter spilled material with high-pressure water streams.
Dike fire-control water for later disposal.

*Spill or Leak
Stop leak if you can do it without risk.
Small dry spills: With clean shovel, place material into clean, dry container
    and cover loosely; move containers from spill area.
Small spills: Take up with sand or other noncombustible absorbent material
    and place into containers for later disposal.
Large spills: Dike far ahead of liquid spill for later disposal.
    Cover powder spill with plastic sheet or tarp to minimize spreading.

*First Aid
   In case of contact with material, immediately flush eye with running water for
      at least 15 min. Wash skin with soap and water.
   Remove and isolate contaminated clothing and shoes at the site.

▼   ▼   ▼   ▼   ▼   ▼   ▼   ▼   ▼   ▼   ▼   ▼   ▼   ▼

--- IDENTIFIERS ---

**Chemical Name and Synonyms: BUTADIENE HOMOPOLYMER
containing Acrylate Functional Groups**

Trade Name and Synonyms: Chemlink 5000 Oligomer

Chemical Family: Acrylated Functional Groups on Butadiene

Manufacturer: Arco Chemical Co., West Chester, Pennsylvania

UN ID Number: NA

--- INGREDIENTS AND COMPOSITION DATA ---

Material	Cas No.	%	TLV
Acrylate terminated polybutadiene	NA	99+	Not established
Catalyst	NA	<0.1	Not established
Inhibitor (BHT)	NA	<0.1	Not established

--- PHYSICAL DATA ---

Boiling Point (°F): NA

Specific Gravity (H_2O = 1): 0.91 (25°C/20°C)

Vapor Pressure (mm Hg): NA

Percent Volatile by Volume (%): Negligible

Vapor Density (air = 1): NA

Freezing Point: NA

Evaporation Rate: NA

pH: NA

Solubility in Water: Negligible

Melting Point/Pour or Congealing Point: NA

Viscosity: 5000 CPS at 25°C (Brookfield)

Appearance/Color/Odor: Clear yellow, viscous liquid with penetrating hydrocarbon odor.

--- FIRE AND EXPLOSION HAZARDS ---

Flash Point: 300°F (COC)

Flammability Limits: NA

Autoignition Temperature: NA

Extinguishing Media: Foam, dry chemical, carbon dioxide, water spray/fog.

Special Fire Fighting Procedures: Do not enter confined spaces without proper protection. Heat may cause decomposition, build pressure, and rupture containers, spreading fire and increasing risk of serious injury. Fight fire from safe distance and protected location. Apply aqueous extinguishing media carefully to prevent steam explosion from contact with hot liquid. Avoid contact with hot liquid to prevent serious burns.

Unusual Fire and Explosion Hazards: On exposure to high temperature, this material may decompose, releasing toxic and flammable gases, which when mixed with air in proper proportions may ignite and explode. The gases released are often heavier than air and may travel long distances before reaching a source of ignition and flash back to the vapor source.

Hazardous Products of Combustion and Decomposition: Incomplete combustion will generate highly poisonous carbon monoxide, perhaps other toxic vapors, and a minor amount of metal oxide dust.

Hazardous Polymerization: This product can polymerize when exposed to high heat or energy sources that deactivate the inhibitor.

--- HEALTH HAZARD DATA ---

Effects of Overexposure: Note the following *CAUTION* statements provided by the manufacturer:

CAUTION: Moderate Skin hazard. May cause sensitization reaction. This material may be absorbed through the skin.

CAUTION: Moderate eye hazard.

CAUTION: Slight inhalation hazard.

CAUTION: Slight ingestion hazard.

The primary hazard with this product is by direct liquid contact with the skin or eyes.

This product may cause a potentially serious sensitization reaction in susceptible persons in addition to eye and skin reaction.

Emergency Response and First Aid:
Skin: Immediately remove and discard contaminated clothing. Wash affected areas with soap and water. If sticky, use waterless cleaner first. Prompt action is essential. Seek medical attention if irritation persists.
Note: Following extensive, repeated and/or prolonged contact, this material may cause skin irritation leading to dermatitis or more serious skin disorders.
In liquid or solution form, this material may be absorbed through intact skin and produce serious toxic effects if exposure is prolonged or extensive.
Eyes: In case of eye contact, immediately flush eyes with clean, low-pressure water for at least 15 min, occasionally lifting eyelids. Obtain immediate medical attention.
Note: Upon direct contact, this material may cause moderate eye irritation, but no irreversible damage.
Inhalation: If overcome by exposure, immediately move victim to fresh air. Keep victim quiet. Administer oxygen or artificial respiration as needed. Obtain medical attention immediately.

Note: Exposure to excessive concentration of this material in either vapor or mist form may result in toxic effects.

Ingestion: If a large quantity of this material is swallowed, administer warm water (quart) and obtain medical attention immediately. Do not induce vomiting because aspiration into lungs will increase the risk of chemical pneumonia.

Note: This material may be toxic if swallowed in large quantity.

Cancer Information: None listed under OSHA/NTP/IARC.

--- TOXICITY DATA ---

No information reported.

--- REACTIVITY DATA ---

Stability and Conditions to Avoid: Elevated temperatures, radiation sources (ultraviolet light, etc.), oxygen-free atmosphere (deactivates polymerization inhibitor).

Materials to Avoid (Incompatibility): Peroxides and other strong oxidizers; strong acids.

Hazard Decomposition Products: Fumes, smoke, carbon monoxide, and carbon dioxide, in the case of incomplete combustion.

Hazardous Polymerization: This product will polymerize.

--- SPILL AND EMERGENCY RESPONSE ---

For Spills: Material can polymerize, causing it to release heat and become progressively thicker, leading ultimately to a solid form. If heat is not dissipated, closed containers may become hot, overpressured, and rupture. Evacuate unnecessary personnel. Extinguish ignition sources including electric supply. Shut off leak if possible without undue risk. Notify fire/pollution/water supply authorities. Blanket spill with water spray/fog or foam.

On land, dike or impound large spill for recycle or recovery and prevent drainage to public waters. For small spill, absorb onto inert solids and transfer into suitable containers for transport or disposal. On water, material is insoluble and will float. Contain with booms, skim, or adsorb. If solid polymer forms, use solids removal equipment. Assure that any closed containers used to store spill cleanup residue are properly vented to prevent overpressure if residue continues to react and generate heat. The National Response Center should be notified at 800-424-8802 if this material is released to the environment.

Waste Disposal Method: Monomers are self-reacting chemicals that polymerize, liberating heat. Unless properly inhibited, such polymerization may result in formation of rigid or rubbery solids. In storage, this process may generate sufficient heat to raise internal temperature and pressure and rupture a container or vessel. Check weekly to confirm inhibitor polymer content. If inhibitor is below recommended level, take immediate steps to add supplemental inhibitor and mix well to make certain it is effective. Store in tightly closed or properly vented containers away from heat, sparks, open flame, or strong oxidizing agents.

Do not store under inert atmosphere (deactivates inhibitor). The manufacturer recommends using the material within 6 months of purchase.

Empty Containers: Keep containers closed when not in use. Do not handle or store near heat, sparks, flame, or strong oxidants.

--- SPECIAL PROTECTION AND PRECAUTIONS ---

Respiratory Protection: If this material is handled at elevated temperature or under mist-forming conditions, NIOSH/MSHA-approved self-contained breathing apparatus operated in positive-pressure mode should be used.

Type of Ventilation Required: Use only in well-ventilated area.

Eye Protection Requirements: Eye protection, such as chemical splash goggles and/or face mask, must be worn when any possibility exists for eye contact due to splashing or spraying liquid. Contact lenses should not be worn.

Protective Gloves and Other: When handling this material, protective clothing including gloves, apron, sleeves, boots, and head and face protection must be worn. This equipment must be cleaned thoroughly after each use.

Use good personal hygiene practices. Wash hands before eating, drinking, smoking, or using toilet facilities. Take a shower after work, using plenty of soap and water.

Storage and Handling Precautions: Containers should be stored in a cool, dry, well-ventilated area. Store away from flammable materials, sources of heat, flame, sparks, and foodstuffs. Exercise due caution to prevent damage to or leakage from the container. Avoid any conditions that might tend to create a dust explosion. Maintain good housekeeping practices to minimize dust buildup. *Note:* This product has a 6-month shelf life.

Other Precautions: Avoid breathing oil mist. Remove oil-soiled clothing and launder before reuse. Discard oil-soaked shoes. Wash skin thoroughly with soap and water after handling.

--- REGULATORY INFORMATION ---

Transportation Incident Information: This product is unregulated by DOT at present.

DOT Identification Number: Not applicable.

▼   ▼   ▼   ▼   ▼   ▼   ▼   ▼   ▼   ▼   ▼   ▼   ▼

--- IDENTIFIERS ---

Chemical Name and Synonyms: COUMARONE-INDENE RESIN

Trade Name and Synonyms: Cumar P-10

Chemical Family: Thermoplastic Resin

Manufacturer: Neville Chemical Co., Pittsburgh, Pennsylvania

UN ID Number: NA

--- INGREDIENTS AND COMPOSITION DATA ---

Material	Cas No.	%	TLV
Cumar P-10	NA	100	Not established

--- PHYSICAL DATA ---

Boiling Point (°F): NA

Specific Gravity (H_2O = 1): 1.04

Vapor Pressure (mm Hg): NA

Percent Volatile by Volume (%): Nonvolatile

Vapor Density (air = 1): NA

Freezing Point: NA

Evaporation Rate: NA

pH: NA

Solubility in Water: Negligible

Melting Point/Pour or Congealing Point: NA

Viscosity: NA

Appearance/Color/Odor: Light straw colored material with mild odor.

--- FIRE AND EXPLOSION HAZARDS ---

Flash Point: 450°F (COC)

Flammability Limits: NA

Autoignition Temperature: NA

Extinguishing Media: Chemical foam, water

Special Fire Fighting Procedures: Wear protective clothing and mask.

Unusual Fire and Explosion Hazards: None reported.

Hazardous Products of Combustion and Decomposition: Carbon dioxide

Hazardous Polymerization: Will not occur.

--- HEALTH HAZARD DATA ---

Effects of Overexposure: Manufacturer recommends avoiding skin and eye contact and to avoid excessive inhalation of dust. This material may cause skin irritation to people with allergic conditions.

Emergency Response and First Aid:
Skin: Flush all affected areas with plenty of water for 15 min. Remove and clean any contaminated clothing and shoes. Seek medical attention if skin irritation occurs.
Eyes: Flush the eyes with plenty of running water for at least 15 min. Hold the eyelids apart during the flushing to ensure rinsing of the surface of the eye and lids with water. Seek medical attention if eye irritation occurs.
Inhalation: There are no effects or recommendations reported.
Ingestion: Consult physician immediately.

Cancer Information: None listed under OSHA/NTP/IARC.

--- TOXICITY DATA ---

No information reported.

--- REACTIVITY DATA ---

Stability and Conditions to Avoid: This product is stable but should be stored away from open flames.

Materials to Avoid (Incompatibility): Excessive heat and open flames.

Hazard Decomposition Products: Fumes, smoke, carbon monoxide, and carbon dioxide, in the case of incomplete combustion. This material has similar properties to that of any organic chemical.

Hazardous Polymerization: Will not occur.

--- SPILL AND EMERGENCY RESPONSE ---

For Spills: Use any inert absorbent such as sand, earth, or vermiculite. The product is described as being slippery. Use proper safety equipment during cleanup. Dispose of in a manner that is consistent with local, state, and federal regulations. After absorbent is applied, the area can be washed off with detergent.

Waste Disposal Method: Assure conformity with applicable disposal regulations. Dispose of absorbed material at an approved waste disposal site or facility. Waste material may be incinerated under conditions that meet federal, state, and local environmental control regulations. At present this is not classified as a RCRA hazardous waste. The manufacturer recommends mixing with more flammable solvent and incinerating.

Empty Containers: Keep containers closed when not in use. Do not handle or store near heat, sparks, flames, or strong oxidants.

--- SPECIAL PROTECTION AND PRECAUTIONS ---

Respiratory Protection: Normally not needed.

Type of Ventilation Required: No recommendations are given by manufacturer. Local exhaust systems should use explosion-proof motors.

Eye Protection Requirements: Use splash goggles or face shield when eye contact may occur.

Protective Gloves and Other: Use polyethylene protective gloves.

Storage and Handling Precautions: Containers should be stored in a cool, dry, well-ventilated area. Store away from flammable materials, sources of heat, flame, sparks, and foodstuffs. Exercise due caution to prevent damage to or leakage from the container. Avoid any conditions that might tend to create a dust explosion. Maintain good housekeeping practices to minimize dust buildup. *Note:* This product has a 1 yr shelf life.

Other Precautions: Avoid breathing oil mist. Remove oil-soiled clothing and launder before reuse. Discard oil-soaked shoes. Wash skin thoroughly with soap and water after handling.

--- REGULATORY INFORMATION ---

Transportation Incident Information: This product is unregulated by DOT at present.

DOT Identification Number: Not applicable.

▼   ▼   ▼   ▼   ▼   ▼   ▼   ▼   ▼   ▼   ▼   ▼   ▼

--- IDENTIFIERS ---

Chemical Name and Synonyms: **PETROLEUM HYDROCARBON FAMILY, BLEND**

Trade Name and Synonyms: Escorez 2000 Series

Chemical Family: Petroleum Hydrocarbon Resin

Manufacturer: Exxon Chemical Co., Houston, Texas

UN ID Number: NA

--- INGREDIENTS AND COMPOSITION DATA ---

Material	Cas No.	%	TLV
Escorez 2000	68527-25-3	100	Not established

--- PHYSICAL DATA ---

Boiling Point (°F): NA

Specific Gravity (H₂O = 1): 1.02

Vapor Pressure (mm Hg): NA

Percent Volatile by Volume (%): NA

Vapor Density (air = 1): NA

Freezing Point: 65 to 100°C (150 to 212°F)

Evaporation Rate: NA

pH: NA

Solubility in Water: Insoluble

Melting Point/Pour or Congealing Point: NA

Viscosity: NA

Appearance/Color/Odor: Pale yellow to amber solid, flakes, or molten liquid.

--- FIRE AND EXPLOSION HAZARDS ---

Flash Point: 430°F (COC)

Flammability Limits: NA

Autoignition Temperature: NA

Extinguishing Media: Chemical foam, water

Special Fire Fighting Procedures: Wear protective clothing and mask.

Unusual Fire and Explosion Hazards: None reported.

Hazardous Products of Combustion and Decomposition: Carbon dioxide

Hazardous Polymerization: Will not occur.

--- HEALTH HAZARD DATA ---

Effects of Overexposure: Material is described by manufacturer as posing the following hazards:
Eye contact: Slightly irritating, but does not injure eye tissue.
Skin contact: Frequent or prolonged contact may irritate. The material is described as having a low order of toxicity.
Inhalation: Negligible hazard at ambient temperature (-18 to 38°C; 0 to 100°F). Dust may be irritating to eyes and respiratory tract.
Ingestion: Minimal toxicity.

The manufacturer recommends the following occupational exposure limit (OEL): 10 mg/m^3 total dust for nuisance dust.

Emergency Response and First Aid:
Skin: Flush all affected areas with plenty of water for 15 min. Remove and clean any contaminated clothing and shoes. Seek medical attention if skin irritation occurs.

Eyes: Flush the eyes with plenty of running water for at least 15 min. Hold the eyelids apart during the flushing to ensure rinsing of the surface of the eye and lids with water. Seek medical attention if eye irritation occurs.

Inhalation: There are no effects or recommendations reported.

Ingestion: Consult physician immediately. Manufacturer notes that first aid is normally not required.

Cancer Information: None listed under OSHA/NTP/IARC.

--- TOXICITY DATA ---

No information reported.

--- REACTIVITY DATA ---

Stability and Conditions to Avoid: Do not heat above 200°C.

Materials to Avoid (Incompatibility): Excessive heat and open flames.

Hazard Decomposition Products: Fumes, smoke, carbon monoxide, and carbon dioxide, in the case of incomplete combustion. This material has similar properties to that of any organic chemical.

Hazardous Polymerization: Will not occur.

--- SPILL AND EMERGENCY RESPONSE ---

For Spills:

Land spill: Eliminate sources of ignition. Prevent additional discharge of material if possible to do so without hazard. For small spills, implement cleanup procedures; for large spills, implement cleanup procedures and, if in public area, keep public away and advise authorities. Also, if this product is subject to CERCLA reporting, notify the National Response Center.

Recover spilled material and place in suitable containers for recycle or disposal.

Consult an expert on disposal of recovered material and ensure conformity to local disposal regulations.

Water spill: Prevent additional discharge of material, if possible to do so without hazard. Advise authorities if floating material enters a watercourse or sewer. If possible, try to contain floating material.

Skim from surface.

Consult an expert on disposal of recovered material and ensure conformity to local disposal regulations.

Waste Disposal Method: Assure conformity with applicable disposal regulations. Dispose of absorbed material at an approved waste disposal site or facility. Waste material may be incinerated under conditions that meet federal, state, and local environmental control regulations. At present this is not classified as a RCRA hazardous waste. The manufacturer recommends mixing with more flammable solvent and incinerating.

Empty Containers: Keep containers closed when not in use. Do not handle or store near heat, sparks, flames, or strong oxidants.

--- SPECIAL PROTECTION AND PRECAUTIONS ---

Respiratory Protection: Where concentrations in air may exceed the limits given by the manufacturer (OEL), work practice or other means of exposure reduction are not adequate. NIOSH/MSHA-approved respirators may be necessary to prevent overexposure by inhalation.

Type of Ventilation Required: No specific recommendations are given by manufacturer. Local exhaust systems should use explosion-proof motors.

Eye Protection Requirements: Use splash goggles or face shield when eye contact may occur.

Protective Gloves and Other: Use polyethylene protective gloves. For open systems where contact is likely, wear safety glasses with side shields, long sleeves, and chemical-resistant gloves.

Storage and Handling Precautions: Containers should be stored in a cool, dry, well-ventilated area. Store away from flammable materials, sources of heat, flame, sparks, and foodstuffs. Exercise due caution to prevent damage to or leakage from the container. Avoid any conditions that might tend to create a dust explosion. Maintain good housekeeping practices to minimize dust buildup. This material should be stored under ambient temperature conditions.
The recommended storage and transport pressure for this product is atmospheric.
Note that loading and unloading temperature recommended is ambient.
Note that this material can pose an electrostatic accumulation hazard. It should be handled using proper grounding procedures.

Other Precautions: Avoid breathing oil mist. Remove oil-soiled clothing and launder before reuse. Discard oil-soaked shoes. Wash skin thoroughly with soap and water after handling.

--- REGULATORY INFORMATION ---

Department of Transportation (DOT)
DOT Hazard Class: Not regulated
DOT Identification Number: Not available

TSCA: This product is listed on the TSCA inventory as UVCB (Unknown, Variable Composition, or Biological) Chemical at CAS Registry Number 68527-25-3.

CERCLA: If this product is accidentally spilled, it is not subject to any special reporting under the requirements of the Comprehensive Environmental Response, Compensation, and Liability Act (CERCLA). The manufacturer recommends contacting local authorities to determine if there may be local reporting requirements.

SARA TITLE III: Under the provisions of Title III, Sections 311/312 of the Superfund Amendments and Reauthorization Act, this product is classified into the following hazard category: **not hazardous.**

▼   ▼   ▼   ▼   ▼   ▼   ▼   ▼   ▼   ▼   ▼   ▼   ▼   ▼

--- IDENTIFIERS ---

Chemical Name and Synonyms: 4,4'-DI(A,A-DIMETHYLBENZYL) DIPHENYLAMINE

Trade Name and Synonyms: Naugard 445

Chemical Family: Aromatic Amine

Manufacturer: Uniroyal Chemical Co., Middlebury, Connecticut

UN ID Number: NA

--- INGREDIENTS AND COMPOSITION DATA ---

Material	Cas No.	%	TLV
Product	10081-67-1	100	Not hazardous

--- PHYSICAL DATA ---

Boiling Point (°F): NA

Specific Gravity (H₂O = 1): 1.14

Vapor Pressure (mm Hg): NA

Percent Volatile by Volume (%): Low

Vapor Density (air = 1): NA

Freezing Point: NA

Evaporation Rate: NA

pH: NA

Solubility in Water: Soluble in most organic compounds.

Melting Point: 203°F (95°C)

Viscosity: NA

Appearance/Color/Odor: White to off white powder with a characteristic odor.

--- FIRE AND EXPLOSION HAZARDS ---

Flash Point: 531°F (276.7°C) (COC)

Flammability Limits: NA

Autoignition Temperature: 569°F (298°C) fire point

Extinguishing Media: Water spray, dry chemical.

Special Fire Fighting Procedures: Protect against inhalation of combustion products.

Unusual Fire and Explosion Hazards: May form explosive dust-air mixtures.

Hazardous Products of Combustion and Decomposition: Oxides of carbon and nitrogen under burning conditions.

Hazardous Polymerization: Will not occur.

--- HEALTH HAZARD DATA ---

Effects of Overexposure: No acute or chronic health hazards have been identified. There are no known medical conditions aggravated by exposure to this material.

Emergency Response and First Aid:
Skin: Flush all affected areas with plenty of water for 15 min. Remove and clean any contaminated clothing and shoes. Seek medical attention if skin irritation occurs.
Eyes: Flush the eyes with plenty of running water for at least 15 min. Hold the eyelids apart during the flushing to ensure rinsing of the surface of the eye and lids with water. Seek medical attention if eye irritation occurs.
Inhalation: Remove victim to fresh air.
Ingestion: Consult physician immediately.

Cancer Information: None listed under OSHA/NTP/IARC.

--- TOXICITY DATA ---

Animal Tests:
Oral toxicity: LD50 (rats) 10 g/kg
Irritation: eye (rabbits), negative; skin (rabbits), negative
Mutagenicity: Ames salmonella, negative
No significant adverse effects have been noted with long-term production and use of this material.

--- REACTIVITY DATA ---

Stability and Conditions to Avoid: This product is stable but should be stored away from open flames. Product is stable at ambient temperatures and pressures.

Materials to Avoid (Incompatibility): Strong oxidizing agents.

Hazard Decomposition Products: Oxides of carbon and nitrogen under burning conditions.

Hazardous Polymerization: Will not occur.

--- SPILL AND EMERGENCY RESPONSE ---

For Spills: Sweep or vacuum up. Shovel into secure containers for proper disposal. Avoid creating dust. Use personal protective equipment as outlined below.

Waste Disposal Method: Assure conformity with applicable disposal regulations. Dispose of absorbed material at an approved waste disposal site or facility. Waste material may be incinerated under conditions that meet federal, state, and local environmental control regulations. At present this is not classified as a RCRA hazardous waste. The manufacturer recommends mixing with more flammable solvent and incinerating.

Empty Containers: Keep containers closed when not in use. Do not handle or store near heat, sparks, flames, or strong oxidants.

--- SPECIAL PROTECTION AND PRECAUTIONS ---

Respiratory Protection: In absence of adequate ventilation, use NIOSH-certified dust cartridge respirator.

Type of Ventilation Required: No recommendations are given by manufacturer.

Eye Protection Requirements: Use splash goggles or face shield when eye contact may occur.

Protective Gloves and Other: Use polyethylene protective gloves.

Storage and Handling Precautions: Containers should be stored in a cool, dry, well-ventilated area. Store away from flammable materials, sources of heat, flame, sparks, and foodstuffs. Exercise due caution to prevent damage to or leakage from the container. Avoid any conditions that might tend to create a dust explosion. Maintain good housekeeping practices to minimize dust buildup. *Note:* This product has a 1 yr shelf life.

Other Precautions: Provide sufficient ventilation to minimize dust exposure. Avoid dust accumulation on building or equipment surfaces.

--- REGULATORY INFORMATION ---

Transportation Incident Information: This product is unregulated by DOT at present.

DOT Identification Number: Not applicable.

▼　▼　▼　▼　▼　▼　▼　▼　▼　▼　▼　▼　▼　▼

--- IDENTIFIERS ---

Name: *N*-OCTADECYLAMINE
Synonyms: Adogenen 142; Alamine 7; Armeen 118D; *n*-Octadecylamine; Oktadecylamin (Czech); Stearylamine
CAS: 124-30-1; **RTECS:** RG4150000
Formula: $C_{18}H_{39}N$; **Mol Wt:** 269.58
WLN: Z18
Chemical Class: Aliphatic amine

See other identifiers listed below under Regulations.

--- PROPERTIES ---

Boiling Point: NA
Melting Point: 328.16 to 330.16 K; 55 to 57°C; 131 to 134.6°F
Flash Point: >383.16 K; >110°C; >230°F
Autoignition: NA
UEL: NA
LEL: NA
Vapor Density: No data
Specific Gravity: No data

Reactivity with Water: No data on water reactivity
Reactivity with Common Materials: No data
Stability during Transport: No data
Neutralizing Agents: No data
Polymerization Possibilities: No data

Toxic Fire Gases: None reported other than possible unburned vapors
Odor Detected at (ppm): Unknown
Odor Description: No data
100% Odor Detection: No data

--- REGULATIONS ---

DOT hazard class: Not given
Packaging exceptions: 173.
Nonbulk packaging: 173.
Bulk packaging: 173.

STCC Number: Not listed

Clean Water Act Sect. 307: No
Clean Water Act Sect. 311: No
Clean Air Act: Not listed
EPA Waste Number: None
CERCLA Ref: Not listed
RQ Designation: Not listed
SARA TPQ Value: Not listed
SARA Sect. 312 Categories:
   Acute toxicity: irritant.

U.S. Postal Service Mailability: Not given

--- TOXICITY DATA ---

Short-term Toxicity: Unknown

Long-term Toxicity: Unknown

Conc IDLH: Unknown

NIOSH REL: Not given

ACGIH TLV: Not listed
ACGIH STEL: Not listed

OSHA PEL: Not in Table Z-1-A

MAK Information: Not listed

Carcinogen: N; **Status:** See below

Carcinogen Lists: *IARC*: Not listed; *MAK*: Not listed; *NIOSH*: Not listed; *NTP*: Not listed; *ACGIH*: Not listed; *OSHA*: Not listed.

LD50 Value: orl-rat LD50:2395 mg/kg

Other Species Toxicity Data: (Source: NIOSH RTECS 1992)
orl-rat LD50:2395 mg/kg
orl-mus LD50:3 g/kg
ipr-mus LD50:250 mg/kg

Reproductive Toxicity (1992 RTECS): This chemical has no known mammalian reproductive toxicity.

--- INITIAL INCIDENT RESPONSE ---

No DOT Guide information for this product.

▼   ▼   ▼   ▼   ▼   ▼   ▼   ▼   ▼   ▼   ▼   ▼   ▼   ▼

--- IDENTIFIERS ---

Chemical Name and Synonyms: DIPHENYLAMINE-ACETONE CONDENSATE

Trade Name and Synonyms: BLE 25

Chemical Family: Amine

Manufacturer: Uniroyal Chemical Co., Naugatuck, Connecticut

UN ID Number: NA

--- INGREDIENTS AND COMPOSITION DATA ---

Material	Cas No.	%	TLV
Product	9003-79-6	100	Not established

--- PHYSICAL DATA ---

Boiling Point (°F): >170°C at 20 mm Hg

Specific Gravity (H₂O = 1): 1.08 to 1.10

Vapor Pressure (mm Hg): NA

Percent Volatile by Volume (%): Low at 70°C

Vapor Density (air = 1): NA

Freezing Point: NA

Evaporation Rate: NA

pH: NA

Solubility in Water: Insoluble in water; soluble in acetone and benzene.

Melting Point: NA

Viscosity: 25 to 50 poises at 30°C

Appearance/Color/Odor: Dark brown liquid. No odor description given by manufacturer.

--- FIRE AND EXPLOSION HAZARDS ---

Flash Point: >230°F (TCC)

Flammability Limits: NA

Autoignition Temperature: NA

Extinguishing Media: Water spray, foam, carbon dioxide.

Special Fire Fighting Procedures: Protect against inhalation of combustion products.

Unusual Fire and Explosion Hazards: None reported by manufacturer.

Hazardous Products of Combustion and Decomposition: Oxides of carbon and nitrogen under burning conditions.

Hazardous Polymerization: Will not occur.

--- HEALTH HAZARD DATA ---

Effects of Overexposure: No acute or chronic health hazards have been identified. There are no known medical conditions aggravated by exposure to this material.

Emergency Response and First Aid:
Skin: Flush all affected areas with plenty of water for 15 min. Remove and clean any contaminated clothing and shoes. Seek medical attention if skin irritation occurs.
Eyes: Flush the eyes with plenty of running water for at least 15 min. Hold the eyelids apart during the flushing to ensure rinsing of the surface of the eye and lids with water. Seek medical attention if eye irritation occurs.
Inhalation: Remove victim to fresh air.
Ingestion: Consult physician immediately.

Cancer Information: None listed under OSHA/NTP/IARC.

--- TOXICITY DATA ---

Animal Effects:
Oral toxicity: LD50 (rats), 2300 mg/kg
Dermal irritation (rabbit): slight
Eye irritation (rabbit): slight
No significant adverse effects have been noted with long-term production and use of this material.

--- REACTIVITY DATA ---

Stability and Conditions to Avoid: This product is stable but should be stored away from open flames. Product is stable at ambient temperatures and pressures.

Materials to Avoid (Incompatibility): Strong mineral acids and oxidizing agents.

Hazard Decomposition Products: Oxides of carbon and nitrogen under burning conditions.

Hazardous Polymerization: Will not occur.

--- SPILL AND EMERGENCY RESPONSE ---

For Spills: Absorb on industrial absorbent. Sweep up. Put into containers for proper disposal. Use proper personal protective equipment.

Waste Disposal Method: Assure conformity with applicable disposal regulations. Dispose of absorbed material at an approved waste disposal site or facility. Waste material may be incinerated under conditions that meet federal, state, and local environmental control regulations. At present this is not classified as a RCRA hazardous waste. The manufacturer recommends mixing with more flammable solvent and incinerating.

Empty Containers: Keep containers closed when not in use. Do not handle or store near heat, sparks, flame, or strong oxidants.

--- SPECIAL PROTECTION AND PRECAUTIONS ---

Respiratory Protection: In the absence of adequate ventilation, use NIOSH-certified dust cartridge respirator.

Type of Ventilation Required: No recommendations are given by manufacturer.

Eye Protection Requirements: Use splash goggles or face shield when eye contact may occur.

Protective Gloves and Other: Use polyethylene protective gloves.

Storage and Handling Precautions: Containers should be stored in a cool, dry, well-ventilated area. Store away from flammable materials, sources of heat, flame, sparks, and foodstuffs. Exercise due caution to prevent damage to or leakage from the container. Avoid any conditions that might tend to create a dust explosion. Maintain good housekeeping practices to minimize dust buildup. *Note:* This product has a 1 yr shelf life.

Other Precautions: Provide sufficient ventilation to minimize dust exposure. Avoid dust accumulation on building or equipment surfaces.

--- REGULATORY INFORMATION ---

Transportation Incident Information: This product is unregulated by DOT at present.

DOT Identification Number: Not applicable.

▼ ▼ ▼ ▼ ▼ ▼ ▼ ▼ ▼ ▼ ▼ ▼ ▼

--- IDENTIFIERS ---

Name: *N*-ISOPROPYL-*N*'-PHENYL-*P*-PHENYLENE-DIAMINE
Synonyms: Cyzone; Elastozone 34; *N*-Fenyl-*N*'-Isopropyl-*p*-Fenylendiamin (Czech); Flexzone 3C; 4-Isopropylaminodiphenylamine; *N*-Isopropyl-*N*'-Fenyl-*p*-Fenylendiamin (Czech); *N*-Isopropyl-*N*'-Phenyl-*p*-Phenylenediamine; NCI-C56304; Nonox ZA; *N*-Phenyl-*N*'-Isopropyl-*p*-Phenylenediamine; *N*-2-Propyl-*N*'-Phenyl-*p*-Phenylenediamine; Santoflex 36
CAS: 101-72-4; **RTECS:** ST2650000
Formula: $C_{15}H_{18}N_2$; **Mol Wt:** 226.35
WLN: 1YMR DMR
Chemical Class: Aromatic amine

See other identifiers listed below under Regulations.

--- PROPERTIES ---

Boiling Point: NA
Melting Point: NA
Flash Point: NA
Autoignition: NA
UEL: NA
LEL: NA
Vapor Density: No data
Specific Gravity: No data

Reactivity with Water: No data on water reactivity
Reactivity with Common Materials: No data
Stability during Transport: No data
Neutralizing Agents: No data
Polymerization Possibilities: No data

Toxic Fire Gases: None reported other than possible unburned vapors
Odor Detected at (ppm): Unknown
Odor Description: No data
100% Odor Detection: No data

--- REGULATIONS ---

DOT hazard class: Not given
Packaging exceptions: 173.
Nonbulk packaging: 173.
Bulk packaging: 173.

STCC Number: Not listed

Clean Water Act Sect. 307: No
Clean Water Act Sect. 311: No
Clean Air Act: Not listed
EPA Waste Number: None
CERCLA Ref: Not listed
RQ Designation: Not listed
SARA TPQ Value: Not listed
SARA Sect. 312 Categories:
   Acute toxicity: irritant.

U.S. Postal Service Mailability: Not given

--- TOXICITY DATA ---

Short-term Toxicity: Unknown

Long-term Toxicity: Unknown

Conc IDLH: Unknown

NIOSH REL: Not given

ACGIH TLV: Not listed
ACGIH STEL: Not listed

OSHA PEL: Not in Table Z-1-A

MAK Information: Not listed

Carcinogen: N; **Status:** See below

Carcinogen Lists: *IARC*: Not listed; *MAK*: Not listed; *NIOSH*: Not listed; *NTP*: Not listed; *ACGIH*: Not listed; *OSHA*: Not listed.

LD50 Value: orl-rat LD50:720 mg/kg

Other Species Toxicity Data: (Source: NIOSH RTECS 1992)
orl-rat LD50:720 mg/kg
orl-mus LD50:1122 mg/kg

Reproductive Toxicity (1992 RTECS): This chemical has no known mammalian reproductive toxicity.

--- PROTECTION AND FIRST AID ---

Recommended Respiration Protection Source: NIOSH Pocket Guide (85-114)
NIOSH (*N*-Isopropyl-*N'*-Phenyl-*p*-Phenylene-Diamine) -
100 ppm: Any supplied-air respirator. Substance reported to cause eye irritation or damage may require eye protection. Any self-contained breathing apparatus.
250 ppm: Any supplied-air respirator operated in a continuous-flow mode. Substance reported to cause eye irritation or damage may require eye protection.
300 ppm: Any self-contained breathing apparatus with a full facepiece. Any supplied-air respirator with a full facepiece.
Emergency or Planned Entry in Unknown Concentrations or IDLH Conditions: Any self-contained breathing apparatus with a full facepiece and operated in a pressure-demand or other positive-pressure mode. Any supplied-air respirator with a full facepiece and operated in a pressure-demand or other positive-pressure mode in combination with an auxiliary self-contained breathing apparatus operated in a pressure-demand or other positive-pressure mode.
Escape: Any air-purifying full facepiece respirator (gas mask) with a chin-style or front- or back-mounting canister. Any appropriate escape-type self-contained breathing apparatus.

▼ ▼ ▼ ▼ ▼ ▼ ▼ ▼ ▼ ▼ ▼ ▼ ▼ ▼

--- IDENTIFIERS ---

Name: SELENIUM, TETRAKIS(DIETHYLDITHIOCARBAMATO)-
Synonyms: Ethyl Selenac; Ethyl Seleram; Selenium Diethyldithiocarbamate;
Tetrakis(Diethylcarbamodithioato-S, S')Selenium; Tetrakis(Diethyldithiocar-
bamato)Selenium
CAS: 5456-28-0; **RTECS:** VT0700000
Formula: $C_{20}H_{40}N_4S_8$ .Se; **Mol Wt:** 672.08
WLN: 2N2

See other identifiers listed below under Regulations.

--- PROPERTIES ---

Boiling Point: NA
Melting Point: NA
Flash Point: NA
Autoignition: NA
UEL: NA
LEL: NA
Vapor Density: No data
Specific Gravity: No data

Reactivity with Water: No data on water reactivity
Reactivity with Common Materials: No data
Stability during Transport: No data
Neutralizing Agents: No data
Polymerization Possibilities: No data

Toxic Fire Gases: Oxides of nitrogen
Odor Detected at (ppm): Unknown
Odor Description: No data
100% Odor Detection: No data

--- REGULATIONS ---

DOT hazard class: Not given
Packaging exceptions: 173.
Nonbulk packaging: 173.
Bulk packaging: 173.

STCC Number: Not listed

Clean Water Act Sect. 307: Yes
Clean Water Act Sect. 311: No
Clean Air Act: CAA '90 by category
EPA Waste Number: D010
CERCLA Ref: Not listed
RQ Designation: Not listed
SARA TPQ Value: Not listed
SARA Sect. 312 Categories: No category
Listed in SARA Sect. 313: Yes
De Minimus Concentration: 1.0%

U.S. Postal Service Mailability: Not given

--- TOXICITY DATA ---

Short-term Toxicity: Unknown

Long-term Toxicity: Unknown

Conc IDLH: Unknown

NIOSH REL: Not given

ACGIH TLV: TLV = 0.2 mg/m^3 as selenium
ACGIH STEL: As selenium

OSHA PEL: Transitional Limits: PEL = 0.2 mg/m^3; Final Rule Limits: TWA = 0.2 mg/m^3

MAK Information: Not listed

Carcinogen: N; **Status:** See below
References: Animal indefinite IARC** 9, 245, 75; Animal indefinite IARC** 12, 107, 76

Carcinogen Lists: *IARC*: Not classified as to human carcinogenicity or probably not carcinogenic to humans; *MAK*: Not listed; *NIOSH*: Not listed; *NTP*: Not listed; *ACGIH*: Not listed; *OSHA*: Not listed.

LD50 Value: No LD50 in RTECS 1992

Reproductive Toxicity (1992 RTECS): This chemical has no known mammalian reproductive toxicity.

▼   ▼   ▼   ▼   ▼   ▼   ▼   ▼   ▼   ▼   ▼   ▼   ▼   ▼

--- IDENTIFIERS ---

Name: HVA-2
MSDS Number: HVA001
Manufacturer/Distributor: Du Pont Polymers, Wilmington, Delaware
Chemical Family: *N,N'-m*-Phenylene Dimaleimide
CAS Number: 3006-93-7

--- COMPONENTS ---

Material	Cas No.	%
N,N'-m-Phenylenebis-maleimide	3006-93-7	100

No skin or eye irritation tests have been run specifically on HVA-2. Tests on a closely similar product show it to be a primary skin irritant and capable of causing eye damage.

--- PHYSICAL DATA ---

Melting Point: 195°C

Specific Gravity: 1.44

Percent Volatiles: NA

Solubility in Water: Negligible

Color: Yellow to light brown

Odor: Essentially none

Form: Powder

--- HAZARDOUS REACTIVITY ---

Stability at Room Temperature: Stable.

Materials to Avoid: None are known.

Conditions to Avoid: None are known.

Hazardous Decomposition Products: Carbon monoxide and nitrogen oxides.

Polymerization: Will not occur.

--- FIRE AND EXPLOSION HAZARDS ---

Flash Ignition Temperature: Above 208°C.

Method: Open cup.

Unusual Fire, Explosion Hazards: No unusual hazards known.

Hazardous Combustion Products: Carbon monoxide, nitrogen oxides.

Extinguishing Media: Water, carbon dioxide, foam, dry chemical.

Special Fire Fighting Instructions: Use self-contained breathing apparatus if exposed to fumes.

--- HEALTH HAZARD DATA ---

Before using HVA-2, read the bulletin on the safe handling of this material.

Acute or Immediate Effects:
Routes of Entry and Symptoms:
Ingestion: The LD50 of HVA in rats is 2025 mg/kg body weight. This classifies the material as being slightly toxic. Ingestion is not a probable route of exposure.
Skin: No data are available on HVA-2. Tests on a similar product showed that it is a strong primary irritant to guinea pig skin. Results are predicted to be similar for HVA-2. HVA-2 is probably not a skin sensitizer.

Eye: No data are available on HVA-2. Tests on a closely similar product showed it is an eye irritant capable of causing severe and irreversible damage. Treat HVA-2 as if it had the same reaction.   Avoid all eye contact with HVA-2.

Inhalation: There are no data on the inhalation toxicity of HVA-2. In view of the fact that it may be a strong skin irritant and strong eye irritant, dust from HVA-2 should not be inhaled.

Chronic Effects: HVA-2 is not mutagenic in the activated or inactivated Ames salmonella test.

Medical Conditions Aggravated by Exposure: Skin dermatitis problems will be aggravated by exposure to HVA-2.

Carcinogenicity: None of the components in this material is listed by IARC, NTP, OSHA, or ACGIH as a carcinogen.

Exposure Limits: TLV (ACGIH), none established; PEL (OSHA), none established.

Safety Precautions: See First Aid and Protection Information sections.

--- FIRST AID ---

If exposed to fumes from overheating or combustion, move to fresh air. Consult a physician.

Wash skin with soap and plenty of water.

Flush eyes with water. Consult a physician

--- PROTECTION INFORMATION ---

Generally Applicable Control Measures and Precautions:
Ventilation: During handling and use of HVA-2, local ventilation must be used to prevent breathing of dust or fumes.

Personal Protective Equipment:
Eye: Eye protection is required to prevent dust from getting in the eyes.
Skin: Impermeable gloves are required to prevent skin contact.
Respirator: Required if local ventilation is not adequate.

--- DISPOSAL INFORMATION ---

Spill, Leak, or Release:
Note: Review Fire and Explosion Hazards and Safety Precautions sections before proceeding with cleanup. Use appropriate personal protective equipment during cleanup.

Material should be vacuumed up or swept up.
Exercise caution to not create a dust hazard. Wear appropriate protective equipment.

Aquatic toxicity: No data are available. Material is probably toxic to fish.

Waste Disposal: Landfill or incinerate in accordance with federal, provincial, and local regulations.

--- STORAGE CONDITIONS ---

Store in a cool dry place. Keep containers closed to prevent contamination.

--- TITLE III HAZARD CLASSIFICATIONS ---

Section 313 Supplier Notification: This product contains no known toxic chemicals subject to the reporting requirements of Section 313 of the Emergency Planning and Community Right-to-Know Act of 1986 and of 40 CFR 372.

--- ADDITIONAL INFORMATION AND REFERENCES ---

NA = Not applicable
NE = Not established
\# = New or revised information in this section when # is in right margin.

State Right-to-Know Laws: No substances on the state hazardous substances list, for the states indicated below, are used in the manufacture of products on this Material Safety Data Sheet, with the exceptions indicated. While we do not specifically analyze these products or the raw materials used in their manufacture for substances on various state hazardous substances lists, to the best of our knowledge the products on this Material Safety Data Sheet contain no such substances except for those specifically listed below:

Substances on the Pennsylvania Hazardous Substances List Present at a Concentration of 1% or More: None known.

Substances on the Pennsylvania Special Hazardous Substances List Present at a Concentration of 0.01% or More: None

Nonhazardous Ingredients Present at a Concentration of 3% or More Required to be Listed by Pennsylvania: Since this product contains no hazardous substances as defined by the Pennyslvania R-T-K Regulations, a MSDS is not required by law.

Warning: Substances Known to the State of California to Cause Birth Defects or Other Reproductive Harm: None known.

Substances on the New Jersey Workplace Hazardous Substance List Present at a Concentration of 1% or More (0.1% for Substances Identified as Carcinogens, Mutagens, or Teratogens): None known.